Solar Energy and Photoenergy Systems

Solar Energy and Photoenergy Systems

Edited by Anderson Sheilds

Syrawood
PUBLISHING HOUSE

New York

Published by Syrawood Publishing House,
750 Third Avenue, 9th Floor,
New York, NY 10017, USA
www.syrawoodpublishinghouse.com

Solar Energy and Photoenergy Systems
Edited by Anderson Sheilds

Cataloging-in-Publication Data

Solar energy and photoenergy systems / edited by Anderson Sheilds.
    p. cm.
Includes bibliographical references and index.
ISBN 978-1-68286-613-9
1. Solar energy. 2. Photovoltaic power generation. I. Sheilds, Anderson.
TJ810 .S65 2018
621.47--dc23

TABLE OF CONTENTS

PREFACE

Photoenergy is the process and science of converting light into energy, particularly into electrical energy by using materials that are semiconductors and use the photovoltaic effect. One of the most successful and widely used form of photovoltaics is solar energy derived by planting solar panels in direct sunlight and harnessing the sunlight into electricity. This book presents researches and studies performed by experts across the globe on the various aspects of photoenergy. Most of the topics introduced in this book cover new techniques and the applications of the subject. It is appropriate for students seeking detailed information in this area as well as for experts.

This book is the end result of constructive efforts and intensive research done by experts in this field. The aim of this book is to enlighten the readers with recent information in this area of research. The information provided in this profound book would serve as a valuable reference to students and researchers in this field.

At the end, I would like to thank all the authors for devoting their precious time and providing their valuable contribution to this book. I would also like to express my gratitude to my fellow colleagues who encouraged me throughout the process.

Editor

Monitoring, Diagnosis, and Power Forecasting for Photovoltaic Fields

S. Daliento,[1] **A. Chouder,**[2] **P. Guerriero,**[1] **A. Massi Pavan,**[3] **A. Mellit,**[4] **R. Moeini,**[5] **and P. Tricoli**[5]

[1]*Department of Electrical Engineering and Information Technology (DIETI), University of Naples Federico II, Via Claudio 21, 80125 Naples, Italy*
[2]*Department of Génie Electrique, Faculty of Technologies, University of M'Sila, BP 166, Ichbelia, M'Sila, Algeria*
[3]*Department of Engineering and Architecture, University of Trieste, Piazzale Europa 1, 34127 Trieste, Italy*
[4]*Renewable Energy Laboratory, Faculty of Sciences and Technology, Jijel University, Jijel, Algeria*
[5]*School of Electronic, Electrical and Systems Engineering, University of Birmingham, Gisbert Kapp Building, Edgbaston, Birmingham B15 2TT, UK*

Correspondence should be addressed to P. Guerriero; pierluigi.guerriero@unina.it

Academic Editor: Wilfried G. J. H. M. Van Sark

A wide literature review of recent advance on monitoring, diagnosis, and power forecasting for photovoltaic systems is presented in this paper. Research contributions are classified into the following five macroareas: (i) electrical methods, covering monitoring/diagnosis techniques based on the direct measurement of electrical parameters, carried out, respectively, at array level, single string level, and single panel level with special consideration to data transmission methods; (ii) data analysis based on artificial intelligence; (iii) power forecasting, intended as the ability to evaluate the producible power of solar systems, with emphasis on temporal horizons of specific applications; (iv) thermal analysis, mostly with reference to thermal images captured by means of unmanned aerial vehicles; (v) power converter reliability especially focused on residual lifetime estimation. The literature survey has been limited, with some exceptions, to papers published during the last five years to focus mainly on recent developments.

1. Introduction

The photovoltaic (PV) market has witnessed over the last years a remarkable growth as a result of various stimulating factors: the significant cost reduction of the PV modules on the market and the changes on support policies. These factors have made the return on investment of a photovoltaic system more interesting. However, like other industrial processes, a photovoltaic system may be subject during its fabrication to defects and anomalies leading to a reduction of the overall system performance or even to total unavailability. These negative consequences will obviously reduce the productivity of a PV system and therefore its profit. Thus, proper early fault detection and real-time diagnostic are crucial not only for lowering cost and time maintenance, but also to avoid energy loss, damage to equipment, and safety hazards.

Several research papers and institutional body reports, such as IEA, have underlined the low yields of PV systems due to faulty components, especially the DC section (i.e., PV cells/modules and MPPT) [1–9]. Broadly speaking, faults of PV arrays are categorized as cracks in the cells, delamination, hot spots, dirt accumulation, modules mismatches, short circuit of modules, junction box faults, caused by damaged connections, corrosion of the connections, open circuit, short circuit, and MPPT faults [10]. Obviously, this is not an exhaustive list and many other faults can be found in the literature [11, 12].

The basic approach for the detection of unexpected power losses of PV systems uses analytical redundancy, which is a comparison between the monitored electrical quantities (output power, voltage, and current) and their counterparts obtained from a reference model. An alarm is triggered when predetermined differences are reached [13–16]. The reference model is often based on the one diode model, whose parameters are determined either by the manufacturer's datasheet or by automatic extraction methods [17]. However, these methods are only effective for the detection and diagnosis of a grouped set of faults but not for an individual location of each defect [18]. Moreover, irradiation and temperature measurements are essential requirements for this approach [15].

Another fault detection and diagnostic method is based on hardware redundancy, in which several similar subsystems undertake the same task. By collecting and analyzing each subsystem's data, abnormalities can be detected. Since monitoring of electrical parameters usually produces a large amount of data, artificial intelligence and data mining are adopted as well. Similar models are adopted for power forecasting, which is important for both monitoring purposes and the management of the utility grid.

Recently, thanks to the widespread diffusion of unmanned aerial vehicles (UAV), thermal analysis is becoming a cost-effective alternative to electrical monitoring. Also the reliability of power converters plays a key role in the correct operations of PV.

The paper is organized as follows: Section 2 describes the monitoring techniques based on the measurement of electrical parameters; Section 3 focuses on the analysis of measured data with artificial intelligence and data mining algorithms; Section 4 is devoted to power forecasting; Section 5 discusses aerial thermal analysis; and Section 6 deals with the reliability of power converters. Conclusions are drawn in Section 7.

2. Electrical Methods

This section presents the recent trends for monitoring and diagnosis (M&D), based on electrical parameters directly acquired from the solar field. In principle, the performance analysis based on such parameters is straightforward, because it is based on the comparison between measurements and predictions. Unfortunately, the large number of unpredictable conditions, which affect the performance of solar panels, poses a serious challenge to the definition of a reliable target for the expected outputs.

The first step towards a suitable monitoring system is the definition of what should be measured, how it can be measured, and how measurements can be handled. The question about what should be measured introduces the first trade-off among possible monitoring/diagnosis approaches, depending on how PV subsystems are grouped. Indeed, the overall performance of a PV system depends on the performance of each subsystem, where the individual subsystem is the single solar cell forming the solar panel. A more pervasive measurement system increases the accuracy of M&D at the expense of an increased cost. Therefore, a rough classification of M&D electrical techniques can be based on the "level of granularity" (LoG). The lowest LoG corresponds to the monitoring of the solar field as a whole. In this case, only the instantaneous output power generated by the PV field, at either the DC side or the AC side, is measured

and then converted into the energy yield of plant. In this case, the widely adopted figure of merit is the Performance Ratio (PR) defined, according to the IEC 61724 [26], as the ratio between the measured instantaneous power, P_i, (or the measured cumulated energy) and the nominal power of the solar field, P_{nom}, (or the cumulated energy produced at nominal power rate), corrected by taking into account the actual instantaneous irradiance G_i. With respect to the irradiance at STC ($1\,kW/m^2$),

$$PR = \frac{P_i}{P_{nom}} \frac{G_{STC}}{G_i}. \tag{1}$$

The main drawback of adopting (1) as figure of merit is that, as well pointed out in [27], the power yield largely depends on the working temperature. In order to take into account thermal effects, an improved version (2) has been proposed:

$$PR(T) = \frac{P_i}{P_{nom} + \beta \cdot \Delta T} \frac{G_{STC}}{G_i}, \tag{2}$$

where β is the temperature coefficient for the power generated (it is always a negative number) and ΔT is the temperature increment with respect to 25°C.

Usually, if PR is lower than 1, the solar system is underperforming. However, the adoption of P_{nom} in both (1) and (2) does not take into account numerous factors leading to a deviation of the actual performance of the solar field from the nominal target, even though all its components operate correctly. In order to overcome this issue, [27] proposes an improvement of both (1) and (2) by replacing the nominal power with a reference power provided by a detailed model of the solar field. The model works with the same environmental conditions of the real system but with ideal solar panels, thus defining a "relative error" as figure of merit.

The limit of complex models, like those presented in [27], is that their effectiveness is based on the reliability of the model and the capability of extracting from its suitable parameters.

An opposite view is presented in [28], as the model used to predict the PV system electricity production has a low complexity.

For all the cases, the expected ac power P_{ac}, adopted in place of P_{nom} in (1), is evaluated as

$$P_{ac} = G_i \left(a_1 + a_2 G_i + a_3 \log(G_i) \right) \left(1 + a_4 (\Delta T) \right), \tag{3}$$

where a_1, a_2, a_3, and a_4 are fitting parameters.

The method proposed in [29] is also very simple. The *sophisticated verification (SV) method* has the following expression for the PR:

$$PR(T) = \frac{P_i}{P_{nom}(G_i/G_{STC})} = \frac{P_{nom}(G_i/G_{STC}) - L}{P_{nom}(G_i/G_{STC})}, \tag{4}$$

where L is a loss term (explicated in [29]) which includes various effects (temperature, shading mismatches, etc.). The measured energy is then plotted as a function of the irradiance and, according to (4), any deviation from the straight line is attributed to some form of malfunctioning.

A similar approach is also proposed in [30] with reference to a solar field with thin film panels. In this paper array losses are defined as

$$L_{\text{Array}} = P_{\text{nom}} \frac{G_i}{G_{\text{STC}}} - V_{\text{pv}} \cdot I_{\text{pv}}, \quad (5)$$

where V_{pv} and I_{pv} are the measured operating voltage and current of the solar field, respectively.

Once a model for evaluating losses has been selected, another issue common to many papers is data transmission, with the main options being WiFi, GSM, and Power Line Communication (PLC). In [31] it is observed that a drawback of GSM [32] is its high operating cost, as the user needs to pay for the data transmission service; thus, the ZigBee protocol is proposed. In [33] wireless communication is exploited to monitor the effect of dust in solar field installed in the desert. Methods based on the monitoring at the array level have the drawback of not being suitable for locating faulted components. This is a very important issue, because maintenance costs strongly depend on the ability to undertake focused interventions. Therefore, the number of monitored parameters needs to be necessarily increased and move towards a higher LoG. A commonly adopted solution is to use the same figures of merit but refer to a single string rather than the entire system. Actually, the availability of string electrical performance can be also exploited to skip the need for weather information. The comparison among the strings allows the direct identification of the faulty ones. This approach is well illustrated in [34], where currents between strings are compared. In that paper, it was observed that fault detection is difficult when failures occur in multiple strings; the disambiguation is carried out by combining the figures of merit with the evaluation of the standard Performance Ratio.

The definition of a more suitable figure of merit can be found in [35], where an inferential tool, returning information about the operation of the PV field, is presented. After initial training, the software defines one or more reference strings that are used in place of the nominal power for the definition of the expected Performance Ratio. A simple method for defining a reference string was presented in [36, 37]. In those papers the instantaneous power generated by the best performing string in a large solar field is assumed as the target for all other strings with the same orientation. This approach has the advantage of being absolutely independent on weather conditions, irradiance, and temperature and does not require any training of the software. Moreover it allows a fast localization of faulty strings and a reliable estimation of energy losses attributable to each string.

A different approach for analyzing string data is proposed in [38], where a given dataset of observed string currents and voltages and their respective low-pass-filtered time derivatives are analyzed a posteriori for the determination of the probabilities of a restricted set of possible fault that could have caused that dataset.

All the techniques listed so far are based on the comparison of the instantaneous power with a set yield target. An alternative method, which can be found in [39], proposes plotting the whole I-V curve of a single string using information from the inverter. Indeed, the inverter control needs to measure instantaneous voltage and current in order to track the maximum power point. Moreover, as pointed out in [40], some commercial inverters carry out a periodic scan of the entire I-V curve in order to distinguish the global maximum from local ones, in case of mismatch among the modules.

The measurement of the whole I-V curves of single strings, compared with a tailored model, is also proposed in [41] to recognize six categories of faults, including shadow effects, bypass diode fault, cell fault, module fault, and so on.

A possible issue, which is often misrecognized when dealing with string level monitoring techniques, is the possible occurrence of reverse currents in parallel connected string. Paper [42] shows the ineffectiveness of usual rules (3σ) used to recognize underperforming strings, since the current of faulty string always lies between the upper and lower bound of the 3σ rule. In order to overcome this problem, a machine-learning local outlier factor (LOF) is defined. This factor provides a quantitative approach to identify the faulty strings (called outliers) by defining a density-based outlier detection rule. This rule is based on the fact that the density around an outlier is significantly different from the density of its neighbors.

An improved fault location capability can be attained by pushing the monitoring at the individual solar panel level. It is obvious that in this case a pervasive sensor network is needed, so that the cost of the system can be justified by higher revenues coming from most effective maintenance strategies. The main issues to be faced when a single panel monitoring system is adopted are the power supply of sensors, data communication, and data management [43], as their effectiveness depends on the monitored parameters. The best performing option [44] consists in measuring the entire I-V curve of each solar panel. This solution is relatively simple to implement in distributed conversion systems [45–49], where each solar panel has its own dc/dc converter that can be properly controlled to plot or estimate the I-V characteristic [50]; other measurements are much more complicated, because they would require the temporary disconnection of individual solar panels from the string.

A more widely adopted solution consists in the measurement of the operating voltage and the operating current of the solar panel to calculate the instantaneous power generated. This approach requires reduced hardware for sensing electrical parameters, while it could be demanding for the power supply, depending on both the adopted communication system and the sampling rate of measurements. For example, [51] proposes GSM, but no details are given about power supply. A possible alternative to GSM is PLC, which exploits the existing dc wiring; in this case it is essential to avoid the fact that signals travel through the solar panels (that are series connected with the dc power cable). To this end, a bypass low impedance path must be provided. Reference [52] proposes connecting a capacitor in parallel with each solar panel, while [53, 54] propose a more effective LC filter. PLC is also suggested in [55] to make a fire proof protection system. A heartbeat is sent through the power line to a microcontroller mounted on each solar panel. If the heartbeat is lost, either because the power line is opened on purpose or broken by fire, the microcontroller trips a series switch to open the circuit and a parallel switch to short circuit the solar panel.

The main drawback of a single panel monitoring system is that the operating current is the same for all the series connected solar panels and depends on the string operating point, which is fixed by the centralized converter. The consequence is that the operating power cannot be considered as a diagnostic measurement for each individual panel. From this point of view, the most advanced single panel monitoring system is described in [56–59]. In addition to the measurement of operating voltage and current, this system uses both the open circuit voltage, V_{oc}, and short-circuit current, I_{sc}, which are unique to each solar panel. Moreover, V_{oc} is an indirect measurement of the temperature while I_{sc} is directly related to the irradiance, so that no specific sensors are needed for these parameters. In order to carry out the aforementioned measurements, the system physically disconnects the solar panel from the string for about 20 ms, thanks to a series solid state switch. The power of the electronic board is drawn from the solar panel. Since, during the measurement of I_{sc}, the output voltage is zero, a supercapacitor is used as a backup storage to allow the continuous controllability of the circuit. Lastly a cheap WiFi communication system is adopted for data transmission.

A common problem encountered by monitoring systems is the large amount of data to be analyzed. In the next section approaches adopted artificial intelligence and data mining are described.

3. Artificial Intelligence and Data Mining

As PV array characteristics are highly nonlinear, the presence of underperformance and faults within the system can lead to uncorrelated effects. Hence, more sophistication and refinement of the algorithms and methods for fault detection and diagnostic are required. One active research area is on the use of artificial intelligence and data mining, which are primarily based on the concept of a knowledge database. These methods can be split into three categories [60–65]: signal processing methods, classification methods, and inference methods. The main idea of signal processing methods is to extract some features of the measured signals, which can be attributed to a particular state of health of the PV system. The most commonly used methods are wavelet transform techniques [66] and Fast Fourier Transform (FFT) [67]. The classification methods are instead based on artificial intelligence, where knowledge is built from an available dataset. As the amount of labeled data is quite large, supervised learning algorithms can learn the characteristics of the system and make the prediction after training. A number of supervised learning models addressing fault detection and diagnostic in PV systems have been proposed in the literature. For instance, artificial neural networks (ANN) have been proposed for PV systems working under partial shading conditions [68]; for the monitoring and supervision of health status of a PV system in [69]; and for short-circuit fault detection of PV arrays in [70]. In other works, Bayesian Neural Network (BNN) and regression polynomial models have been proposed to predict the soiling effects on large-scale PV arrays [71]. Data mining methods for

fault detection and isolation in PV systems are also proposed in the literature such as decision-tree method, K-nearest neighbor, and support vector machine (SVM).

4. Forecasting of Power Production in Photovoltaic Plants

The forecasting of the power generated by a PV plant is a key activity for supporting the monitoring of PV fields [72, 73]. Usually, it makes use of either solar radiation measurements made on module plane or solar radiation data taken from meteorological service providers [74], with the aim being to calculate reference values for energy yield.

Depending on the task required, forecasting techniques can have different timescale: the very short (up to one hour) and short (up to 6 hours) time scales belong to intraday forecasts, while longer forecasts have time scales of one or more days. With reference to the spatial extension, forecasting can be related to a single plant or, for regional models [75, 76], a cluster of plants.

As PV systems are greatly influenced by weather conditions, such as solar irradiance and air temperature, accurate models are required for a reliable prediction of their power generation. Many approaches have been developed over the last decades and a good review of them is presented [73, 77]. Models for forecasting the power generated by a PV plant can be broadly classified into three categories. A first type of techniques is based on Numerical Weather Prediction (NPW) for the forecast of meteorological parameters such as the solar irradiance and the air temperature. These parameters are used as the input of a model of the PV system to forecast the power generated. Another approach is based on statistical modelling of the historical record of the power generated. This approach includes regressive (e.g., autoregressive, AR; autoregressive moving average, ARMA; autoregressive integrated moving average, ARIMA, etc.) and Artificial Intelligence (AI) models (e.g., artificial neural networks, ANN; support vector machine, SVM; adaptive neurofuzzy inference system, ANFIS, etc.). Finally, a third way to forecast the power generated by a PV system combines physical and statistical modelling, called hybrid technique [78–80]. Hybrid techniques are usually applied when some of the data required by physical or statistical methods are missing and can also be used for improving the accuracy of the forecasting activity [73]. The three different approaches have different temporal capability: most of the techniques produce short-term predictions, while NPW-based methods are better suited for long-term predictions of up to 15 days [81].

Among the different approaches presented in the literature, where the main challenge is the design of cost-effective models working for different PV technologies, locations, and working conditions, physical, regressive, and ANN-based models are the most applied techniques accounting for 50% of the reviewed literature [73].

Physical approaches, which represent 11% of the used techniques [73], use models of the PV system to generate the forecasts, while the major research attempts are spent on solar irradiance forecasting [75]. NWP modelling is based on the physical state and on the dynamic motion of atmosphere [75, 77, 82–84]. One of the latest works in this field can be found in [85], where the solar irradiance is forecasted on a day-ahead and intraday basis by means of a model provided by the European Center for Medium Range Wheatear Forecasting (ECMWF model). NPW modelling is also used to calculate the temperature that, together with the solar irradiance, usually is the main input of any physical model for the calculation of the power generated [86].

Statistical modelling does not require any information regarding the system to model and use data to predict the future behavior of the plant [73]. These approaches include a number of different types of time series regression models [87], accounting for 14% of the total forecasting techniques [73]. The mostly used techniques are ARIMA-based models because of their generality [88]. As reported in [89], ARMAX models, which use exogenous inputs, give the best results for this type of modelling.

In a similar way, techniques based on artificial intelligence do not require any information regarding the system and include a number of different approaches: artificial neural networks, fuzzy logic, evolutionary algorithms, expert systems, and others [90]. The most used AI-based techniques use ANN, representing 24% of the total, and they can be classified as follows.

(i) A first type of ANN-based model estimates the power generated by the PV plant starting from the instantaneous working conditions of solar irradiance and temperature [91–93]. The working conditions can come either from sensors mounted on the field or from NWP-based models.

(ii) Other ANN-based models take as an input the current and the past values of the output power [81, 94–96]. These models directly forecast the power output without any additional meteorological parameters.

(iii) A third type of ANN-based models is a combination of the first two types [97–100].

5. Aerial Thermal Analysis

As it is widely known, the degradation of long-term performance and overall reliability of PV plants can drastically reduce expected revenues. It should be considered that medium- and large-size plants are composed by thousands of modules, with each one potentially affected by the following main types of faults:

(i) optical degradation or fault: bubbles, delamination, discoloration of the encapsulant, and front cover (i.e., glass) fracture;

(ii) electrical mismatches: cell cracks/fractures, breakage of interconnection ribbons, poor soldering, snail tracks, shunts, shading;

(iii) nonclassified: potential induced degradation (PID), defective/short-circuited bypass diodes, short-circuited modules or strings, and junction box failure.

Standard monitoring approaches, that is, electrical string monitoring, only ensure power losses detection in a portion of the PV field, while the accurate localization of faulty modules requires strings disassembling, visual inspection, and/or electrical and thermographic analysis. Unfortunately the above-mentioned techniques are time demanding, cause undesired stops of the energy generation, and often require laboratory instrumentation, thus resulting in cost effectiveness only in case of catastrophic faults. Moreover, it should be noted that PV plants are often located in inaccessible places, for example, rooftops, thus making any intervention dangerous. As a consequence, the safety of operation deeply impacts on the maintenance costs.

The introduction of diagnostic techniques provides on one hand rapid detection and effective classification over a large number of faults, but on the other hand they limit monitoring and diagnostics costs. This requires in the majority of PV fields cost-effective O&M.

In the recent literature [19–25, 37, 101–106], a new nondestructive diagnostic approach uses unmanned aerial systems (UAS) equipped with thermal and/or visual cameras to inspect PV fields and automatic tools for image processing and fault detection and classification. The main challenges of this approach are

(a) positioning;

(b) individual module identification;

(c) defect detection;

(d) defect classification.

The critical aspect of PV module automatic identification in infrared images (point (b)) has been studied in [107]. Unfortunately, the small 5% error obtained by means of manual camera increases to 30% when a drone carries out the measurement from a flying altitude of 20 m.

An automatic defects detection and classification procedure is proposed in [108] for a cell-level analysis. First, the variance and the mean value of the temperature of each pixel of the photograph are calculated for each PV cell. Nonuniform cells, that is, cells exhibiting a large variance in their temperature distribution, are discarded and separately analyzed. Uniform cells are classified into light, medium, and strong hot spot according to their mean temperature. Subsequently, hot cells are classified as a function of the most common defects.

A simple and effective method to identify the frames of PV modules from thermal images is proposed in [109]. This approach relies on the assumption that the solar cells have temperature higher than the metallic frame and there is a sharp transition between the two regions. Moreover, the proposed procedure is capable of classifying the defects by means of the thermal gradient analysis carried out at cell level. The obtained results are promising, but the applicability of aerial inspections is still limited by strict requirements in terms of thermal camera resolution and/or low flight height.

Even though defects in solar panels often cause an increase of the surface temperature, sometimes the poor resolution of the thermal cameras hinders an accurate defect classification. In [20], a double stage procedure is proposed:

in the first step UAV (in the following also referred to as drone) equipped with a thermal camera provides a preliminary thermal analysis to detect and classify large-size defects, while a subsequent visual inspection with a second drone, equipped by a HD photo camera, provides small-size defects classification. In particular, the first stage detects defective points in the PV field and classifies them depending on their shape and location. The following faults can be identified: (a) interconnection issues, that is, entire module warming; (b) defective bypass diodes, internal short circuits, cell mismatch, and snail trails, that is, isolated hot spots or "patchwork pattern"; (c) partial shadowing and cracks, that is, hot spots and/or polygonal patches. Additional information is obtained by reducing the distance between PV arrays and the UAV, thus allowing a more accurate visual analysis. Indeed, visual images can validate the detection and the classification of defects and failures like browning, bubbles, cracked cell, burning, corrosion, cell or module breakages, white spots, snail tails, discoloration, broken interconnections, solder bond failures, and dirty points.

Nevertheless, some critical aspects have to be addressed for an accurate aerial visual inspection, as reported in [21]:

(i) the vertical photography (axis of the camera perpendicular to the ground) should produce a sort of a map of the PV field, where the objects are slightly affected by perspective issues;

(ii) overlap between two consecutive pictures has to be ensured;

(iii) a given flight height must be respected depending on the specific faults to be detect;

(iv) the stability of the drone has to be ensured by flying without wind (wind speed < 3 m/s) and sunny, cloudless, and clear sky;

(v) there is a need to find suitable flight trajectory to minimize the reflection of objects located near the modules and the sunlight.

In [22, 104] images acquired by a light UAV produced an IRT map, that, is thermal orthophotoplan, of the investigated PV installation by means of aerotriangulation methods. In particular, both photogrammetry techniques and global positioning system (GPS) receivers are employed to ensure correct positioning, while an image postprocessing procedure based on Canny edge approach allows highlighting hot spot of photovoltaic modules. Unfortunately no automatic classification tool is suggested, but auxiliary diagnostic measurements (e.g., IRT, I-V characterization, and EL) validate fault detection and qualitatively classify the analyzed results into a specific fault type, corresponding to a specific thermal pattern, I-V characteristic, and EL pattern. Automated diagnostic tools based on aerial thermal analysis are proposed in [23–25, 37].

The system described in [23, 24] uses a three-step procedure: (i) undertaking a raw preliminary defects detection; (ii) selecting faulty modules from the thermal image according to health index; (iii) carrying out accurate defects detection and classification. In the first step the thermal image is converted

into grayscale and digital filters (namely, rectangular average, rectangular ideal, and Gaussian filters) are applied to the frequency domain. Subsequently, a high-pass filter evidences the hot area in the panel. According to the assumption that a hot area suggests a fault, all the panels showing a high percentage of hot area with respect to the global one are selected for further analyses and their frames are accurately extracted by means of a Laplace filter. In the third step, the Decision Support Center evaluates the defect and failure type and proposes the best solution for the specific plant by comparing actual performance and its monitored history. Nevertheless, there are still no studies tackling the accurate description of the algorithms executed by the Decision Support Center.

An effective statistical data-driven approach is adopted in [25], where the identification of individual modules consists of the following steps: (1) normalization, (2) thresholding, (3) orientation estimation of the photovoltaic modules, and finally (4) correction and refinement. Moreover, in the proposed pipeline, all data corresponding to the detected photovoltaic modules within an infrared image are processed to obtain four sets of features. Suited statistical test highlights outliers, thus suggesting temperature abnormalities caused by module defects. Then, major temperature abnormalities are classified accurately into three main groups: overheated modules, hot spots, and overheated substrings. The method reaches high accuracy level, but the classification is still poor and generic. Table 1 reports information regarding drones and cameras adopted to implement the techniques discussed above.

Results obtained in [23–25, 37] suggest that drone-assisted diagnostic is going to achieve an important role in O&M of PV plants thanks to its effectiveness in terms of detection and localization. Moreover, even though these techniques still require high resolution cameras often costly and heavy, as indicated in Table 1, today the market has on offer a growing number of light electric drones with high payload, sophisticated navigation systems mainly based on GPS receivers, and extended flight time. Nevertheless, an accurate classification of defects is still a challenge.

6. Converter Reliability

Among possible faults in photovoltaic systems, those associated with dc/ac power converters are the most dramatic. In fact, this occurrence completely stops the energy generation. Although a malfunction of the dc/ac power converter is easy to detect (even though reference data from inverter manufacturers might be needed), it is not so easy to fix, as it often requires the location and the replacement of damaged devices inside the case. Therefore, the challenge is to prevent failures by estimating the residual lifetime (RLT) of the power modules. RLT algorithms are different in terms of type of applications and module characteristics. In the following, different methods to estimate the expected RLT of IGBT modules based on accelerated life tests are summarized. First, temperature cycling tests are introduced to correlate RLT to temperature stresses. Then, methods for the estimation of the

TABLE 1: Drones and cameras employed in aerial inspections.

| Reference | Drone | | Camera | | | | | | |
	Model	Propulsion	Type	Pixels	Range	Temperature range	Accuracy	Thermal sensitivity	Weight
[19]	dji S1000	Electric	GoPRo Hero 3	1920 × 1080	—	—	—	—	76 g
			Optris PI 450 + recorder	382 × 288	7.5–13 μm	−20°C to 100°C 0°C to 250°C 150°C to 900°C	±2°C or ±2%.	0.040 K	380 g
[20]	Nimbus EosXi	Gasoline	ThermoteknixMicroCAM 640	640 × 480	8–12 μm	—	—	0.060 K	74 g
	Nimbus PLP-610	Electric	Nikon1 V1 HD	3906 × 2606	—	—	—	—	383 g
[21]	Nimbus PLP-610	Electric	Nikon1 V1 HD	3906 × 2606	—	—	—	—	383 g
[22]	Condor AY 704	Electric	Optris PI 450	382 × 288	7.5–13 μm	−20°C to 100°C 0°C to 250°C 150°C to 900°C	±2°C or ±2%.	0.040 K	320 g
[23] [24]	Nimbus PLP-610	Electric	Flir A35	320 × 256		−40°C to 160°C −40°C to 550°C	±5°C or ±5%	0.05 K	200 g
[25]	DaVinci Copters ScaraBot X8	Electric	GoPRo Hero 3+	1920 × 1080	—	—	—	—	76 g
			Optris PI 450	382 × 288	7.5–13 μm	−20°C to 100°C 0°C to 250°C 150°C to 900°C	±2°C or ±2%.	0.04 K	320 g

junction temperature (JT) and the temperature humidity bias (THB) tests are presented.

Power modules consist of materials with different thermal expansion coefficients and are subjected to temperature swings due to the variability of the load [110–112]. These stresses lead to a degradation of the module integrity and development of faults, such as heal cracking [113], bond wire lift off (BWLO) [114], and solder fatigue [115]. Corrosion has received recently more attention as IGBT modules are packaged in plastic cases that do not normally offer sufficient moisture resistance.

Due to temperature swings and moisture penetrations, the actual RLT of IGBT modules can be significantly different from the manufacturer's predictions. Thus, these factors should be taken into account to correct online the expected RTL. Condition monitoring systems (CMS) can give an important contribution to this problem, minimizing the risk of failure of IGBTs. CMS gather real-time data on temperature swings and moisture penetration during the operations of converters and use dedicated algorithms to correlate the measurements and update the prediction. These algorithms are based on accelerated life tests, where overstress conditions (high temperature, high temperature cycling, high power cycling, humidity, etc.) are applied to the module to understand in short time the effects of these stresses on the external characteristics and parameters of IGBTs. The CMS will then monitor the stresses in normal conditions and calculate the associated residual lifetime in the correct time scale using the acceleration factor used for the test.

Temperature cycling tests (TCT) refer to the power cycling of IGBT modules at high and low temperatures when the modules are on-state and off-state, respectively. During these tests, both the temperature difference ΔT_j of the junction and the mean value of the junction temperature (JT) need to be collected. The RLT model is obtained from the Palmgren-Miner rule [116]. In this model, the lifetime consumption (LC) is calculated as the ratio of total cycle numbers and number of cycles to failure. The former is obtained from a counting algorithm, like the Rain Flow [117]; the latter is calculated from the Coffin-Manson law [118], which physically models each fault. The RLT is finally estimated by time history (time duration when fault occur) over LC. LCs equal to one or greater than one can be symptoms of being close to failure or occurrence of failure, respectively. The most challenging issue is that TCT rely on JT. Measurement of JT is significantly difficult in a direct way, because of the difficulty in accessing the junction of the module [119]. Different alternative methods have been investigated by researchers to estimate JT, as explained hereinafter.

The JT can be directly measured by temperature sensors [120] and IR cameras [121]. However, due to the slow response of temperature sensors and dependency on the point of installation, the direct method is not recommended. Moreover IR cameras allow only average measurements of relatively large areas of the module and are also not easy to calibrate for the entire range of temperature variations.

The estimation of the JT is preferred to the direct measurement, as it does not require the modification of the internal structure of the IGBT [7]. The current estimation methods of JT are based on the measurement and analysis of thermal sensitive electrical parameters (TSEPs). These can be used to obtain the static (e.g., on-state collector-emitter voltage and collector current) and dynamic (e.g., gate parameters and turn-on and turn-off delay times) characteristics of the IGBT under examination. The improvement of models and algorithms for the analysis of indirect measurements of JT is now bringing the indirect methods to the same accuracy levels of their direct counterparts, even when a detection of JT at fast sampling rate is required.

Kuhn and Mertens [122] and Brown et al. [123] have demonstrated that turn-on and turn-off delay times are positively correlated to JT variations. However, the measurement of these times is challenging for the high sampling rate required to achieve a good accuracy. A second negative aspect of this method is that an increase of the off-state to on-state time (t_{on}) and the on-state to off-state time (t_{off}) can also be caused by a degradation of the gate-oxide characteristics, which increases the gate-emitter voltage (V_{GE}).

Denk and Bakran [124] and Baker et al. [125] showed that the internal gate resistance (R_{Gi}) of IGBTs has a negative correlation with JT, so that the value of gate current changes and, hence, the voltage across the external gate resistance linearly increase by a variation of JT. Variation of this voltage has been considered as a TSEP, useful to estimate the JT. The value of the proposed TSEP is accurately measured when the gate voltage is negative; that is, the IGBT is off. However, this method requires an external high frequency sinusoidal voltage signal that has to be superimposed on the negative gate voltage during the off-state time, using auxiliary MOSFET.

Barlini et al. [118] showed that the rates of change of the collector-emitter voltage (V_{CE}) and the collector current (I_C) are linearly related to the JT. However, this relation has been verified only for MOSFETs and it is of difficult practical implementation, because it requires the measurement of time derivatives with high sampling rates.

The measurement of the static V_{CE} and I_C characteristic of IGBTs has been investigated as another method to estimate the JT [126]. However, this method is not recommended, as the estimated JT is affected by BWLO and the method becomes unreliable if this fault occurs.

As mentioned above, RLT of IGBT modules can be adversely affected by moisture, since the plastic package is not hermetically sealed. Moisture can lead to corrosion of the aluminum inside the module and, hence, faults. Zorn and Kaminski [127] showed that the increase of moisture causes a decrease of avalanche voltage and, above a certain level of humidity, a surge in the leakage current. This leads to temperature stress, especially at the junction the IGBT. Therefore, moisture can significantly decrease RLT of IGBT modules.

The correlation between moisture levels and RLT can be obtained from THB tests. Generally, a THB test is conducted in relative humidity of 85% and 85°C to assess the moisture resistance of the IGBT package. The test voltage applied to the IGBT is used to regulate the leakage current. This voltage should be carefully selected, because if the leakage current heats the module, the moisture evaporates and the degradation effect of humidity becomes less evident [128]. As shown by Zorn and Kaminski [129], the problem is particularly delicate when the test voltage is above the bias voltage (typically 90% of the nominal voltage), because the leakage current noticeably increases and makes the degradation effect more observable, albeit the higher temperature accelerates the evaporation of moisture.

7. Conclusions

This paper has presented a literature survey on reliability issues of photovoltaic fields. The main aspects of the subject have been covered by reviewing papers dealing with data acquisition, data management, and modelling. Tradeoffs among high sensitivity, pervasiveness, hardware requirements, effectiveness, and costs have been pointed out. The abundance of high quality works is an indicator of the relevance of the problem for the scientific community; nevertheless it also evidenced that many issues are still debated and need a more in-depth investigation.

Competing Interests

The authors declare that they have no competing interests.

Acknowledgments

Dr. A. Mellit would like to thank the ICTP, Trieste (Italy), for providing the materials and the computers facilities.

References

[1] S. E. Forman, "Performance of experimental terrestrial photovoltaic modules," *IEEE Transactions on Reliability*, vol. 31, no. 3, pp. 235–245, 1982.

[2] C. Baltus, J. Eikelboom, and R. Van Zolingen, "Analytical monitoring of losses in pv systems," in *Proceedings of the 14th European Photovoltaic Solar Energy Conference*, Barcelona, Spain, July 1997.

[3] D. L. King, W. E. Boyson, and J. A. Kratochvil, "Analysis of factors influencing the annual energy production of photovoltaic systems," in *Proceedings of the 29th IEEE Photovoltaic Specialists Conference*, New Orleans, La, USA, May 2002.

[4] G. Petrone, G. Spagnuolo, R. Teodorescu, M. Veerachary, and M. Vitelli, "Reliability issues in photovoltaic power processing systems," *IEEE Transactions on Industrial Electronics*, vol. 55, no. 7, pp. 2569–2580, 2008.

[5] A. Chouder and S. Silvestre, "Analysis model of mismatch power losses in pv systems," *Journal of Solar Energy Engineering, Transactions of the ASME*, vol. 131, no. 2, Article ID 024504, 5 pages, 2009.

[6] M. A. Quintana, D. L. King, T. J. McMahon, and C. R. Osterwald, "Commonly observed degradation in field-aged photovoltaic modules," in *Proceedings of the 29th IEEE Photovoltaic Specialists Conference*, pp. 1436–1439, May 2002.

[7] P. Guerriero, F. Di Napoli, V. D'Alessandro, and S. Daliento, "Accurate maximum power tracking in photovoltaic systems affected by partial shading," *International Journal of Photoenergy*, vol. 2015, Article ID 824832, 10 pages, 2015.

[8] G. Cipriani, V. Di Dio, L. P. Di Noia et al., "A PV plant simulator for testing MPPT techniques," in *Proceedings of the 4th International Conference on Clean Electrical Power: Renewable Energy Resources Impact (ICCEP '13)*, pp. 483–489, Alghero, Italy, June 2013.

[9] G. Brando, A. Dannier, and R. Rizzo, "A sensorless control of H-bridge multilevel converter for maximum power point tracking in grid connected photovoltaic systems," in *Proceedings of the International Conference on Clean Electrical Power (ICCEP '07)*, May 2007.

[10] M. A. Munoz, M. C. Alonso-García, N. Vela, and F. Chenlo, "Early degradation of silicon PV modules and guaranty conditions," *Solar Energy*, vol. 85, no. 9, pp. 2264–2274, 2011.

[11] S. Daliento and L. Lancellotti, "3D analysis of the performances degradation caused by series resistance in concentrator solar cells," *Solar Energy*, vol. 84, no. 1, pp. 44–50, 2010.

[12] P. Guerriero, V. D'Alessandro, L. Petrazzuoli, G. Vallone, and S. Daliento, "Effective real-time performance monitoring and diagnostics of individual panels in PV plants," in *Proceedings of the 4th International Conference on Clean Electrical Power: Renewable Energy Resources Impact (ICCEP '13)*, pp. 14–19, Alghero, Italy, June 2013.

[13] D. Stellbogen, "Use of PV circuit simulation for fault detection in PV array fields," in *Proceedings of the 23rd IEEE Photovoltaic Specialists Conference*, pp. 1302–1307, May 1993.

[14] H. Haeberlin and C. Beutler, "Normalized representation of energy and power for analysis of performance and on-line error detection in PV-systems," in *Proceedings of the 13th EU PV Conference on Photovoltaic Solar Energy Conversion*, Nice, France, 1995.

[15] A. Drews, A. C. de Keizer, H. G. Beyer et al., "Monitoring and remote failure detection of grid-connected PV systems based on satellite observations," *Solar Energy*, vol. 81, no. 4, pp. 548–564, 2007.

[16] S. Silvestre, M. A. D. Silva, A. Chouder, D. Guasch, and E. Karatepe, "New procedure for fault detection in grid connected PV systems based on the evaluation of current and voltage indicators," *Energy Conversion and Management*, vol. 86, pp. 241–249, 2014.

[17] G. T. Klise and J. S. Stein, "Mode used to assess the performance of photovoltaic systems," Tech. Rep., Sandia National Laboratories, 2009.

[18] A. Chouder and S. Silvestre, "Automatic supervision and fault detection of PV systems based on power losses analysis," *Energy Conversion and Management*, vol. 51, no. 10, pp. 1929–1937, 2010.

[19] U. Muntwyler, E. Schüpbach, and M. Lanz, "Infrared (IR) drone for quick and cheap PV inspection," in *Proceedings of the 31st European Photovoltaic Solar Energy Conference and Exhibition*, 5CO.15.6, pp. 1804–1806, September 2015.

[20] P. B. Quater, F. Grimaccia, S. Leva, M. Mussetta, and M. Aghaei, "Light Unmanned Aerial Vehicles (UAVs) for cooperative inspection of PV plants," *IEEE Journal of Photovoltaics*, vol. 4, no. 4, pp. 1107–1113, 2014.

[21] S. Leva, M. Aghaei, and F. Grimaccia, "PV power plant inspection by UAS: correlation between altitude and detection of defects on PV modules," in *Proceedings of the 15th IEEE International Conference on Environment and Electrical Engineering (EEEIC '15)*, pp. 1921–1926, IEEE, Rome, Italy, June 2015.

[22] J. A. Tsanakas, D. Chrysostomou, P. N. Botsaris, and A. Gasteratos, "Fault diagnosis of photovoltaic modules through image processing and Canny edge detection on field thermographic measurements," *International Journal of Sustainable Energy*, vol. 34, no. 6, pp. 351–372, 2015.

[23] M. Aghaei, F. Grimaccia, C. A. Gonano, and S. Leva, "Innovative automated control system for PV fields inspection and remote control," *IEEE Transactions on Industrial Electronics*, vol. 62, no. 11, pp. 7287–7296, 2015.

[24] M. Aghaei, A. Gandelli, F. Grimaccia, S. Leva, and R. E. Zich, "IR real-time analyses for PV system monitoring by digital image processing techniques," in *Proceedings of the 1st International Conference on Event-Based Control, Communication and Signal Processing (EBCCSP '15)*, pp. 1–6, Krakow, Poland, June 2015.

[25] S. Dotenco, M. Dalsass, L. Winkler et al., "Automatic detection and analysis of photovoltaic modules in aerial infrared imagery," in *Proceedings of the IEEE Winter Conference on Applications of Computer Vision (WACV '16)*, pp. 1–9, New York, NY, USA, March 2016.

[26] IEC 61724, https://webstore.iec.ch/preview/info_iec61724% IEC 61724, 7Bed1.0%7Den.pdf.

[27] F. Bizzarri, A. Brambilla, L. Caretta, and C. Guardiani, "Monitoring performance and efficiency of photovoltaic parks," *Renewable Energy*, vol. 78, pp. 314–321, 2015.

[28] R. Platon, J. Martel, N. Woodruff, and T. Y. Chau, "Online fault detection in PV systems," *IEEE Transactions on Sustainable Energy*, vol. 6, no. 4, pp. 1200–1207, 2015.

[29] A. Tahri, T. Oozeki, and A. Draou, "Monitoring and evaluation of photovoltaic system," *Energy Procedia*, vol. 42, pp. 456–464, 2013.

[30] R. Hariharan, M. Chakkarapani, G. Saravana Ilango, and C. Nagamani, "A method to detect photovoltaic array faults and partial shading in PV systems," *IEEE Journal of Photovoltaics*, vol. 6, no. 5, pp. 1278–1285, 2016.

[31] F. Shariff, N. A. Rahim, and W. P. Hew, "Zigbee-based data acquisition system for online monitoring of grid-connected photovoltaic system," *Expert Systems with Applications*, vol. 42, no. 3, pp. 1730–1742, 2015.

[32] M. Gagliarducci, D. A. Lampasi, and L. Podestà, "GSM-based monitoring and control of photovoltaic power generation," *Measurement*, vol. 40, no. 3, pp. 314–321, 2007.

[33] F. Touati, M. A. Al-Hitmi, N. A. Chowdhury, J. A. Hamad, and A. J. R. San Pedro Gonzales, "Investigation of solar PV performance under Doha weather using a customized measurement and monitoring system," *Renewable Energy*, vol. 89, pp. 564–577, 2016.

[34] M. Baba, T. Shimakage, and N. Takeuchi, "Examination of fault detection technique in PV systems," in *Proceedings of the 35th International Telecommunications Energy Conference 'Smart Power and Efficiency' (INTELEC '13)*, pp. 431–434, Hamburg, Germany, 2013.

[35] L. Cristaldi, M. Faifer, G. Leone, and S. Vergura, "Reference strings for statistical monitoring of the energy performance of photovoltaic fields," in *Proceedings of the 5th International Conference on Clean Electrical Power (ICCEP '15)*, pp. 591–596, Taormina, Italy, June 2015.

[36] P. Guerriero, G. Vallone, M. Primato et al., "A wireless sensor network for the monitoring of large PV plants," in *Proceedings of the 2014 International Symposium on Power Electronics, Electrical Drives, Automation and Motion (SPEEDAM '14)*, pp. 960–965, Sorrento, Italy, June 2014.

[37] P. Guerriero, F. Di Napoli, and S. Daliento, "Real time monitoring of solar fields with cost/revenue analysis of fault fixing," in *Proceedings of the IEEE 16th International Conference on*

Environment and Electrical Engineering (EEEIC '16), Florence, Italy, June 2016.

[38] S. Ben-Menahem and S. C. Yang, "Online photovoltaic array hot-spot Bayesian diagnostics from streaming string-level electric data," in *Proceedings of the 38th IEEE Photovoltaic Specialists Conference (PVSC '12)*, pp. 2432–2437, June 2012.

[39] M. Davarifar, A. Rabhi, A. Hajjaji, E. Kamal, and Z. Daneshifar, "Partial shading fault diagnosis in PV system with discrete wavelet transform (DWT)," in *Proceedings of the 3rd International Conference on Renewable Energy Research and Applications (ICRERA '14)*, pp. 810–814, Milwaukee, Wis, USA, October 2014.

[40] S. Spataru, D. Sera, T. Kerekes, and R. Teodorescu, "Diagnostic method for photovoltaic systems based on light I–V measurements," *Solar Energy*, vol. 119, pp. 29–44, 2015.

[41] W. Chine, A. Mellit, A. M. Pavan, and V. Lughi, "Fault diagnosis in photovoltaic arrays," in *Proceedings of the 5th International Conference on Clean Electrical Power (ICCEP '15)*, pp. 67–72, Taormina, Italy, June 2015.

[42] Y. Zhao, F. Balboni, T. Arnaud, J. Mosesian, R. Ball, and B. Lehman, "Fault experiments in a commercial-scale PV laboratory and fault detection using local outlier factor," in *Proceedings of the 40th IEEE Photovoltaic Specialist Conference (PVSC '14)*, pp. 3398–3403, Denver, Colo, USA, June 2014.

[43] T. Hu, M. Zheng, J. Tan, L. Zhu, and W. Miao, "Intelligent photovoltaic monitoring based on solar irradiance big data and wireless sensor networks," *Ad Hoc Networks*, vol. 35, pp. 127–136, 2015.

[44] P. Papageorgas, D. Piromalisb, T. Valavanis, S. Kambasis, T. Iliopoulou, and G. Vokas, "A low-cost and fast PV I-V curve tracer based on an open source platform with M2M communication capabilities for preventive monitoring," *Energy Procedia*, vol. 74, pp. 423–438, 2015.

[45] M. Coppola, P. Guerriero, F. Di Napoli, S. Daliento, D. Lauria, and A. Del Pizzo, "A PV AC-module based on coupled-inductors boost DC/AC converter," in *Proceedings of the International Symposium on Power Electronics, Electrical Drives, Automation and Motion (SPEEDAM '14)*, pp. 1015–1020, June 2014.

[46] M. Coppola, P. Guerriero, F. Di Napoli et al., "Modulation technique for grid-tied PV multilevel inverter," in *Proceedings of the International Symposium on Power Electronics, Electrical Drives, Automation and Motion (SPEEDAM '16)*, pp. 923–928, 2016.

[47] P. Guerriero, M. Coppola, F. Di Napoli et al., "Three-phase PV CHB inverter for a distributed power generation system," *Applied Sciences*, vol. 6, no. 10, article 287, 2016.

[48] M. Coppola, F. D. Napoli, P. Guerriero et al., "Maximum power point tracking algorithm for gridtied photovoltaic cascaded hbridge inverter," *Electric Power Components and Systems*, vol. 43, no. 8–10, pp. 951–963, 2015.

[49] M. Coppola, S. Daliento, P. Guerriero, D. Lauria, and E. Napoli, "On the design and the control of a coupled-inductors boost dc-ac converter for an individual PV panel," in *Proceedings of the 21st International Symposium on Power Electronics, Electrical Drives, Automation and Motion (SPEEDAM '12)*, pp. 1154–1159, Sorrento, Italy, June 2012.

[50] B. Ando, S. Baglio, A. Pistorio, G. M. Tina, and C. Ventura, "Sentinella: smart monitoring of photovoltaic systems at panel level," *IEEE Transactions on Instrumentation and Measurement*, vol. 64, no. 8, pp. 2188–2199, 2015.

[51] F. Shariff, N. A. Rahim, and H. W. Ping, "Photovoltaic remote monitoring system based on GSM," in *Proceedings of the IEEE Conference on Clean Energy and Technology (CEAT '13)*, pp. 379–383, November 2013.

[52] F. J. Sánchez-Pacheco, P. J. Sotorrío-Ruiz, J. R. Heredia-Larrubia, F. Pérez-Hidalgo, and M. S. De Cardona, "PLC-based PV plants smart monitoring system: field measurements and uncertainty estimation," *IEEE Transactions on Instrumentation and Measurement*, vol. 63, no. 9, pp. 2215–2222, 2014.

[53] F. Di Napoli, P. Guerriero, V. D'Alessandro, and S. Daliento, "A power line communication on DC bus with photovoltaic strings," in *Proceedings of the 3rd Renewable Power Generation Conference (RPG '14)*, IET, Naples, Italy, September 2014.

[54] J. Han, I. Lee, and K. Sang-Ha, "User friendly Monitoring System for residential PV System Based on Low cost Power Line Communication," *IEEE Transactions on Consumer Electronics*, vol. 61, no. 2, pp. 175–180, 2015.

[55] F. Di Napoli, P. Guerriero, V. d'Alessandro, and S. Daliento, "Single-panel voltage zeroing system for safe access on PV plants," *IEEE Journal of Photovoltaics*, vol. 5, no. 5, pp. 1428–1434, 2015.

[56] P. Guerriero, F. Di Napoli, G. Vallone, V. D'Alessandro, and S. Daliento, "Monitoring and diagnostics of PV plants by a wireless self-powered sensor for individual panels," *IEEE Journal of Photovoltaics*, vol. 6, no. 1, pp. 286–294, 2015.

[57] V. D'Alessandro, P. Guerriero, S. Daliento, and M. Gargiulo, "Accurately extracting the shunt resistance of photovoltaic cells in installed module strings," in *Proceedings of the 3rd International Conference on Clean Electrical Power: Renewable Energy Resources Impact (ICCEP '11)*, pp. 164–168, June 2011.

[58] P. Guerriero, V. D'Alessandro, L. Petrazzuoli, G. Vallone, and S. Daliento, "Effective real-time performance monitoring and diagnostics of individual panels in PV plants," in *Proceedings of the 4th International Conference on Clean Electrical Power: Renewable Energy Resources Impact (ICCEP '13)*, pp. 14–19, June 2013.

[59] M. Gargiulo, P. Guerriero, S. Daliento et al., "A novel wireless self-powered microcontroller-based monitoring circuit for photovoltaic panels in grid-connected systems," in *Proceedings of the 2010 International Symposium on Power Electronics, Electrical Drives, Automation and Motion (SPEEDAM '10)*, pp. 164–168, Pisa, Italy, June 2010.

[60] H. Braun, S. T. Buddha, V. Krishnan et al., "Signal processing for fault detection in photovoltaic arrays," in *Proceedings of the IEEE International Conference on Acoustics, Speech, and Signal Processing (ICASSP '12)*, pp. 1681–1684, March 2012.

[61] C. R. Griesbach, "Fault-tolerant solar array control using digital signal processing for peak power tracking," in *Proceedings of the 31st Intersociety Energy Conversion Engineering Conference (IECEC '96)*, pp. 260–265, Washington, DC, USA, August 1996.

[62] H. Braun, S. T. Buddha, V. Krishnan et al., "Signal processing for solar array monitoring, fault detection, and optimization," *Synthesis Lectures on Power Electronics*, 95 pages, 2012.

[63] Y. Zhao, L. Yang, B. Lehman, J.-F. De Palma, J. Mosesian, and R. Lyons, "Decision tree-based fault detection and classification in solar photovoltaic arrays," in *Proceedings of the 27th Annual IEEE Applied Power Electronics Conference and Exposition (APEC '12)*, pp. 93–99, Orlando, Fla, USA, February 2012.

[64] W. Chine, A. Mellit, V. Lughi, A. Malek, G. Sulligoi, and A. Massi Pavan, "A novel fault diagnosis technique for photovoltaic systems based on artificial neural networks," *Renewable Energy*, vol. 90, pp. 501–512, 2016.

[65] A. Al-Amoudi and L. Zhang, "Application of radial basis function networks for solar-array modelling and maximum power-point prediction," *IEE Proceedings: Generation, Transmission and Distribution*, vol. 147, no. 5, pp. 310–316, 2000.

[66] I.-S. Kim, "On-line fault detection algorithm of a photovoltaic system using wavelet transform," *Solar Energy*, vol. 126, pp. 137–145, 2016.

[67] J. A. Momoh and R. Button, "Design and analysis of aerospace DC arcing faults using fast fourier transformation and artificial neural network," in *Proceedings of the IEEE Power Engineering Society General Meeting*, pp. 788–793, Ontario, Canada, July 2003.

[68] D. D. Nguyen, B. Lehman, and S. Kamarthi, "Performance evaluation of solar photovoltaic arrays including shadow effects using neural network," in *Proceedings of the IEEE Energy Conversion Congress and Exposition (ECCE '09)*, pp. 3357–3362, September 2009.

[69] D. Riley and J. Johnson, "Photovoltaic prognostics and heath management using learning algorithms," in *Proceedings of the 38th IEEE Photovoltaic Specialists Conference (PVSC '12)*, pp. 1535–1539, Austin, Tex, USA, June 2012.

[70] Syafaruddin, E. Karatepe, and T. Hiyama, "Controlling of artificial neural network for fault diagnosis of photovoltaic array," in *Proceedings of the 16th International Conference on Intelligent System Applications to Power Systems (ISAP '11)*, pp. 1–6, IEEE, September 2011.

[71] A. Massi Pavan, A. Mellit, D. De Pieri, and S. A. Kalogirou, "A comparison between BNN and regression polynomial methods for the evaluation of the effect of soiling in large scale photovoltaic plants," *Applied Energy*, vol. 108, pp. 392–401, 2013.

[72] A. Mellit, A. Massi Pavan, and V. Lughi, "Short-term forecasting of power production in a large-scale photovoltaic plant," *Solar Energy*, vol. 105, pp. 401–413, 2014.

[73] J. Antonanzas, N. Osorio, R. Escobar, R. Urraca, F. Martinez-de-Pison, and F. Antonanzas-Torres, "Review of photovoltaic power forecasting," *Solar Energy*, vol. 136, pp. 78–111, 2016.

[74] V. d'Alessandro, F. Di Napoli, P. Guerriero, and S. Daliento, "An automated high-granularity tool for a fast evaluation of the yield of PV plants accounting for shading effects," *Renewable Energy*, vol. 83, pp. 294–304, 2015.

[75] E. Lorenz, T. Scheidsteger, J. Hurka, D. Heinemann, and C. Kurz, "Regional PV power prediction for improved grid integration," *Progress in Photovoltaics: Research and Applications*, vol. 19, no. 7, pp. 757–771, 2011.

[76] J. G. Da Silva Fonseca, T. Oozeki, H. Ohtake, T. Takashima, and K. Ogimoto, "Regional forecasts of photovoltaic power generation according to different data availability scenarios: a study of four methods," *Progress in Photovoltaics: Research and Applications*, vol. 23, no. 10, pp. 1203–1218, 2015.

[77] M. Q. Raza, M. Nadarajah, and C. Ekanayake, "On recent advances in PV output power forecast," *Solar Energy*, vol. 136, pp. 125–144, 2016.

[78] M. Bouzerdoum, A. Mellit, and A. M. Pavan, "A hybrid model (SARIMA-SVM) for short-term power forecasting of a small-scale grid-connected photovoltaic plant," *Solar Energy*, vol. 98, pp. 226–235, 2013.

[79] P. Ramsami and V. Oree, "A hybrid method for forecasting the energy output of photovoltaic systems," *Energy Conversion and Management*, vol. 95, pp. 406–413, 2015.

[80] D. P. Larson, L. Nonnenmacher, and C. F. M. Coimbra, "Day-ahead forecasting of solar power output from photovoltaic plants in the American Southwest," *Renewable Energy*, vol. 91, pp. 11–20, 2016.

[81] M. Ding, L. Wang, and R. Bi, "An ANN-based approach for forecasting the power output of photovoltaic system," *Procedia Environmental Sciences C*, vol. 11, pp. 1308–1315, 2011.

[82] S. Pelland, J. Remund, J. Kleissl, T. Oozeki, and K. De Brabandere, "Photovoltaic and solar forecasting: state of the art," *IEA PVPS Task*, vol. 14, pp. 1–36, 2013.

[83] S. Pelland, G. Galanis, and G. Kallos, "Solar and photovoltaic forecasting through post-processing of the Global Environmental Multiscale numerical weather prediction model," *Progress in Photovoltaics: Research and Applications*, vol. 21, no. 3, pp. 284–296, 2013.

[84] J. Zhang, B.-M. Hodge, J. Simmons et al., "Baseline and target values for PV forecasts: toward improved solar power forecasting," in *Proceedings of the IEEE Power and Energy Society General Meeting (PESGM '15)*, Denver, Colo, USA, July 2015.

[85] P. Lauret, E. Lorenz, and M. David, "Solar forecasting in a challenging insular context," *Atmosphere*, vol. 7, no. 2, p. 18, 2016.

[86] E. Lorenz, J. Kühnert, B. Wolff, A. Hammer, O. Kramer, and D. Heinemann, "PV power predictions on different spatial and temporal scales integrating PV measurements, satellite data and numerical weather predictions," in *Proceedings of the 29th European Photovoltaic Solar Energy Conference and Exhibition (EUPVSEC '14)*, pp. 22–26, Amsterdam, The Netherlands, September 2014.

[87] P. Alan, *Forecasting with Dynamic Regression Models*, John Wiley & Sons, 1991.

[88] G. E. Box, G. M. Jenkins, G. C. Reinsel, and G. M. Ljung, *Time Series Analysis: Forecasting and Control*, John Wiley & Sons, New York, NY, USA, 2015.

[89] Y. Li, Y. Su, and L. Shu, "An ARMAX model for forecasting the power output of a grid connected photovoltaic system," *Renewable Energy*, vol. 66, pp. 78–89, 2014.

[90] A. Mellit and S. A. Kalogirou, "Artificial intelligence techniques for photovoltaic applications: a review," *Progress in Energy and Combustion Science*, vol. 34, no. 5, pp. 574–632, 2008.

[91] L. A. Fernandez-Jimenez, A. Muñoz-Jimenez, A. Falces et al., "Short-term power forecasting system for photovoltaic plants," *Renewable Energy*, vol. 44, pp. 311–317, 2012.

[92] A. Mellit and A. M. Pavan, "A 24-h forecast of solar irradiance using artificial neural network: application for performance prediction of a grid-connected PV plant at Trieste, Italy," *Solar Energy*, vol. 84, no. 5, pp. 807–821, 2010.

[93] F. Almonacid, C. Rus, L. Hontoria, and F. J. Muñoz, "Characterisation of PV CIS module by artificial neural networks. A comparative study with other methods," *Renewable Energy*, vol. 35, no. 5, pp. 973–980, 2010.

[94] E. Izgi, A. Öztopal, B. Yerli, M. K. Kaymak, and A. D. Şahin, "Short-mid-term solar power prediction by using artificial neural networks," *Solar Energy*, vol. 86, no. 2, pp. 725–733, 2012.

[95] P. Mandal, S. T. S. Madhira, J. Meng, and R. L. Pineda, "Forecasting power output of solar photovoltaic system using wavelet transform and artificial intelligence techniques," *Procedia Computer Science*, vol. 12, pp. 332–337, 2012.

[96] A. Mellit and S. Shaari, "Recurrent neural network-based forecasting of the daily electricity generation of a Photovoltaic power system," in *Proceedings of the Ecological Vehicle and Renewable Energy (EVER '09)*, pp. 26–29, Monte-Carlo, Monaco, March 2009.

[97] T. T. Teo, T. Logenthiran, and W. L. Woo, "Forecasting of photovoltaic power using extreme learning machine," in *Proceedings of the IEEE Innovative Smart Grid Technologies—Asia (ISGT ASIA '15)*, November 2015.

[98] C. Chen, S. Duan, T. Cai, and B. Liu, "Online 24-h solar power forecasting based on weather type classification using artificial neural network," *Solar Energy*, vol. 85, no. 11, pp. 2856–2870, 2011.

[99] S. Daliento, L. Mele, E. Bobeico, L. Lancellotti, and P. Morvillo, "Analytical modelling and minority current measurements for the determination of the emitter surface recombination velocity in silicon solar cells," *Solar Energy Materials and Solar Cells*, vol. 91, no. 8, pp. 707–713, 2007.

[100] S. Daliento and L. Mele, "Approximate closed-form analytical solution for minority carrier transport in opaque heavily doped regions under illuminated conditions," *IEEE Transactions on Electron Devices*, vol. 53, no. 11, pp. 2837–2839, 2006.

[101] H. Denio III, "Aerial solar Thermography and condition monitoring of photovoltaic systems," in *Proceedings of the 38th IEEE Photovoltaic Specialists Conference (PVSC '12)*, pp. 613–618, Austin, Tex, USA, June 2012.

[102] P. Guerriero, F. Di Napoli, M. Coppola, and S. Daliento, "A new bypass circuit for hot spot mitigation," in *Proceedings of the International Symposium on Power Electronics, Electrical Drives, Automation and Motion (SPEEDAM '16)*, pp. 1067–1072, June 2016.

[103] N. Tyutyundzhiev, F. Martinez-Moreno, J. Leloux, and L. Narvarte, "Equipment and procedures for on-site testing of PV plants and BIPV," in *Proceedings of the 29th European Photovoltaic Solar Energy Conference and Exhibition (PVSEC '14)*, pp. 3499–3503, 2014.

[104] J. A. Tsanakas, G. Vannier, A. Plissonnier, D. L. Ha, and F. Barruel, "Fault diagnosis and classification of large-scale photovoltaic plants through aerial orthophoto thermal mapping," in *Proceedings of the 31st European Photovoltaic Solar Energy Conference and Exhibition 2015*, pp. 1783–1788, Hamburg, Germany, August 2016.

[105] J. R. Martinez-De Dios and A. Ollero, "Automatic detection of windows thermal heat losses in buildings using UAVS," in *Proceedings of the World Automation Congress (WAC '06)*, Budapest, Hungary, June 2006.

[106] F. Grimaccia, M. Aghaei, M. Mussetta, S. Leva, and P. B. Quater, "Planning for PV plant performance monitoring by means of unmanned aerial systems (UAS)," *International Journal of Energy and Environmental Engineering*, vol. 6, no. 1, pp. 47–54, 2015.

[107] G. Leotta, P. M. Pugliatti, A. D. Stefano, F. Aleo, and F. Bizzarri, "Post processing technique for thermo-graphic images provided by drone inspections," in *Proceedings of the 31st European Photovoltaic Solar Energy Conference and Exhibition (EU PVSEC '15)*, pp. 1799–1803, Hamburg, Germany, 2015.

[108] S. Vergura, F. Marino, and M. Carpentieri, "Processing infrared image of PV modules for defects classification," in *Proceedings of the International Conference on Renewable Energy Research and Applications (ICRERA '15)*, pp. 1337–1341, IEEE, Palermo, Italy, November 2015.

[109] P. Guerriero, G. Cuozzo, and S. Daliento, "Health diagnostics of PV panels by means of single cell analysis of thermographic images," in *Proceedings of the IEEE 16th International Conference on Environment and Electrical Engineering (EEEIC '16)*, Florence, Italy, June 2016.

[110] N. I. Tziavos, H. Hemida, N. Metje, and C. Baniotopoulos, "Grouted connections on offshore wind turbines: a review," *Proceedings of the Institution of Civil Engineers—Engineering and Computational Mechanics*, vol. 169, no. 4, pp. 183–195, 2016.

[111] L. Piegari and R. Rizzo, "A control technique for doubly fed induction generators to solve flicker problems in wind power generation," in *Proceedings of the 1st International Power and Energy Conference (PECon '06)*, pp. 19–23, November 2006.

[112] G. Brando, A. Danmer, A. Del Pizzo, and R. Rizzo, "A generalized modulation technique for multilevel converters," in *Proceedings of the International Conference on Power Engineering, Energy and Electrical Drives (POWERENG '07)*, April 2007.

[113] B. Ji, V. Pickert, B. Zahawi, and M. Zhang, "In-situ bond wire health monitoring circuit for IGBT power modules," in *Proceedings of the 6th IET International Conference on Power Electronics, Machines and Drives (PEMD '12)*, pp. 1–6, March 2012.

[114] V. N. Popok, K. B. Pedersen, P. K. Kristensen, and K. Pedersen, "Comprehensive physical analysis of bond wire interfaces in power modules," *Microelectronics Reliability*, vol. 58, pp. 58–64, 2016.

[115] A. Morozumi, K. Yamada, T. Miyasaka, S. Sumi, and Y. Seki, "Reliability of power cycling for IGBT power semiconductor modules," *IEEE Transactions on Industry Applications*, vol. 39, no. 3, pp. 665–671, 2003.

[116] R. Bayerer, T. Herrmann, T. Licht, J. Lutz, and M. Feller, "Model for power cycling lifetime of IGBT modules-various factors influencing lifetime," in *Proceedings of the 5th International Conference on Integrated Power Systems (CIPS '08)*, pp. 1–6, VDE, Nuremberg, Germany, March 2008.

[117] M. Denk and M.-M. Bakran, "Comparison of counting algorithms and empiric lifetime models to analyze the load-profile of an IGBT power module in a hybrid car," in *Proceedings of the 3rd International Electric Drives Production Conference (EDPC '13)*, pp. 1–6, IEEE, October 2013.

[118] D. Barlini, M. Ciappa, M. Mermet-Guyennet, and W. Fichtner, "Measurement of the transient junction temperature in MOSFET devices under operating conditions," *Microelectronics Reliability*, vol. 47, no. 9-11, pp. 1707–1712, 2007.

[119] R. Moeini, P. Tricoli, and H. Hemida, "Increasing the reliability of wind turbines using condition monitoring of semiconductor devices: a review," in *Renewable Power Generation*, IET, 2016.

[120] R. Schmidt and U. Scheuermann, "Using the chip as a temperature sensor—the influence of steep lateral temperature gradients on the Vce(T)-measurement," in *Proceedings of the 13th European Conference on Power Electronics and Applications (EPE '09)*, pp. 5–11, Barcelona, Spain, September 2009.

[121] W. Brekel, T. Duetemeyer, G. Puk, and O. Schilling, "Time resolved in situ Tvj measurements of 6.5 kV IGBTs during inverter operation," in *Proceedings of the PCIM Europe Conference*, pp. 808–813, Nuremberg, Germany, May 2009.

[122] H. Kuhn and A. Mertens, "On-line junction temperature measurement of IGBTs based on temperature sensitive electrical parameters," in *Proceedings of the 13th European Conference on Power Electronics and Applications (EPE '09)*, 10, 1 pages, September 2009.

[123] D. W. Brown, M. Abbas, A. Ginart, I. N. Ali, P. W. Kalgren, and G. J. Vachtsevanos, "Turn-off time as an early indicator of insulated gate bipolar transistor latch-up," *IEEE Transactions on Power Electronics*, vol. 27, no. 2, pp. 479–489, 2012.

[124] M. Denk and M.-M. Bakran, "An IGBT driver concept with integrated real-time junction temperature measurement," in

Proceedings of the International Exhibition and Conference for Power Electronics, Intelligent Motion, Renewable Energy and Energy Management (PCIM '14), pp. 214–221, May 2014.

[125] N. Baker, S. Munk-Nielsen, M. Liserre, and F. Iannuzzo, "Online junction temperature measurement via internal gate resistance during turn-on," in *Proceedings of the IEEE 16th European Conference on Power Electronics and Applications (EPE '14-ECCE Europe)*, pp. 1–10, August 2014.

[126] S. Bęczkowski, P. Ghimre, A. R. Vega, S. Munk-Nielsen, and P. Th, "Online Vce measurement method for wear-out monitoring of high power IGBT modules," in *Proceedings of the 15th European Conference on Power Electronics and Applications (EPE '13)*, pp. 1–7, IEEE, Lille, France, 2013.

[127] C. Zorn and N. Kaminski, "Acceleration of temperature humidity bias (THB) testing on IGBT modules by high bias levels," in *Proceedings of the 27th IEEE International Symposium on Power Semiconductor Devices and IC's (ISPSD '15)*, pp. 385–388, May 2015.

[128] N. Flourentzou, V. G. Agelidis, and G. D. Demetriades, "VSC-based HVDC power transmission systems: an overview," *IEEE Transactions on Power Electronics*, vol. 24, no. 3, pp. 592–602, 2009.

[129] C. Zorn and N. Kaminski, "Temperature humidity bias (THB) testing on IGBT modules at high bias levels," in *Proceedings of the 8th International Conference on Integrated Power Systems (CIPS '14)*, pp. 1–7, 2014.

PV-Powered CoMP-Based Green Cellular Networks with a Standby Grid Supply

Abu Jahid,[1] **Abdullah Bin Shams,**[2] **and Md. Farhad Hossain**[3]

[1]*Department of Electrical, Electronic and Communication Engineering, Military Institute of Science and Technology,*
 Dhaka 1216, Bangladesh
[2]*Department of Electrical and Electronic Engineering, Islamic University of Technology, Gazipur 1704, Bangladesh*
[3]*Department of Electrical and Electronic Engineering, Bangladesh University of Engineering and Technology,*
 Dhaka 1000, Bangladesh

Correspondence should be addressed to Md. Farhad Hossain; mfarhadhossain@eee.buet.ac.bd

Academic Editor: Md. Rabiul Islam

This paper proposes a novel framework for PV-powered cellular networks with a standby grid supply and an essential energy management technique for achieving envisaged green networks. The proposal considers an emerging cellular network architecture employing two types of coordinated multipoint (CoMP) transmission techniques for serving the subscribers. Under the proposed framework, each base station (BS) is powered by an individual PV solar energy module having an independent storage device. BSs are also connected to the conventional grid supply for meeting additional energy demand. We also propose a dynamic inter-BS solar energy sharing policy through a transmission line for further greening the proposed network by minimizing the consumption from the grid supply. An extensive simulation-based study in the downlink of a Long-Term Evolution (LTE) cellular system is carried out for evaluating the energy efficiency performance of the proposed framework. System performance is also investigated for identifying the impact of various system parameters including storage factor, storage capacity, solar generation capacity, transmission line loss, and different CoMP techniques.

1. Introduction

Due to the recent unprecedented growth in the number of subscribers and diverse data applications, mobile cellular network operators are deploying a higher number of BS in their infrastructure resulting in an exponential increase in energy consumption. Such growth of energy demand in the cellular network industry is exerting enormous detrimental effect on both the economical and the environmental aspects [1–3]. Recent studies suggest that around 50% of the operating expenditure (OPEX) of a cellular system attributes to the energy cost for running the network. On the other hand, it has been reported that the information and communication technology (ICT) industry contributes about 2–2.5% of total carbon emissions and this is expected to increase every year

with the exponential growth of mobile traffic [4, 5]. Moreover, this increased energy consumption in cellular networks places heavy burden on the electric grid. Therefore, with the escalating awareness of global warming and energy costs for operating cellular networks, green communications have received considerable attention among the telecommunication researchers and have led to an emerging trend to improve the energy efficiency (EE) of the overall system [4, 6]. Powering the cellular networks as much as possible by renewable energy sources is potentially the best alternative for reducing and even completely phasing out the consumption from the conventional grid supply leading to improved EE and decreased carbon footprint. The most popular among the renewable energy sources are solar, wind, and hydropower. A step towards green communication requires the renewable energy source to be easily integratable with the existing cellular networks. It must also be economical, widely available, and modular and should occupy a smaller area so that it can easily be installed at the vicinity of the BSs. From this perspective, the most feasible and lucrative renewable energy source is photovoltaic (PV) cells.

1.1. Photovoltaic Power Plants. As an efficient way to utilize solar energy, PV power plant has received increased attention all over the world due to the fossil fuel crisis and its associated environmental pollution. The tremendous growth of energy consumption around the world has led to an increase in operating cost and global warming. Solar energy offers attractive solutions to reduce carbon footprints and mitigate the global climate change. Burning of nonrenewable energy sources like fossil fuel produces greenhouse gases, whereas PV-based power plants and industries have no such detrimental effect on the environment. Since renewable solar energy is derived from resources which are regenerative, it does not emit carbon. In accordance with the growing trend of PV power plant, solar energy can also be used for a variety of purposes, such as solar-powered BSs for green communications, solar irrigation, solar cold storage, and solar boat systems [7]. Many emerging economies [7–9] have an excellent solar resource and have adopted policies to encourage the development of the solar industry to realize the benefits of PV technology. This can generate positive impact on their economies, as well as on the local and global environment, and improve energy security. Chandel et al. [8] analyzed the potential and cost-effectiveness of solar PV power plant of 2.5 MW for meeting the energy demand of the garment zone in India. A viability analysis [9] of 1 MW PV power plants was conducted in Serbia by taking different types of solar modules to find out the best possibilities of generating high electricity.

1.2. PV Solar Energy for Green Communications. BSs in the radio access network (RAN) of cellular mobile networks are the most energy hungry equipment amounting around 60%–80% of the total consumption [10–12], whereas the accumulated energy requirement for user equipment (UE) is around 1% [13]. On the other hand, cellular network data traffic is expected to increase approximately by a factor of ten every five years resulting in a tremendous pressure on energy demand [14]. Thus, this unprecedented growth in energy consumption exerts a detrimental impact on the environment in terms of carbon footprints [4, 5, 15]. Therefore, energy-efficient resource management system in RANs has become the center of focus of the researchers from both academia and industry. This trend has motivated the interest of researchers in an innovative research area called "green communications" concentrating on the environmental effects of cellular networks.

Being inspired for curving down the energy costs, telecom operators have started the deployment of renewable energy sources, such as solar panels, for improving EE of RAN infrastructure. The objective of green cellular communications is to reduce the overall nonrenewable energy consumption leading to improved EE and higher economic benefits.

Conventional design approaches focus on optimizing the quality of service (QoS) parameters such as cell coverage, capacity, and throughput with no consideration on the EE aspect of cellular networks. However, with the introduction of green communication paradigm, designed networks must also maintain the same level of QoS while improving EE [11, 16]. On the other hand, the intermittency and spatial randomness of renewable energy generation can severely degrade the system performance of large-scale cellular networks, and hence, it is a fundamental design issue to utilize the harvested energy to sustain traffic demands of users in the network [17]. As a consequence, though renewable energy sources are being deployed in BSs, a provision of conventional grid energy is still required to mitigate for the variability of the renewable energy generation.

Considering the aforementioned concerns, envisioning BSs to be powered by hybrid supplies combining solar energy sources with on-grid sources has become a promising alternative stimulating the proposed work in this paper. In such cellular networks, the primary energy source for BSs is the solar energy. If enough green energy is not available, BSs draw energy from the grid supply for serving its associated UEs. The focus of such green networking is to maximize the usage of solar energy while minimizing the conventional grid energy utilization. The optimal use of solar energy over a period of time depends on the proper energy management techniques integrated into the network operation.

1.3. Coordinated Multipoint (CoMP) Transmission Technique. Spectral efficiency (SE) and EE are considered as the prime performance metrics for planning and operation of next-generation cellular networks. SE is a key performance parameter defined as the overall throughput per unit bandwidth. On the other hand, coordinated multipoint (CoMP) transmission has been widely discussed as a promising candidate for future LTE-Advanced (LTE-A) cellular systems [18, 19]. In a cellular network with CoMP, multiple BSs coordinate among themselves for serving a UE in the best possible way. Thus, CoMP has the potential to improve the network performance in terms of interference management, cell-edge throughput, and overall SE as well as EE [20–22]. The downlink CoMP can be categorized into three types based on data availability at multipoint: joint transmission (JT), dynamic point selection (DPS), and coordinated scheduling/coordinated beamforming (CS/CB), which are outlined by 3GPP [20, 23]. In the DPS technique, BS offering the highest SINR is dynamically selected for serving a UE. In contrast, under the JT technique, multiple coordinating BSs transmit data simultaneously to a UE. On the other hand, in the CS/CB technique, signal is transmitted from only one BS by employing beamforming, which is achieved through proper scheduling among the coordinated BSs for avoiding intercell interference.

1.4. Contributions. This paper proposes and explores the potential of different approaches for improving the EE of CoMP transmission-based future cellular networks. The main contributions of this paper can be summarized as follows:

(i) This paper proposes a novel framework for improving the EE of the CoMP-based next-generation cellular networks by employing a hybrid power supply for BSs. Under the proposed framework, PV solar modules work as the primary energy source, while conventional grid power is proposed as the standby source for running BSs in case of insufficient solar energy for serving the UEs with no interruption. The proposed hybrid energy usage scheme is then investigated for both DPS and JT CoMP transmission technique-based cellular networks, which has not been reported yet in literature.

(ii) Then, a technique for maximizing the green energy utilization (i.e., minimizing the consumption from a conventional grid supply) is developed, while BSs are still powered by hybrid sources with the proposed energy usage scheme as outlined above. Therefore, a heuristic policy for sharing green energy (i.e., solar) among the BSs is proposed. For enabling the inter-BS energy transfer, neighboring BSs are proposed to be connected through a resistive lossy transmission line. The proposed energy sharing scheme is also integrated in and investigated for both the DPS and the JT CoMP-based cellular networks. To the best of our knowledge, we are the first to propose a green energy sharing technique for CoMP-based cellular systems.

(iii) Tempo-spatial cellular traffic diversity as well as solar energy generation variability plays a significant role in developing effective green networking techniques. On the other hand, intercell interference, wireless channel propagation model including shadow fading, and BS power consumption model are the other major factors that can affect any system performance. All these factors are taken into consideration in this proposed research and thus makes the network scenario near realistic.

(iv) Extensive simulations are carried out for investigating the energy usage analysis of the proposed framework in terms of various performance metrics such as EE, energy consumption indicator (ECI), and on-grid energy savings. Simulations are carried out considering both temporal traffic diversity over 24 hours and spatial traffic diversity over the entire network area.

(v) The impact of various system parameters including solar storage capacity, storage factor, transmission line loss, solar generation capacity, and CoMP techniques on the performance metrics is demonstrated and critically analyzed. Furthermore, system performance is also compared with that of the existing hybrid non-CoMP-based cellular system having no energy sharing.

The rest of the paper is organized as follows. Section 2 presents a thorough study on the related works. A detailed discussion of the system model along with the network layout and green energy model is outlined in Section 3. In addition, the energy consumption model for macrocell BS and the formulation of performance metrics are also presented in the same section. Section 4 presents the user association policy and the proposed algorithms. Section 5 shows the simulation results with insightful analysis, and finally, Section 6 concludes the paper summarizing the key findings.

2. Related Works

Over the last decades, the ever-increasing energy consumption in cellular networks has received intensive attention from regulatory bodies and mobile operators. With the growing awareness of global warming and financial consequences, both researchers and industries have initiated projects to reduce the increasing trend of energy consumption [2, 24, 25]. Switching off some BSs during the low traffic period is the most popular technique for minimizing energy consumption in RAN infrastructure [26–28]. Moreover, the concept of dynamic sectorization of BSs [29] and the traffic-aware intelligent cooperation among BSs [30] have shown remarkable aptitude for improving EE.

In recent years, comprehensive surveys on green cellular networks using various energy saving methods are presented in [4, 12, 31]. Hasan et al. [4] categorized energy saving mechanisms as cooperative networks, adoption of renewable energy resources, deployment of heterogeneous networks, and efficient usage of spectrum. In [31], Xu et al. outlined various distinctive approaches to reduce grid energy consumption in modern cellular networks. The strategies can be broadly classified into energy-efficient hardware design, selectively turning off some components during low-traffic period, optimizing radio transmission process in a physical layer, and powering RANs by renewable energy resources. On the other hand, several research works presented in [13, 32–34] were carried out to improve the EE of the cellular networks with hybrid power supplies. In [13], Peng et al. proposed an energy management technique for cellular networks with the provision of a hybrid energy supply and BS sleep mode. Han and Ansari [33] investigated the optimization of green energy utilization resulting in a significant reduction of conventional grid energy consumption during peak traffic periods. This work did not consider green energy sharing among the BSs. On the other hand, Chia et al. [32] proposed a model for energy sharing between two BSs through a resistive power line, whereas Xu et al. [34] focused on energy harvesting and coordinate transmission-enabled wireless communication by investigating a joint energy and communication cooperative approach. In the proposed paradigm, energy cooperation was implemented by the cellular network operators via signing a contract with the grid operator so that BSs can exchange green energy via the existing grid infrastructure. Besides, the system model has not discussed energy saving issues in this paper. However, none of these papers in [13, 32–34] considered either the JT or the DPS CoMP-based cellular networks for

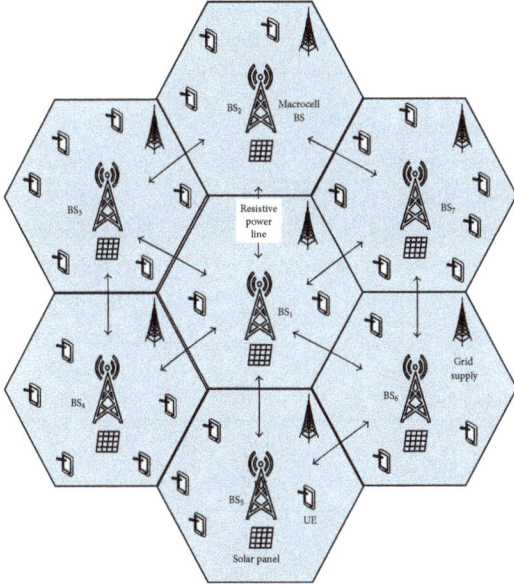

FIGURE 1: A section of the proposed network model with a hybrid energy supply.

investigating the EE performance or developing energy sharing mechanisms.

3. System Model

This section presents the proposed network model and other system components in the context of orthogonal frequency division multiple access- (OFDMA-) based LTE-A cellular systems, which can also be adopted to other standards.

3.1. Network Layout. The downlink of a multi-cell cellular network having a set of N BSs $\mathbb{B} = \{\mathcal{B}_1, \mathcal{B}_2, ..., \mathcal{B}_N\}$ and covering an area $\mathcal{A} = \{\mathcal{A}_1 \cup \mathcal{A}_2 \cup \cdots \cup \mathcal{A}_N\} \subset \mathbb{R}^2$ is considered. Here, $\mathcal{A}_n$ is the coverage area of BS $\mathcal{B}_n$, $n = 1, 2, ..., N$. BSs are assumed to be deployed using omnidirectional antennas in a hexagonal grid layout and orthogonal frequency bands are allocated in a BS resulting in zero intracell interference. On the other hand, universal frequency reuse is considered resulting in intercell interference when same frequency band is allocated in two BSs.

All the BSs in the considered LTE-A cellular network are powered by hybrid supplies, namely, PV solar energy and commercial on-grid energy. PV solar energy is the primary energy source, while the grid supply is the standby one. Each BS has an independent on-site solar energy harvester and energy storage device such as a battery bank. For sharing green solar energy among the BSs, each BS is connected with its neighboring BSs through resistive power lines. A segment of the network layout with seven macrocells is depicted in Figure 1. It is also considered that the proposed network is deployed using either the DPS or the JT CoMP transmission technique. On the other hand, UEs are assumed to be distributed uniformly throughout the network. Furthermore, any BS having no user to serve are switched into low power sleep mode for saving energy.

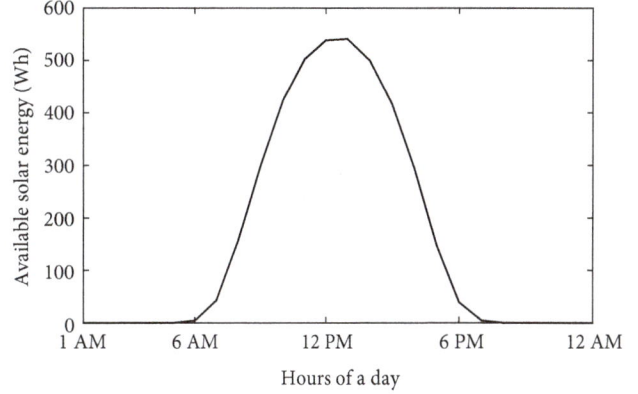

FIGURE 2: Average hourly solar energy generation.

3.2. Link Model. This paper considers a channel model with log-normally distributed shadow fading. For a separation d between transmitter and receiver, path loss in dB can be expressed as

$$\text{PL}(d) = \text{PL}(d_0) + 10n\log\left(\frac{d}{d_0}\right), \quad (1)$$

where $\text{PL}(d_0)$ is the path loss in dB at a reference distance d_0 and n is the path loss exponent. $\text{PL}(d_0)$ can be calculated using the free-space path loss equation.

Thus, the received power in dBm for jth UE at a distance $d = d^{n,j}$ from nth BS $\mathcal{B}_n$ is given by

$$P_r^{n,j} = P_t^{n,j} - \text{PL} + X_\sigma, \quad (2)$$

where $P_t^{n,j}$ is the transmitted power in dBm and X_σ is the amount of shadow fading modeled as a zero-mean Gaussian random variable with a standard deviation σ dB. Then the received SINR $\gamma_{n,j}$ at jth UE from $\mathcal{B}_n$ can be given by.

$$\gamma_{n,j} = \frac{P_r^{n,j}}{\mathcal{I}_{j,\text{inter}} + \mathcal{I}_{j,\text{intra}} + \mathcal{P}_N}, \quad (3)$$

where $\mathcal{I}_{j,\text{inter}}$ is the intercell interference, $\mathcal{I}_{j,\text{intra}}$ is the intracell interference, and $\mathcal{P}_N$ is the additive white Gaussian noise (AWGN) power given by $\mathcal{P}_N = -174 + 10\log_{10}(\Delta f)$ in dBm with Δf as the bandwidth in Hz.

3.3. Solar Energy Generation Model. This paper considers the PV solar panel as the on-site green energy harvester. The solar energy generation profile is nondeterministic and depends on some factors, such as temperature, solar light intensity, panel materials, generation technology, and the geographic location of the solar panel. The daily solar energy generation thus shows temporal dynamics over a period of a day in the given area and exhibits spatial variations with geographical location [7]. Due to the tempo-spatial diversity, the available solar energy may not guarantee the adequate energy supplies for a BS to run for a whole day.

Average hourly solar energy generation profile for a full year in Dhaka city of Bangladesh is shown in Figure 2. Here, the solar energy profile for a particular region is estimated by adopting the System Advisor Model (SAM) [35]. The curve indicates that the green energy generation starts from around 6:00 AM, reaches peak value at noon, and stops at about 6:00 PM. SAM supports various solar power generation technologies. However, without losing the generality, distributed type concentrated solar power (CSP) PV technology with 1 kW solar panel is used for generating the shown curve. On the other hand, though solar batteries such as Ni-Cd, NiMH, Li-ion, and sodium nickel chloride are available for using in solar systems, lead-acid batteries are commonly used in solar-powered BSs. The parameters of the solar generation and storage systems for the considered 1 kW solar panel are summarized in Table 1.

3.4. Solar Energy Storage Model.

For the proposed system, the green energy storage of the nth BS $\mathcal{B}_n$ at time t can be given by

$$s_n(t) = \mu s_n(t-1) + r_n(t) - d_n(t), \tag{4}$$

where s_n is the green energy storage, r_n is the incoming energy from PV solar panel, d_n is the energy demand of the BS, and $0 \le \mu \le 1$ is the storage factor, that is, the percentage of storage energy retained after a unit period of time. For example, $\mu = 0.9$ indicates that 10% of energy will be lost in the storage during the time interval. It is to be noted that the stored energy cannot exceed the maximum storage capacity. Therefore, if the generation is higher than the storage capacity, that amount of energy is considered as wastage.

3.5. BS Power Consumption Model.

It is important to investigate the traffic demand to be served by the BSs in order to analyze the energy consumption of the network. The mobile traffic volume exhibits both temporal and spatial diversity. Mobile users are assumed randomly distributed. It is also assumed that BSs transmit data to all users with the same data rate. Based on internal surveys on operator traffic data within the EARTH project and the Sandvine report [36], the daily traffic demand in the network is characterized by the normalized traffic profile illustrated in Figure 3.

The BSs energy consumption is directly related to the traffic volumes [37]. The energy consumption of BSs can be subdivided into two parts: the static energy consumption and the dynamic energy consumption. Holtkamp et al. [38] approximated the operating power of a BS as a linear function of RF output power P_{MAX} and BS loading parameter x, which can be given by [38]

$$P_{in} = \begin{cases} M_{sec}[P_1 + \Delta_p P_{MAX}(x-1)], & \text{if } 0 < x \le 1 \\ M_{sec} P_{sleep}, & \text{if } x = 0 \end{cases}, \tag{5}$$

where the expression in the square brackets represents the total power requirement for a transceiver (TRX) chain, M_{sec} is the number of sectors in a BS, and P_1 is the maximum power consumption in a sector. The load dependency is accounted for by the power gradient, Δ_p. The loading parameter $x = 1$ indicates a fully loaded system, that is, BS transmitting at full power with all of their LTE resource blocks (RBs)

TABLE 1: Solar panel and storage device parameters.

Parameters	Type (value)
Solar module type	Photovoltaic (distributed)
Generation technology	CSP PV cell
Solar panel capacity	1 kWdc
DC-to-AC ratio	0.9
Array type	Fixed roof mount
Tilt	20 degrees
Azimuth	180 degrees
Storage type	Lead-acid battery
Storage capacity	2000 Wh
Storage factor	0.96

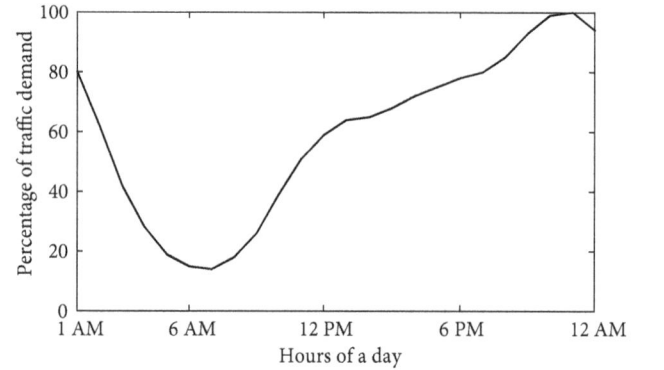

FIGURE 3: Daily traffic profile of a residential area.

occupied, and $x = 0$ indicates idle state. Furthermore, a BS without any traffic load enters into sleep mode with lowered consumption, P_{sleep}. Now P_1 can be expressed as follows [38]:

$$P_1 = \frac{P_{BB} + P_{RF} + P_{PA}}{(1 - \sigma_{DC})(1 - \sigma_{MS})(1 - \sigma_{cool})}, \tag{6}$$

where P_{BB} and P_{RF} are the power consumption of baseband unit and radio frequency transceiver, respectively. Losses incurred by DC-DC power supply, main supply, and active cooling can be approximated by the loss factors σ_{DC}, σ_{MS}, and σ_{cool}, respectively. However, power consumption in the power amplifiers is represented by P_{PA} which depends on the maximum transmission power and power amplifier efficiency η_{PA} and can be given as follows [38]:

$$P_{PA} = \frac{P_{MAX}}{\eta_{PA}(1 - \sigma_{feed})}. \tag{7}$$

BS power consumption model parameters used in this paper are summarized in Table 2.

3.6. Performance Metrics.

This paper evaluates the on-grid energy savings offered by the proposed network models compared to that of the conventional networks powered by grid supply only (i.e., no solar power). The average on-grid energy savings at time t denoted by $E_s(t)$ can be written as

TABLE 2: BS power consumption model parameters [38].

Parameters	Value
BS type	Macro
η_{PA}	0.306
γ	0.15
P_{BB} (W)	29.4
P_{RF} (W)	12.9
σ_{feed}	0.5
σ_{DC}	0.075
σ_{MS}	0.09
σ_{cool}	0.1
Number of sectors, M_{sec}	1
Maximum transmit power, P_{MAX} (dBm)	43
Δ_p	4.2
P_{sleep} (W)	54

$$E_s(t) = \frac{\sum_{n=1}^{N} P_s(n, t)}{\sum_{n=1}^{N} P_{in}(n, t)} \times 100\%, \tag{8}$$

where $P_{in}(n, t) = d_n(t)$ is the required total power in BS $\mathcal{B}_n$ at time t and $P_s(n, t)$ is the green solar power utilized by the BS $\mathcal{B}_n$ for serving its UEs.

On the other hand, EE performance metric of a network given in terms of bits per joule can be defined as the ratio of the total throughput to the total power required for running the network. In this paper, we define the EE metric of the proposed network models with CoMP techniques and hybrid power supply as the ratio of the aggregate throughput of the network to that of the net on-grid power consumed by the network. Total achievable throughput in a network at time t can be calculated by Shanon's capacity formula as follows:

$$R_{total}(t) = \sum_{j=1}^{U} \sum_{n=1}^{N_j} \Delta f \log_2(1 + \gamma_{n,j}), \text{ bps}, \tag{9}$$

where N_j is the number of transmitting BSs for serving jth UE and U is the total number of UEs in the network. Thus, the EE metric denoted as η_{EE} for time t can be written as follows:

$$\eta_{EE}(t) = \frac{R_{total}(t)}{\sum_{n=1}^{N} P_g(n, t)}, \text{ bits/joule}, \tag{10}$$

where $P_g(n, t) = P_{in}(n, t) - P_s(n, t)$ is the on-grid energy consumption in BS $\mathcal{B}_n$ at time t.

An alternative performance metric for evaluating the EE of a BS is the Energy Consumption Index (ECI) defined in [39], which can be given by

$$ECI = \frac{P_{in}}{KPI}, \tag{11}$$

where P_{in} refers to the total input power of a BS, whereas KPI (key performance indicator) indicates the total throughput of the BS. In other words, ECI is the reciprocal of EE, and hence, for the proposed networks, it can be evaluated by taking the inverse of (10), while a lower value of ECI implies better EE and vice versa. ECI is more suitable for better visualization of network behavior when the denominator of (10) becomes zero.

4. User Association and Algorithm

4.1. User Association Policy. The term user association means assigning a UE with a BS for receiving service. Associating users with the closest BS does not always ensure the best SINR due to the randomness of shadow fading. Therefore, user association policy based on the better signal quality (i.e., higher SINR) can support better performance. Therefore, this paper proposes SINR-based user association policy, which is presented for the DPS and JT CoMP-based networks as follows:

(i) *Network with DPS CoMP.* In a network deployed with the DPS CoMP-based transmission technique, one of the available BSs is dynamically selected for serving a UE in the best way. Thus, under the proposed network models with DPS, the BS which provides the highest SINR is selected for associating a UE.

(ii) *Network with JT CoMP.* In JT CoMP-based networks, multiple BSs are dynamically selected for serving a UE. For instance, in a network with 2-BS JT system, two BSs providing the top two SINR values are selected for associating a UE, which jointly transmits data to the particular UE.

4.2. Energy Sharing Algorithm. This section presents the proposed energy management scheme with the provision of green energy sharing. Under the proposed network model, each BS is equipped with a PV solar module with a storage facility, which acts as the primary energy source and can be shared among the neighboring BSs. In case there is not adequate energy stored in the storage of a BS, it seeks solar green energy from the neighboring BSs for supporting continuous service for its users. While seeking solar energy, a BS aims to share through the feasible shortest path for minimizing the power loss in the interconnecting resistive transmission line. This implies that a BS can share solar energy only from the six neighboring BSs placed in the first-tier surrounding it as illustrated in Figure 1. Furthermore, a BS aims to share from the BSs having higher solar energy stored.

On the other hand, a neighboring BS shares the surplus energy from its storage only after fulfilling its own demand. If solar energy is not available from the neighboring BSs, only then energy from the standby grid supply is used. Thus, there can arise two different cases for using energy in the BSs, which are presented as below with respect to the nth BS $\mathcal{B}_n$.

4.2.1. Case I: Sufficient Green Energy in Storage.

If $s_n(t) \geq d_n(t)$, then the nth BS $\mathcal{B}_n$ has sufficient solar energy for serving its UEs, and hence, the BS will be powered using its own stored energy. Thus, there is no need of green energy sharing from other BSs as well as no on-grid energy is consumed. The remaining solar energy in the storage after fulfilling the demand denoted by $g_n(t)$ can be expressed as

$$g_n(t) = s_n(t) - d_n(t). \tag{12}$$

Therefore, after meeting the demand of time t, the available solar energy in the storage of $\mathcal{B}_n$ for the time slot $(t+1)$ can be written as

$$s_n(t+1) = \mu g_n(t) + r_n(t+1) - d_n(t+1), \tag{13}$$

where $r_n(t+1)$ and $d_n(t+1)$ are the generation of solar energy and the total energy demand for time $(t+1)$, respectively.

4.2.2. Case II: Insufficient Green Energy in Storage.

The scenario with $s_n(t) < d_n(t)$ implies that there is not sufficient solar energy stored for powering the BS $\mathcal{B}_n$, and hence, solar energy sharing is required. Hence, $\mathcal{B}_n$ seeks for the additional solar energy from its neighbors, which is the difference between the total energy demand and the solar energy remaining in its own storage. Therefore, the total green energy required to be shared by $\mathcal{B}_n$ at time t denoted by $g_{n,s}(t)$ can be expressed as

$$g_{n,s}(t) = d_n(t) - s_n(t). \tag{14}$$

For sharing solar energy, BS $\mathcal{B}_n$ sorts its neighbors in a descending order of the available solar energy in their respective storages. Let the set of sorted BSs be given by $\mathbb{B}_n = \{\mathcal{B}_{n,1}, \mathcal{B}_{n,2}, ..., \mathcal{B}_{n,M}\}$, where M is the number of neighboring BSs of $\mathcal{B}_n$ for sharing energy and $\mathcal{B}_{n,p}$ has higher storage than $\mathcal{B}_{n,q}$ for $p < q$. Now, if the neighboring BS $\mathcal{B}_{n,1}$ has a shareable solar energy $\geq g_{n,s}(t)$, BS $\mathcal{B}_n$ accepts this amount from $\mathcal{B}_{n,1}$ that fulfills its demand. The sharable solar energy of a BS is the amount that can be shared after fulfilling its own demand. If the sharable energy of $\mathcal{B}_{n,1}$ is $< g_{n,s}(t)$, BS $\mathcal{B}_n$ accepts the amount from $\mathcal{B}_{n,1}$ that it can share.

For the remaining amount of required energy, BS $\mathcal{B}_n$ seeks to share from $\mathcal{B}_{n,2}$ and continues to the next BSs in $\mathbb{B}_n$. Let $\varepsilon_{n,m}$ be the amount of solar energy shared by $\mathcal{B}_n$ from the neighboring BS $\mathcal{B}_{n,m}$. Then the total solar energy received by $\mathcal{B}_n$ from the neighboring BSs can be given by

$$\varepsilon_n(t) = \sum_{m=1}^{M} \alpha_{n,m} \varepsilon_{n,m}(t), \tag{15}$$

where $0 \leq \alpha_{n,m} \leq 1$ is the utilization factor representing the line loss between BS $\mathcal{B}_n$ and its neighbor $\mathcal{B}_{n,m}$, that is, while sharing $\alpha_{n,m} \times 100\%$ of the energy is dissipated as line loss.

If $\varepsilon_n(t) = g_{n,s}(t)$, no on-grid energy is used by BS $\mathcal{B}_n$ at time t. Otherwise, on-grid energy is consumed for powering BS $\mathcal{B}_n$ for serving its UEs. Thus, the conventional grid energy consumption denoted by $c_n(t)$ by $\mathcal{B}_n$ at time t can be given by

TABLE 3: Pseudo code of the proposed energy sharing algorithm for nth BS $\mathcal{B}_n$.

1: **Initialize:** $s_n(t), d_n(t), \mu, \alpha_{n,m}, \varepsilon_{n,m} = 0, \forall n = 1, 2, ..., N;$ $\forall m = 1, 2, ..., M$
2: **If** $s_n(t) \geq d_n(t)$
3:     $P_s(n,t) = d_n(t)$ and $P_g(n,t) = 0$
4:     $g_n(t) = s_n(t) - d_n(t)$
5: **Else**
6:     Coordinate with other BSs for sharing solar energy   Sort the neighboring M BSs with respect to stored energy   i.e., find the set $\mathbb{B}_n = \{\mathcal{B}_{n,1}, \mathcal{B}_{n,2}, ..., \mathcal{B}_{n,M}\}$, $s.t., s_{n,p}(t) \geq s_{n,q}(t)$ for $p < q$
7:     **For** $m = 1 : M$
8:       Calculate $r_{n,s}(t) = g_{n,s}(t) - \sum_{k=0}^{m-1} \alpha_{n,k} \varepsilon_{n,k}(t)$, $\varepsilon_{n,k}(t) = 0$ for $k < 1$
9:       **If** $s_{n,m}(t) - d_{n,m}(t) \geq r_{n,s}(t)$
10:        Share solar energy $\varepsilon_{n,m}(t) = r_{n,s}(t)$ from $\mathcal{B}_{n,m}$
11:       **Else**
12:        Share solar energy $\varepsilon_{n,m}(t) = s_{n,m}(t) - d_{n,m}(t)$ from $\mathcal{B}_{n,m}$
13:        **If** $\sum_{k=1}^{m} \alpha_{n,k} \varepsilon_{n,k}(t) = g_{n,s}(t)$
14:         Stop the algorithm and Go to Step 21
15:        **Else**
16:         $m = m + 1$ and Go to Step 8
17:        **End If**
18:       **End If**
19:     **End For**
20:     **If** $\sum_{m=1}^{M} \alpha_{n,m} \varepsilon_{n,m}(t) = g_{n,s}(t)$
21:       $P_s(n,t) = d_n(t)$ and $P_g(n,t) = 0$
22:       Stop the algorithm
23:     **Else**
24:       $P_s(n,t) = \varepsilon_n(t) = \sum_{m=1}^{M} \alpha_{n,m} \varepsilon_{n,m}(t)$
25:       $P_g(n,t) = d_n(t) - \varepsilon_n(t)$
26:     **End If**
27: **End If**

$$P_g(n,t) = d_n(t) - \varepsilon_n(t). \tag{16}$$

Pseudo codes of the energy sharing algorithm with respect to the nth BS $\mathcal{B}_n$ is presented in Table 3. In the pseudo code, $d_{n,m}(t)$ and $s_{n,m}(t)$ are the total energy demand and the stored solar energy of the mth BS $\mathcal{B}_{n,m} \in \mathbb{B}_n$.

5. Performance Analysis

5.1. Simulation Setup.

This section analyzes the performance of the proposed cellular network framework with BSs powered by PV solar energy and standby grid supply. A MATLAB-based Monte Carlo simulation platform is developed for carrying out extensive simulations. For each data point, results are calculated by averaging over 10,000 independent iterations each with a simulation time of seven days. The network is deployed using a hexagonal grid layout with a cell radius of 1000 m. For comprehensive performance evaluation, intercell interference contributions by the 18 BSs placed in the two surround tiers are taken into consideration. On the other hand, UEs are considered uniformly distributed over the geographical area. Performance of the proposed network model for any nonuniform UE distribution can also be evaluated in a similar way. It is assumed that one UE occupies one RB and equal transmit power over all RBs. Furthermore, under the JT CoMP-based network, it is considered that the top two BSs providing the best SINR values serve a UE simultaneously. A summary of the system parameters of the simulated network are set in reference to the LTE standard [18] as summarized in Table 4. On the other hand, unless otherwise specified, the proposed network models are simulated considering x as a uniform random variable in [0,1] for modeling the spatial variation in traffic generation among the BSs, while equal solar generation is assumed in all the BSs having the solar module parameters as presented in Table 1.

5.2. Result Analysis

5.2.1. SINR and Throughput Analysis. Figure 4 demonstrates the empirical cumulative distribution function (CDF) of received SINR at UEs located throughout the considered network model. The network is considered with fully loaded (i.e., $x = 1$) BSs, which are supplied with hybrid supply with no inter-BS energy sharing. From the figure, a clear distinction in SINR distribution is observed among the JT CoMP-, DPS CoMP-, and non-CoMP-based hybrid systems. The JT CoMP-enabled hybrid system keeps its optimistic nature achieving comparatively stronger SINR among UEs, which ranges around −2 dB to 60 dB. This is because the two BSs offering the highest SINR values simultaneously serve a UE resulting in significantly better signal quality. In contrast, a non-CoMP-based hybrid scheme has the worst SINR performance as it spreads out over a larger range compared to that of the other techniques. SINR performance of the DPS CoMP-based system lies in between the JT and the non-CoMP-based systems as it selects the BS supporting the best signal quality. Thus, from the view point of SINR, the JT CoMP-based hybrid model is a preferred choice compared to the others.

On the other hand, a comparison of the throughput performance over a day among the different hybrid schemes is shown in Figure 5. As seen from the figure, throughput curves clearly follows the given traffic pattern (Figure 3). Also, a clear distinction is observed between the low traffic and the peak traffic times. This is because higher traffic means the allocation of higher number of RBs to the users resulting in higher throughput. Further, during peak traffic arrivals, throughput gap is more significant among the different hybrid systems. On the other hand, it is observed that the JT CoMP-based hybrid system outperforms others in terms of throughput. As throughput is directly related to the signal quality, superior SINR performance of the JT CoMP-based

TABLE 4: Simulation parameters.

Parameters	Value
Resource block (RB) bandwidth	180 kHz
System bandwidth, BW	10 MHz (50 RBs), 600 subcarriers
Carrier frequency, f_c	2 GHz
Duplex mode	FDD
Cell radius	1000 m
BS transmission power	43 dBm
Noise power density	−174 dBm/Hz
Number of antennas	1
Reference distance, d_0	100 m
Path loss exponent, n	3.574
Shadow fading, σ	8 dB
Access technique, DL	OFDMA
Traffic model	Randomly distributed

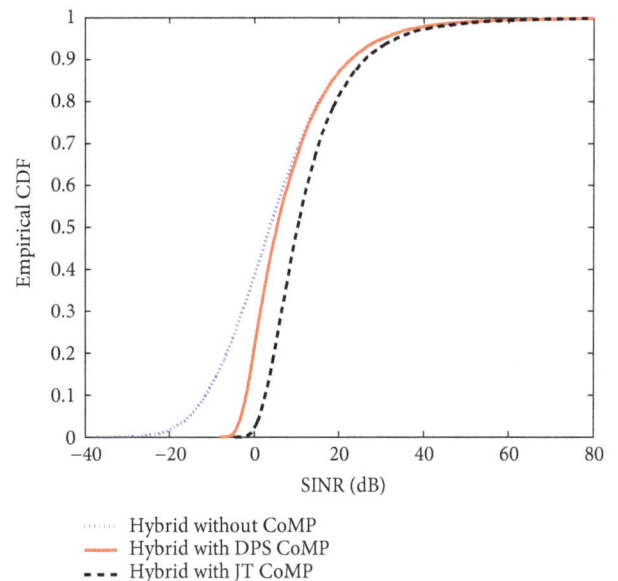

FIGURE 4: Empirical CDF of received SINR among the different hybrid systems for $x = 1$.

model as observed in Figure 4 ensures its relatively higher throughput.

5.2.2. Energy Analysis with Hybrid Supply under No Energy Sharing. Figure 6 compares the temporal variation of energy consumption by a BS in a conventional cellular system to that in the proposed network model with hybrid energy supply without energy sharing option. Here, the conventional scheme implies a cellular system powered by only grid energy with no energy sharing option and no CoMP transmission mechanism. As seen from the figure, on-grid consumption of a BS in the conventional system follows traffic distribution, reaches at peak when traffic demand is the highest, and goes down as the traffic decreases. Use of solar energy in BSs and sharing this energy among BSs is to minimize this grid energy usage. The energy consumption curve for the hybrid model

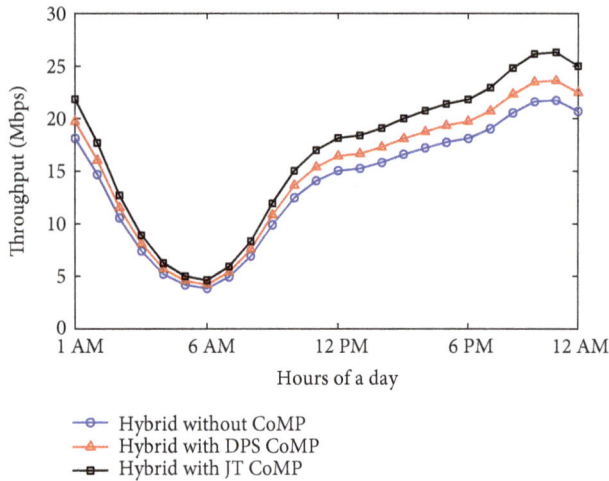

FIGURE 5: Throughput comparison of a single BS among the different hybrid models.

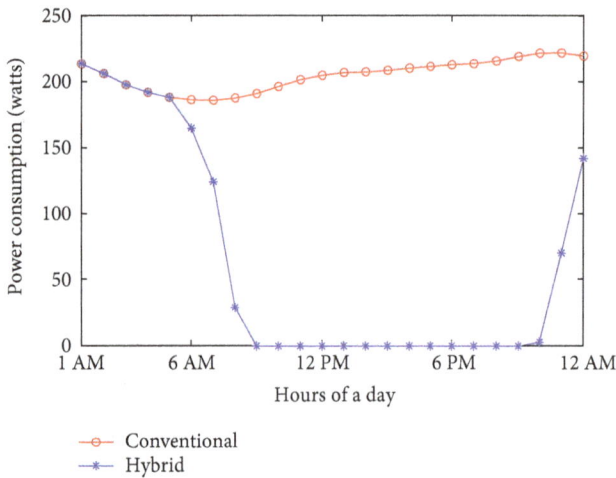

FIGURE 6: Comparison of on-grid power consumption in a single BS between the existing hybrid energy model (no sharing) and the conventional scheme with grid supply only. No CoMP technique is implemented in the network models.

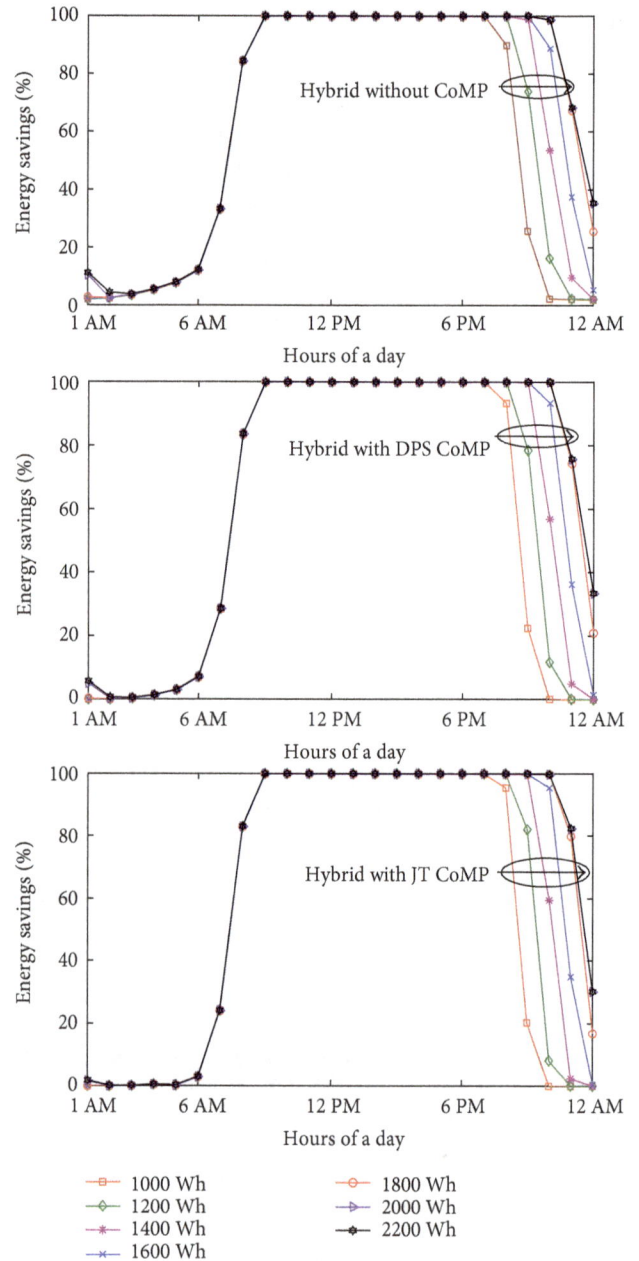

FIGURE 7: Average on-grid savings of a BS for different storage capacities with 1 kW solar module and no energy sharing.

with no sharing option is also presented assuming a storage capacity of 2 kWh. As seen, up to around 5 AM, a BS is completely run by on-grid supply as solar energy is unavailable during this period. After this, on-grid energy consumption gradually decreases with the increase of solar energy availability and becomes zero at 9 AM. Between 9 AM and 9 PM, there is adequate solar energy available for running the BS, and hence, no consumption of conventional energy. During this period, a BS fulfill its demand from its own solar energy storage and stores the surplus energy for future use. As time goes, stored solar energy decreases gradually with the decrease of solar light intensity, and once again, on-grid energy is required to serve its associated users after 9 PM. Thus, due to this temporal dynamics of solar energy generation, the available solar energy is not always sufficient for supplying the BS, and hence, on-grid energy is still required to fully meet the BS demand.

Percentage of energy savings under the different network models with hybrid power supply for various solar storage capacity is demonstrated in Figure 7. Here the network model having hybrid energy supply with no energy sharing and non-CoMP transmission refers to the existing hybrid system. As seen from the figure, during 12 AM to 6 AM, the BSs are mainly run by grid energy, and hence the savings in this period are mainly due to the switching of some BSs having no traffic into sleep mode. Under JT CoMP, energy savings in this period are found almost zero as the probability of BSs to enter into sleep mode is negligible. As the time proceed from 6 AM, solar energy generation increases leading to higher on-grid energy savings, which eventually reaches the peak around noon and following a gradual decrease around the evening. Thus, the hybrid scheme has the potential to reduce the on-grid energy consumption up to 100% for a prolonged period of time as evident from the figure. On the other hand, a significant impact of storage capacity on the energy savings is observed. As seen from the figure, the energy savings region is expanded with the increase of storage capacity resulting in higher savings. For example, energy savings curve for 2 kWh storage capacity lasts longer compared to that for a capacity of 1 kWh. Furthermore, the saving curves for 2 kWh is fully overlapped with that of 2.2 kWh and further increases in the storage capacity have no impact on energy savings as evident from the figure. Thus, the optimal value of storage capacity is 2 kWh. Notably, the energy saving performance follows similar fashion for all the three hybrid models having no significant variation due to CoMP techniques. This implies that for the particular network setting, savings is dependent predominantly on the solar energy generation and storage capacity.

A comparison of ECI performance metric with the three hybrid models is presented in Figure 8. During the low traffic periods in the morning, ECI increases rapidly up to a certain point and then starts to fall beyond that point. The upward trending nature of ECI implies the relatively higher on-grid energy consumption as the available solar energy is almost negligible. With the increase of solar energy generation, the ECI curve is pushed downward. It can be seen that during 9 AM to 9 PM, the ECI curve falls to zero indicating no on-grid energy consumption, that is, maximum EE. With the diminishing sunlight, stored energy also runs out by powering the BSs, and once again after 9 PM, on-grid energy is required to supply the BSs resulting in an upward trend of the ECI curve. On the other hand, since throughput performance of JT CoMP is better than that of DPS CoMP as observed in Figure 5, the JT CoMP-based hybrid system provides superior ECI performance. It can also be seen that the proposed network models have better ECI performance compared to that of the existing hybrid system.

Figure 9 illustrates the variation of average EE and the average on-grid power consumption in a BS under different hybrid schemes with the solar storage capacity. The network is simulated over a period of a week considering 1 kW solar panel in each BS as a green energy harvester. The three power consumption curves decreases in a similar fashion. The tendency of the down trending of the energy consumption curves indicates that the energy drawn from the grid decreases with the increment of storage capacity. The curves eventually reach to their respective constant values, which is also supported by Figure 7. This is because the storage limit reached optimum value for the given solar panel capacity and further increase in storage capacity does not make any significant improvement in reducing on-grid consumption. It is also observed that the grid power consumption is slightly higher under the JT CoMP-based hybrid system as two BSs simultaneously serve a particular UE. On the other hand, the EE curves demonstrate the opposite trend with the increment of storage capacity. However, the JT-based hybrid model shows superior EE performance compared to the DPS-based and the existing non-CoMP-based hybrid model

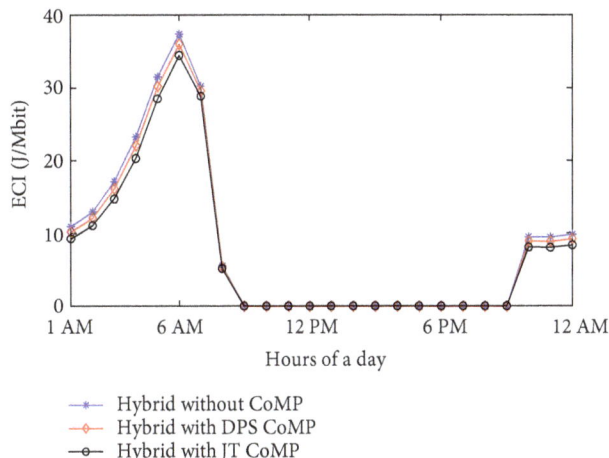

FIGURE 8: ECI comparison among the different hybrid models without sharing.

due to its significantly higher throughput. As expected, EE metric also eventually reaches its peak and remains constant after the optimal storage capacity, which is found equal to 2 kWh.

5.2.3. Energy Analysis with Hybrid Supply under Energy Sharing
(1) Equal Solar Generation Capacity. Figure 10 presents the impact of the resistive loss in the green energy transmission lines on the EE metric under different hybrid models evaluated over a period of one week. The systems are simulated assuming that each BS has an equal solar panel capacity of 1 kW with the optimal storage capacity of 2 kWh. Results for the three different scenarios, namely, hybrid only with no CoMP, hybrid with DPS CoMP, and hybrid with JT CoMP, are shown. From the figure, a clear difference is noticed in the EE metric performance with energy sharing and no sharing techniques of the respective hybrid models. The figure depicts that the EE performance has a decreasing trend with the increase of line loss. For the case of the hybrid model with no CoMP system, up to a certain percentage of line loss (around 55%), the EE performance with energy cooperation among BSs remains better than that of the corresponding non-cooperation-based scheme. Beyond this amount of line loss, energy cooperation degrades the network EE. Further analysis of the figure identifies that the concept of solar energy sharing can improve the EE of the DPS-based CoMP scheme if the line loss is less than 15%, while apparently no positive impact of such cooperation is observed for the JT-based CoMP scheme.

Dependency of EE on the storage factor of batteries over a period of a week is illustrated in Figure 11. The solar panel capacity and the storage capacity are same as those of Figure 10. As shown, all of the curves have a similar pattern reaching their respective peak values at a storage factor of 1. Storage factor indicates percentage of storage energy retained after unit period of time and the higher value of μ provides better EE. For the hybrid case with no CoMP mechanism,

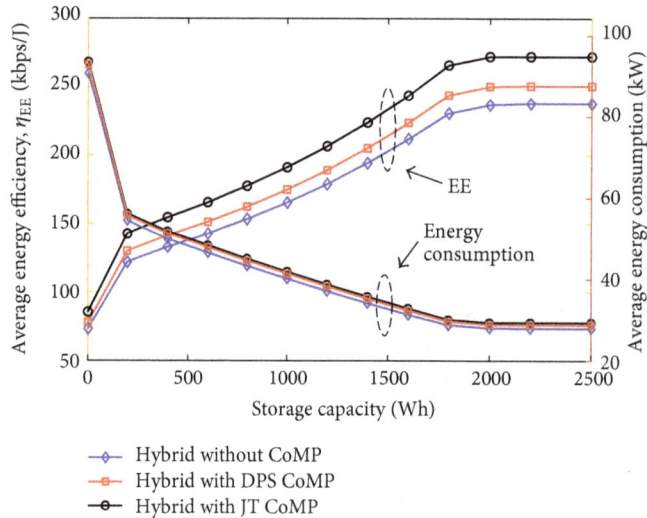

FIGURE 9: Comparison of EE and on-grid power consumption for the three different hybrid schemes without sharing.

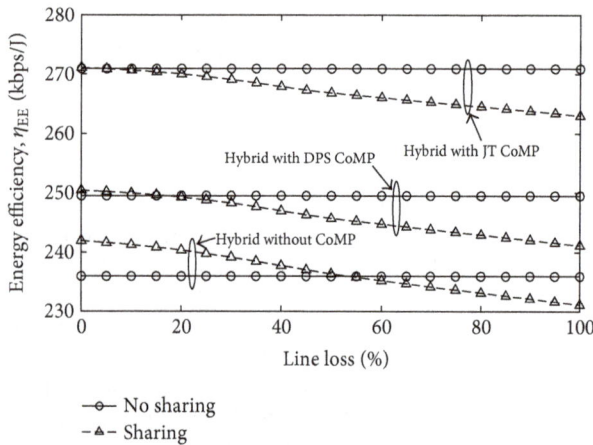

FIGURE 10: EE variation with line loss under the different hybrid scenarios with energy sharing considering equal solar capacity.

energy sharing demonstrates a positive impact on EE with the increase of storage factor μ. However, the best EE is found for the JT-based hybrid model in which the curves of sharing and no sharing are fully overlapping with each other, whereas the EE curves of DPS CoMP lie in between the former two hybrid models.

On the other hand, Figure 12 illustrates the impact of solar panel capacity on the EE performance of the proposed models. With the increase of solar panel capacity, storage capacity is also linearly scaled for guaranteeing no wastage of generated solar energy. As evident from the figure, with the increase of solar capacity (i.e., higher available solar energy), EE of the proposed network models substantially improves, which is mainly due to the increasing use of solar energy for running the BSs. From the figure, we can also determine the minimum solar capacity required for running the BSs 24 hours from solar energy. This is the capacity beyond which EE becomes infinity implying that no on-grid energy is required. As seen in the figure, this capacity is found around 1800 W and 1700 W for no sharing and

sharing schemes, respectively. Once again, the difference between the energy sharing models and the corresponding no sharing models is not that much significant as also observed in Figures 10 and 11. Furthermore, comparison of Figures 10–12 clearly demonstrates that the proposed hybrid models with CoMP techniques and energy sharing mechanism has superior EE performance compared to that of the existing hybrid system.

(2) Spatial Diversity in Solar Energy Generation. On top of the temporal diversity in solar generation in a BS as considered in the previous simulations, this section presents the results by introducing the spatial diversity as well. This phenomenon of spatial diversity in solar energy generation among the BSs is modeled as a uniform random variable distributed in [0,1] multiplied by a constant c_s. Unless otherwise specified, $c_s = 1$ kW is used for the simulations.

Under such scenario, comparison of EE with the resistive line loss evaluated over one week is illustrated in Figure 13. The figure follows the similar fashion of Figure 10. However, the figure depicts significant performance gap between the energy sharing and the corresponding no energy sharing based schemes. As the solar generation now varies from BS to BS, energy cooperation becomes more effective for improving EE by sharing surplus solar energy of some BSs with other BSs having lower amount. For the same reasons, compared to Figure 11, Figure 14 demonstrates a clear impact of energy sharing on the EE with varying storage factor μ. Furthermore, EE gap between the energy sharing and the corresponding no sharing cases is found higher for lower values of μ, which diminishes as μ increases to 1. The lower storage factor indicates lower amount of useful energy stored in the batteries, which stimulates the necessity of sharing solar energy from the neighboring BSs leading to higher gap and vice versa.

Figure 15 presents the variation of EE with the solar module capacity demonstrating the impact of energy sharing and no sharing operation for a network with spatial diversity of

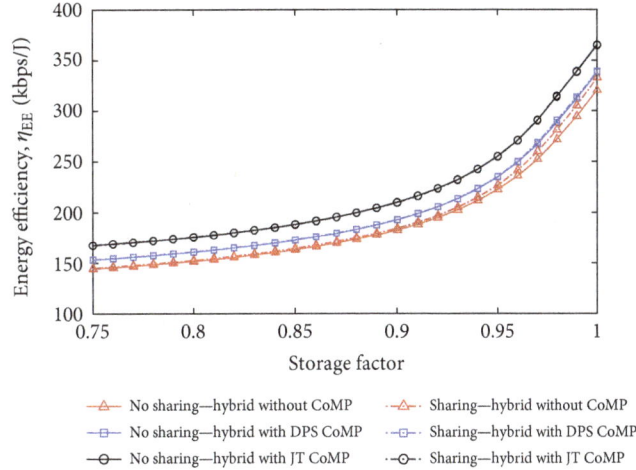

FIGURE 11: Variation of EE with storage factor with energy sharing considering equal solar capacity.

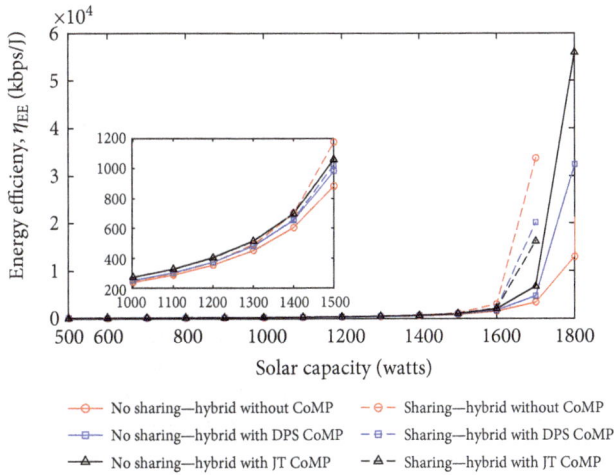

FIGURE 12: EE of the different hybrid schemes with energy sharing under equal solar capacity.

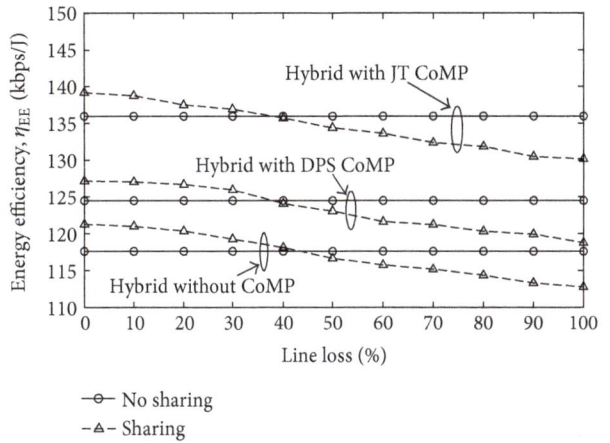

FIGURE 13: EE with line loss for different hybrid scenarios with energy sharing under varying solar capacity.

solar energy generation. Any capacity shown in the x-axis, for instance, 1 kW, implies that the solar energy generation varies among the BSs according to a uniform random variable distributed in [0,1] with $c_s = 1$ kW. As expected, EE of the proposed networks improves with the increase of solar capacity. Furthermore, comparison with the Figure 12 identifies that the positive impact of energy sharing on the EE is more apparent under the case of varying solar energy generation, which can be explained in the same way as presented for Figures 13 and 14. Once again, EE performance of the proposed CoMP-based systems are found significantly better than that of the existing hybrid system as illustrated in Figures 13–15.

6. Conclusion

This paper have proposed a framework for an energy efficient cellular network with hybrid-powered BSs, where a PV solar module acts as the main energy source for a BS and the grid remains as the standby. A solar energy sharing algorithm among the BSs has also been proposed for further greening the cellular networks by minimizing on-grid energy consumption. The proposed framework has been analyzed for both the DPS and the JT CoMP transmission techniques based on future cellular networks. System performance has been evaluated in terms of EE, energy savings, and throughput by comprehensive Monte Carlo simulations under varying system parameters, such as storage capacity, resistive line loss, storage factor, solar generation capacity, and CoMP techniques. Simulation results have shown that EE and energy savings of the proposed hybrid system increase with the increase in the storage capacity but approach to a peak value after a certain optimum capacity beyond which no further improvement is inflicted. Moreover, a continuous increase in EE has been observed for better storage factors, whereas the resistive loss in the transmission lines has been found to have significant deteriorating impact resulting in reduced improvement in EE. On the other hand, the proposed solar energy cooperation among the neighboring BSs

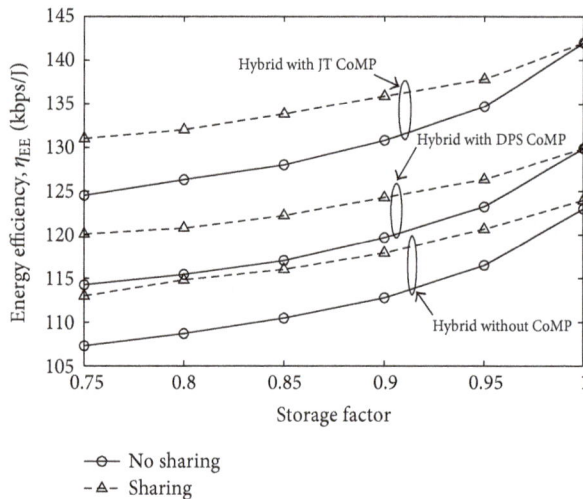

FIGURE 14: EE with storage factor for different hybrid scenarios with energy sharing under varying solar capacity.

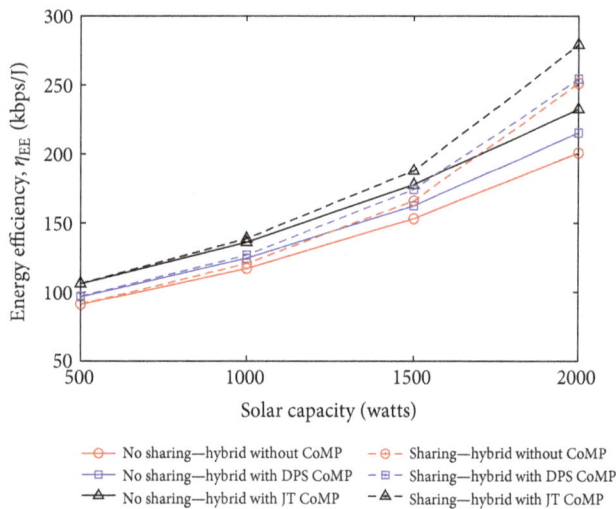

FIGURE 15: EE for different hybrid schemes with energy sharing under varying solar capacity.

have demonstrated a significant improvement in network EE for all the CoMP- and non-CoMP-based networks. It has also been identified that the JT CoMP hybrid system has the best EE performance compared to the DPS CoMP scheme. Moreover, the proposed network models have always been found higher energy efficient than the existing hybrid scheme with no energy sharing and non-CoMP transmission. Furthermore, energy sharing has been figured out more effective for improving EE in networks having spatial diversity in solar energy generation. In summary, the degree of improvement in EE of the proposed hybrid-powered network models with and without the proposed energy sharing mechanism has been found highly dependent on the network scenarios.

Conflicts of Interest

The authors declare that there is no conflict of interest regarding the publication of this paper.

References

[1] K. A. Adamson and C. Wheelock, "Off-Grid Power for Mobile Base Stations," *Technical Report, Navigant Research*, 2013.

[2] A. Feshke, G. Fettweis, J. Malmodin, and G. Biczok, "The global footprint of mobile communications: the ecological and economic perspective," *IEEE Communications Magazine*, vol. 49, no. 8, pp. 55–62, 2011.

[3] R. Mahapatra, Y. Nijsure, G. Kaddoum, N. Ul Hassan, and C. Yuen, "Energy efficiency tradeoff mechanism towards wireless green communications: a survey," *IEEE Communication Surveys and Tutorials*, vol. 18, no. 1, pp. 686–705, 2016.

[4] Z. Hasan, H. Boostanimehr, and V. K. Bhargava, "Green cellular networks: a survey, some, research issues and challenges," *IEEE Communication Surveys and Tutorials*, vol. 13, no. 4, pp. 524–540, 2011.

[5] J. Xu, C. Liu, Y. Yang, X. Ge, and T. Chen, "An overview of energy efficiency analytical models in communication networks," in *IEEE International Conference on Wireless Communications and Signal Processing (WCSP)*, pp. 1–6, Suzhou, China, October 2010.

[6] A. P. Blanzino, C. Chaudet, D. Rossi, and J. Rougier, "A survey of green networking research," *IEEE Communication Surveys and Tutorials*, vol. 14, no. 1, pp. 3–20, 2012.

[7] M. R. Islam, M. R. Islam, and M. R. A. Beg, "Renewable energy resources and technologies practice in Bangladesh," *Renewable and Sustainable Energy Reviews*, vol. 12, no. 2, pp. 299–343, 2008.

[8] M. Chandel, G. D. Agrawal, S. Mathur, and A. Mathur, "Techno-economic analysis of solar photovoltaic power plant for garment zone of Jaipur city," *Elsevier Journal of Case Studies in Thermal Engineering*, vol. 2, no. 1, pp. 1–7, 2013.

[9] T. Pavlovic, D. Milosavljevic, I. Radonjic, L. Pantic, A. Radivojevic, and M. Pavlovic, "Possibility of electricity generation using PV solar plants in Serbia," *Elsevier Journal of Renewable and Sustainable Energy Reviews*, vol. 20, no. 1, pp. 201–218, 2013.

[10] K. Son, H. Kim, Y. Yi, and B. Krishnamachari, "Base station operation and user association mechanisms for energy-delay tradeoffs in green cellular networks," *IEEE Journal on Selected Areas in Communications*, vol. 29, no. 8, pp. 1525–1536, 2011.

[11] R. Bolla, R. Bruschi, F. Davoli, and F. Gucchietti, "Energy efficiency in the future internet: a survey of existing approaches and trends in energy aware fixed network infrastructure," *IEEE Communication Surveys and Tutorials*, vol. 13, no. 2, pp. 223–244, 2011.

[12] M. Ismail, W. Zhuang, E. Serpedin, and K. Qaraqe, "A survey of green mobile networking: from the perspectives of network operators and mobile users," *IEEE Communication Surveys and Tutorials*, vol. 17, no. 3, pp. 1535–1556, 2015.

[13] C. Peng, S. B. Lee, S. Lu, H. Luo, and H. Li, "Traffic-driven power saving in operational 3G cellular networks," in *ACM International Conference on Mobile Computing and Networking (MobiCom)*, pp. 121–132, 2011.

[14] T. Chen, H. Kin, and Y. Yang, "Energy efficiency metrics for green wireless communications," in *IEEE International Conference on Wireless Communications and Signal Processing (WCSP)*, pp. 1–6, Suzhou, China, October 2010.

[15] G. Auer, I. Godor, L. Hevizi et al., "Enablers for energy efficient wireless networks," in *IEEE Vehicular Technology Conference (VTC)—Fall*, pp. 1–5, Ottawa, Canada, September 2010.

[16] Y. Zhang, P. Chowdhury, M. Tornatore, and B. Mukherjee, "Energy efficiency in telecom optical networks," *IEEE Communication Surveys and Tutorials*, vol. 12, no. 4, pp. 441–458, 2010.

[17] L. Cai, H. Poor, Y. Liu, T. Luan, X. Shen, and J. Mark, "Dimensioning network deployment and resource management in green mesh networks," *IEEE Wireless Communications*, vol. 18, no. 5, pp. 58–65, 2011.

[18] 3GPP TR 36.913, "LTE; Requirements for further advancements for evolved universal terrestrial radio access (E-UTRA) (LTE-advanced)," *Technical Report*, Version 10.0.0, Release 10, 2011.

[19] I. F. Akyildiz, D. M. G. Estevez, and E. C. Reyes, "The evolution of 4G cellular systems: LTE-advanced," *Physical Communication*, vol. 3, no. 4, pp. 217–244, 2010.

[20] J. Lee, Y. Kim, H. Lee et al., "Coordinated multipoint transmission and reception in LTE-Advanced systems," *IEEE Communications Magazine*, vol. 50, no. 11, pp. 44–50, 2012.

[21] R. Irmer, H. Droste, P. Marsch et al., "Coordinated multipoint: concepts, performance and field trail results," *IEEE Communications Magazine*, vol. 49, no. 2, pp. 102–111, 2011.

[22] A. Jahid, A. B. Shams, and M. F. Hossain, "Energy cooperation among BS with hybrid power supply for DPS CoMP based cellular networks," in *IEEE Conference on Electrical, Computer and Telecommunication Engineering (ICECTE)*, pp. 1–4, Rajshahi, Bangladesh, 2016.

[23] D. Lee, H. Seo, B. Clerckx et al., "Coordinated multipoint transmission and reception in LTE-Advanced: deployment scenarios and operational challenges," *IEEE Communications Magazine*, vol. 50, no. 2, pp. 148–155, 2012.

[24] White Paper, "Improving energy efficiency, lower CO_2 emission and TCO," *Huawei Energy Efficiency Solution, Huawei Technologies Co. Ltd.*, pp. 1–13, 2011.

[25] Vodafone, "Carbon and energy," http://www.vodafone.com/content/index/ukcorporateresponsibility/greener/carbon+energy.html.

[26] 3GPP TR 36.902 ver. 9.3.1 Rel. 9, "Evolved universal terrestrial radio access network (E-UTRAN); self-configuring and self-optimizing network (SON): use cases and solutions," 2011.

[27] H. Tabassum, U. Siddique, E. Hossain, and M. J. Hossain, "Downlink performance of cellular systems with base station sleeping, user association, and scheduling," *IEEE Transactions on Wireless Communications*, vol. 13, no. 10, pp. 5752–5767, 2014.

[28] G. Cili, H. Yanikomeroglu, and F. R. Yu, "Cell switch off technique combined with coordinated multi-point transmission for energy efficiency in beyond LTE cellular networks," in *IEEE ICC Workshop on Green Communications and Networking*, pp. 5931–5935, Ottawa, Canada, June 2012.

[29] M. F. Hossain, K. S. Munasinghe, and A. Jamalipour, "Energy-aware dynamic sectorization of base stations in multi-cell OFDMA networks," *IEEE Wireless Communications Letters*, vol. 2, no. 6, pp. 587–590, 2013.

[30] M. F. Hossain, K. S. Munasinghe, and A. Jamalipour, "Distributed inter- BS cooperation aided energy efficient load balancing for cellular networks," *IEEE Transactions on Wireless Communications*, vol. 12, no. 11, pp. 5929–5939, 2013.

[31] J. Xu, Y. Zhang, M. Zukerman, and E. K.-N. Yung, "Energy-efficient base stations sleep mode techniques in green cellular networks: a survey," *IEEE Communication Surveys and Tutorials*, vol. 17, no. 2, pp. 803–826, 2015.

[32] Y. K. Chia, S. Sun, and R. Zhang, "Energy cooperation in cellular networks with renewable powered base stations," *IEEE Transactions on Wireless Communications*, vol. 13, no. 12, pp. 6996–7010, 2014.

[33] T. Han and N. Ansari, "On optimizing green energy utilization for cellular networks with hybrid supplies," *IEEE Transactions on Wireless Communications*, vol. 12, no. 8, pp. 3872–3882, 2013.

[34] J. Xu, Y. Guo, and R. Zhang, "CoMP meets energy harvesting: a new communication and energy cooperation paradigm," in *IEEE Global Communications Conference (GLOBECOM)*, pp. 2508–2513, Atlanta, GA, USA, December 2013.

[35] "System advisor model (SAM)," *National Renewable Energy Laboratory (NREL), U.S. Department of Energy*, https://sam.nrel.gov/.

[36] "Mobile internet phenomena report," 2010, http://www.sandvine.com/downloads/documents/2010GlobalInternetPhenomenaReport.pdf.

[37] J. Lorincz, T. Garma, and G. Petrovic, "Measurements and modeling of base station power consumption under real traffic loads," *Sensors*, vol. 12, no. 4, pp. 4281–4310, 2012.

[38] H. Holtkamp, G. Auer, V. Giannini, and H. Hass, "A parameterized base station power model," *IEEE Communications Letters*, vol. 17, no. 11, pp. 2033–2035, 2013.

[39] EARTH, "EARTH deliverable D2.4, most suitable efficiency metrics and utility functions," 2012, https://bscw.ict-earth.eu/pub/bscw.cgi/d70454/EARTH_WP2_D2.4.pdf.

3

Technoeconomic Evaluation for an Installed Small-Scale Photovoltaic Power Plant

Bulent Yaniktepe, Osman Kara, and Coskun Ozalp

Engineering Faculty, Department of Energy Systems Engineering, University of Osmaniye Korkut Ata, Fakiusagi, 80000 Osmaniye, Turkey

Correspondence should be addressed to Bulent Yaniktepe; byaniktepe@osmaniye.edu.tr

Academic Editor: Cheuk-Lam Ho

Solar energy production and economic evaluation are analyzed, in this study, by using daily solar radiation and average temperature data which are measured for 3 years in the Osmaniye province in Turkey. Besides, this study utilizes the photovoltaic- (PV-) based grid connected to a power plant which has an installed capacity of 1 MW investment in electricity production. Economic values show that the net present value (NPV), the first economic method in the research, is about 111941 USD, which is greater than zero. Therefore, the payback year of this investment is approximately 8.3. The second one of these methods, the payback period of the simple payback period (PBP), is 6.27 years. The last method, which is the mean value of the internal rate of return (IRR), is 10.36%. The results of this study show that Osmaniye is a considerable region for the PV investment in electricity production. As a result, investment of a PV system in Osmaniye can be applicable.

1. Introduction

Power production from fossil fuel resources has detrimental effects on the environment considering fossil fuel-induced global climate change and air pollution. The increase in the cost of fossil fuels and soaring consumption of energy resources will both play a critical role in the usage of clean energy in many parts of the world in the near future. In the last decade, the trend of electricity generation from renewable energy has taken the place of conventional fossil fuels throughout the world. Solar energy, one of the main sources of renewable energy, is very convenient for electricity generation. In comparison to fossil fuels, solar energy, as a clean, inexhaustible, and immense energy source, is also one of the most promising renewable energies that present a sustainable alternative electricity generation. Furthermore, the areas where solar energy is used intensively, such as electricity generation, water heating, water pumping in agriculture, lightening, and charging, are application areas in which the use photovoltaic (PV) is easy [1–4]. The usage of PV as a renewable source for reliably producing electricity has grown rapidly for more than 30 years as a result of specific national and transnational incentive programs [5]. Moreover, PV as a usage of utilization of renewable energy sources becomes a popular day-by-day account of high modularity, no requirement for additional resource, not having moving parts, and low-maintenance cost. Another advantage of using PV is that when the growing rate of PV usage increases, the cost of the PV electricity generation decreases steadily [6]. Studies have showed that photovoltaic systems will be broadly used in the future, considering the rapidly decreasing cost of photovoltaic systems. Because price analysis is very important for energy marketing, a review of the cost potential factors on photovoltaic panels was realized, and the expected cost potential of photovoltaic systems was examined considering numerous studies [7].

Researchers have started to carry out economic and technical analyses of a PV power plant to be installed in Turkey. In this context, the producing capacity of grid-connected photovoltaic power plants for 135 locations in Turkey has been investigated [8]. This study has found that although Osmaniye, Dalaman, and Koycegiz have the highest energy generation, Tosya, Gumushane, and Artvin have the lowest energy production. Three separate PV panels for a PV power

TABLE 1: Solar energy potential for subregions in Turkey [16].

	Average radiation (kWh/m^2/year)	Maximum (kWh/m^2/year)	Minimum (kWh/m^2 year)	Sunshine duration period		
				Average (h/year)	Maximum (h/year)	Minimum (h/year)
Southeastern Anatolia	1492	2250	600	3016	408	127
Mediterranean	1453	2112	588	2924	360	102
General Anatolia	1434	2112	504	2712	381	98
Aegean	1407	2028	492	2726	371	96
East Anatolia	1395	2196	588	2694	374	167
Marmara	1144	1992	396	2528	351	88
Black Sea	1086	1704	408	1966	274	84

plant which has the capacity of 3 kWp for Kahramanmaras are investigated to obtain optimal PV type in terms of thin-film, monocrystalline, and polycrystalline silicon PV panels. The investigation concluded that the system consisting of polycrystals is the most suitable type for the region in terms of the shortest breakeven point for the investment [9]. Solar power generation potential in 2009 for Istanbul was evaluated using the parameters of technical measurements. Moreover, the results obtained using different tariffs, such as time-of-use and feed-in tariffs, show that electricity payments of customers can be decreased by using solar PV systems by more than 40% [10].

Economic analyses of electricity generation from solar renewable energy sources in the world have been investigated by several researches. Hrayshat and Al-Soud [11] show that the total PV installed capacity in Jordan is 82 kWp, generating a total of 182.5 MWh of electricity each year. They use a PV system which has a 5 MWp installed capacity for energy production. Also, economic analysis is calculated using monthly mean values of temperature, global solar radiation, and latitude in RETScreen software.

Furthermore, Al-Badi et al. [12] utilize a 5 MW solar PV power plant for 25 locations in Oman, and the mean energy production value from the PV power plant is found to be 7700 MWh for each year. They conclude that the capacity factor varies between 20% and 14% and the cost of electricity is 250 USD/MWh on average. Another study carries out the economic analysis of a 1.2 MW capacity grid-connected photovoltaic (PV) power plant installed at the Colorado State University-Pueblo [13]. Cash flow economic analysis is performed on a 1 MWp PV power station in Farafenni [14]. Quansah et al. [15] develop analytical models to conduct a technical and economic comparison of grid-charged battery-inverter systems (GBIS) and solar PV with battery storage systems (SPVS). In the study, GBIS is compared with an alternative approach that uses SPVS facility and is designed to meet half (50%) of the user's regular load.

As a result, it is the objective of this paper to verify the economic feasibility of implementing PV solar power in Osmaniye and to examine the economic benefit of solar energy. The rest of the paper will be structured according to the following: Section 1 presents Osmaniye and Turkey's annual solar radiation potential. Section 2 identifies the

TABLE 2: Feed-in tariff and incentives, locally manufactured components, and PV [17].

Locally manufactured component	Bonus (USD cent/kWh)
PV panel integration and production	0.8
PV modules	1.3
PV module cells	3.5
Inverter	0.6
Material which focuses radiation on PV module	0.5
Total	13.3 + 6.7 = 20

location of Osmaniye and determines the status of the photovoltaic solar modules. Section 3 examines the economic feasibility of Osmaniye. Section 3 also evaluates the economic viability of solar energy. Finally, Section 4 presents the main conclusions and recommendations.

2. Materials and Methods

2.1. Solar Energy Potential of Osmaniye. The location of Turkey has a considerable solar energy potential in terms of sunshine duration to produce electricity using PV. The yearly average solar radiation is 3.6 kWh/m^2/day, and the total yearly radiation period is approximately 2640 h, which is sufficient to provide adequate energy for solar applications. The technical and economic usages of solar energy in Turkey are 6105 and 305 billion kWh, respectively. Moreover, the solar energy breakdown in Turkey is shown in Table 1 in terms of subregions. As can be seen from Table 1, the highest average radiation values are in the Southeastern Anatolia and Mediterranean regions, which are 1492 and 1453 kWh/m^2/year, respectively [16].

Renewable energy has been one of the hot topics on Turkey's energy agenda. Significant progress has been made in the field of renewable energy starting from 2005, after the enactment of the Law on Utilization of Renewable Energy Resources for the Purpose of Generating Electrical Energy (Renewable Energy Law (REL)). Investments in renewable energy technologies remained limited between 2005 and 2010 due to the lack of secondary legislation and relatively low feed-in tariff prices. The RE Law of 2010 offers

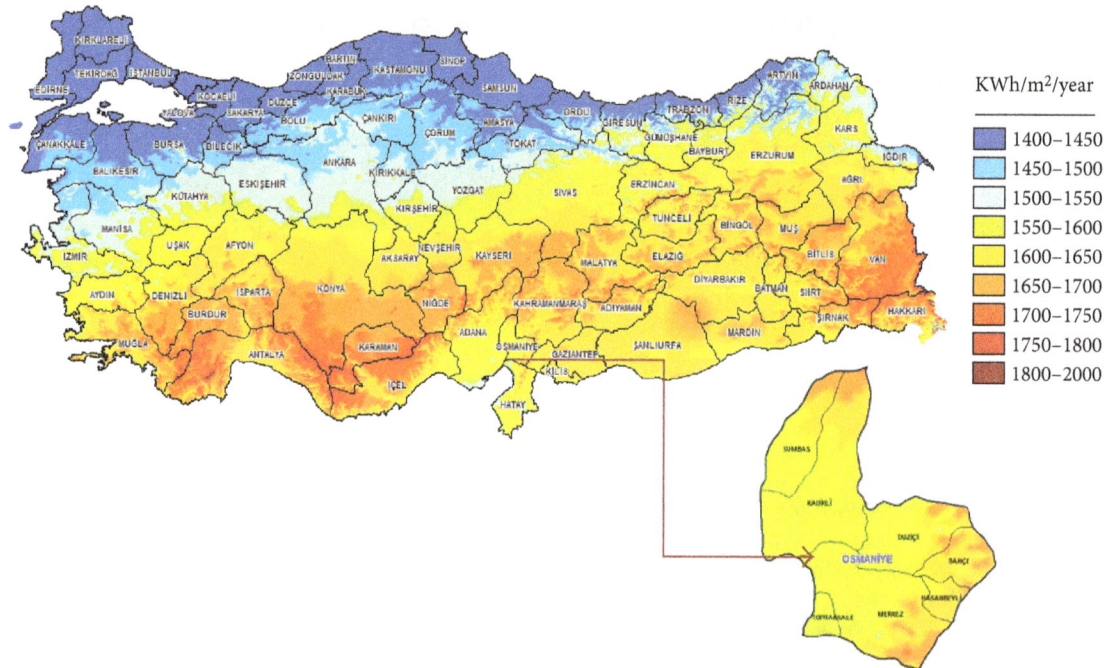

FIGURE 1: Annual solar energy capacity and location of Osmaniye in Turkey [18].

renewable electricity producers higher FIT rate schemes if they use local components in their projects. Currently, PVs dominate the solar market and global PV installed capacity grew by 74% in 2011. Support and incentives for PV investment in REL is given in Table 2. Generous subsidies around the world have been the main drivers of this significant growth. Even though prices of PVs have fallen dramatically over the years, additional reductions are needed for further implementation [17].

Osmaniye, 37.05 north and 36.14 east, is located in the Mediterranean region in Turkey, as shown in Figure 1 [18]. Solar radiation data in this study were measured every 5 minutes from 22 June 2012 to 01 June 2015 by using a meteorological measuring device (Vantage Pro2 Weather Station) located at the building of the Energy Systems Engineering Department in the campus of Osmaniye Korkut Ata University. A meteorological measuring device was mounted 20 m high from ground level. After the data was measured, the solar energy potential was calculated as averaged values of hourly, daily, monthly, and yearly.

As seen from Figure 2(a), the maximum and minimum total solar radiation occurred in 2013 and 2012, respectively. In addition, the highest monthly average daily radiation of 7.3 kWh/m²/day was recorded in the month of June, whereas the lowest monthly average solar radiation was 1.6 kWh/m²/day in December of 2012. It is understood from Figure 1 that monthly averaged values of global solar radiation stay nearly constant every month for each year. As a result, the total value of measured solar radiation based on the year 2013 was about 1.7 MWh/m²/year for one complete year. Hourly measured monthly average daily global solar radiation during the year 2014 is also shown in Figure 2(b). As can be seen from this figure, the highest and

lowest values of monthly averaged total global solar radiation are 695 W/m² and 476 W/m² in 2014. Considering the monthly values, Figure 2(b) also indicates that the daily mean and maximum solar radiation values are generally higher in summer (May-June-July), whereas comparatively, lower values are seen in winter months (November-December-January).

2.2. Solar PV System Description and Economic Feasibility Analysis. Global radiation, sunshine duration, and temperature as measured data are generally needed to produce electricity with a photovoltaic power plant in a selected region. Therefore, values of average daily measurement data from the meteorological measuring device in Osmaniye are given in Table 3. The table shows that the value of monthly measured average radiation is 136892 kWh/m²; the value of annual solar duration is 2959 h, and the value of annual average temperature is 18.3°C. Moreover, it is clear that Osmaniye has a significant solar energy potential to produce electricity.

PV modules can be collocated to form a solar array to provide the specific power at a specified voltage and current. In this context, as boundary conditions of this study, a 250 W peak PV module, which encloses polycrystalline silicon solar cells [20], is considered for obtaining 1 MW installed capacity of a power plant which includes 3986 modules calculated using the technical specifications in Table 4. If the PV system is installed on the roof of the building, the roof will be assumed to be a shed roof of 18° tilt made, covering a surface of approximately 6800 m² for 1 MW installed capacity.

In the economic viability aspect, the net present value (NPV), payback period (PBP) method, and internal rate of return (IRR) are analyzed in this study. Moreover, net present value (NPV) as an economic indicator is defined as the

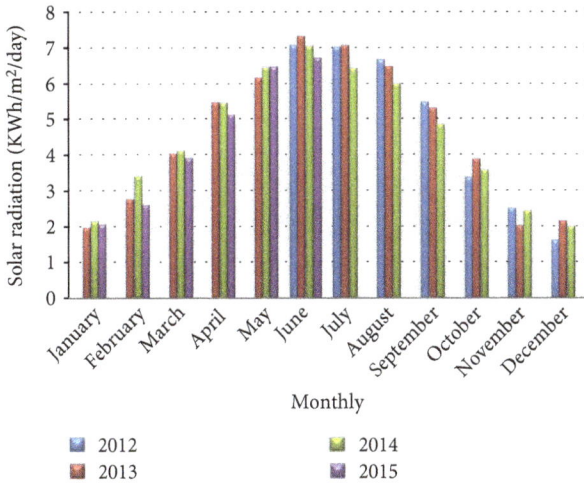

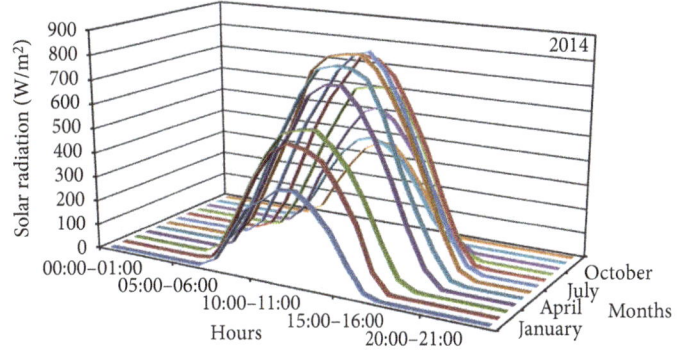

Figure 2: (a) Monthly averages of daily solar radiation throughout 2012–2015. (b) Hourly measured monthly average daily global radiation for 2014.

Table 3: The values of daily average measurement data.

Months	Daily average radiation (kWh/m^2)	Daily average sunshine duration (h) [18]	Daily average temperature ($^{\circ}$C) [19]
January	2.09	4.57	8.6
February	3.10	5.66	9.7
March	4.1	6.76	12.7
April	5.48	7.87	16.9
May	6.31	9.83	21.1
June	7.16	11.39	25.2
July	6.85	11.79	27.8
August	6.40	11.19	28.4
September	5.23	10.15	25.3
October	3.64	7.78	20.4
November	2.35	5.92	13.8
December	1.94	4.24	9.7

Table 4: Technical specifications of a photovoltaic module used [20].

Item description	Item specification	Item description	Item specification
Module number	BYD 250P6-30	Short-circuit current	8.98 A
Efficiency	15.37 percent	Open-circuit current	38 V
Rated power ($P_{\max}$)	250 W	Frame area	1.63 m^2
Voltage at $P_{\max}$	30.40 V	Dimension (mm)	$1640 \times 992 \times 50$
Current at $P_{\max}$	8.22 A	Weight	19.6 kg

value of all future cash flows, calculated at the discount rate. Positive NPV represents an indicator of a potentially feasible project [21]. Sowe et al. [14] also provided the following equation:

$$\sum_{n=0}^{N} \frac{B_n}{(1+i)^n} - \sum_{n=0}^{N} \frac{C_n}{(1+i)^n} = \text{PVB} - \text{PVC}, \quad (1)$$

where B_n = expected benefit at the end of the year n, C_n = expected cost at the end of year n, i = discountrate, n = project duration in years, N = project period, PVB = present value benefit, and PVC = present value cost. The payback period (PBP), which is the second economic method applied in this study, is substantially indicated by the payback period of the project as stated in (2). The project investment

is unacceptable with respect to economics when the PBP presents a high value (long payback periods). PBP does not incorporate the time value of money; moreover, assumptions on discount or interest rates are not required. Otherwise, the shorter PBP indicates the better investment. It is well known that the criterion of PBP value for the availability is higher than the profitability of the PV project. PBP is shown as follows [14, 22]:

$$\text{PBP} = \frac{\text{Initial investment (USD)}}{\text{Annual saving (USD/years)}} = \text{years}. \quad (2)$$

The last applied economic method is the internal rate of return (IRR), calculated by the subtraction of the total present value benefit from the total present value cost. As the IRR is greater than the discount rate, the PV project is considered as satisfactory and worthwhile. The IRR is defined in (3) where i = IRR [14, 23].

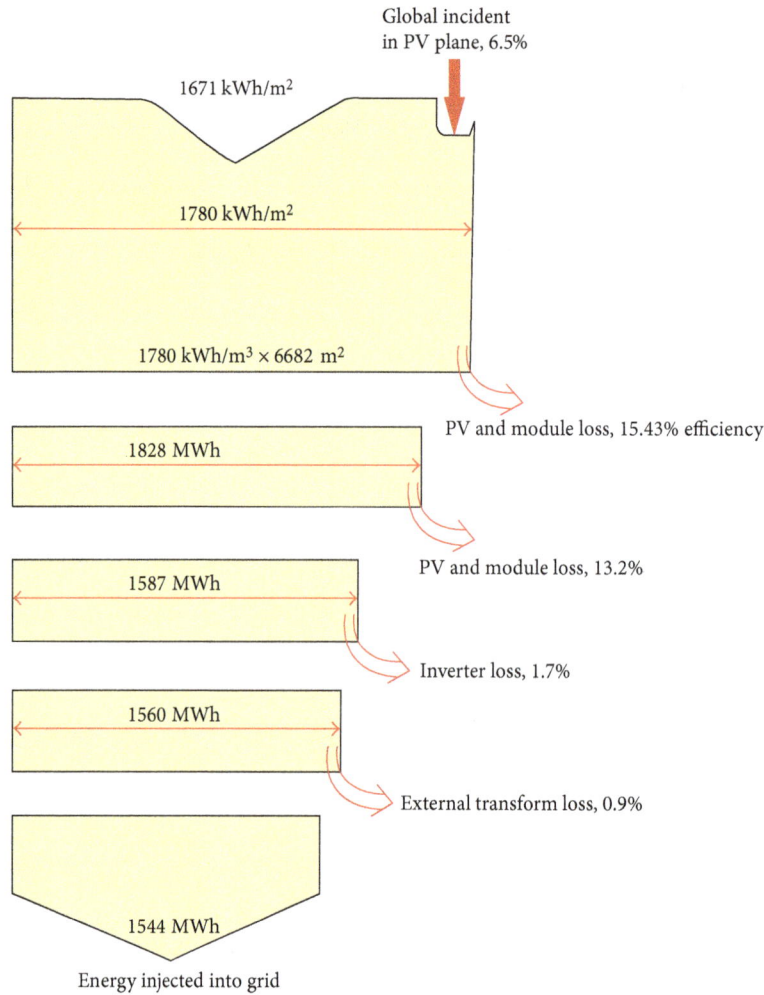

FIGURE 3: Electricity generation of a proposed solar PV system considering losses [24].

TABLE 5: Assumption of various interest rates used in the economic feasibility [25].

Item description	Value
Interest rate (TL)	10.79%
Inflation rate (TL)	2.96%
Discount rate (TL)	7.6%
Photovoltaic panel yield loss	8%
Project life	25 years

$$\sum_{n=0}^{N} \frac{B_n}{(1+i)^n} - \sum_{n=0}^{N} \frac{C_n}{(1+i)^n} = 0. \tag{3}$$

3. Results and Discussions

Total investment cost, life cycle of panel sand inverters, fixed operation, maintenance, repairing and utilizing costs, purchasing price offered by the government, and increasing energy costs were considered while calculating the economic

TABLE 6: Cost and economic assumption of the PV power plant [1].

Item description	Cost ($)	% of total cost
Feasibility study	9012	0.7%
Development cost	6437	0.5%
Engineering cost	5150	0.4%
Solar PV equipment	1125257	87.4%
Balance of power plant	102998	8%
Miscellaneous	38624	3%
Total initial cost	1287479	100
Annual operation and maintenance	5000	Annual
Depreciation	81000	Annual

viability of a proposed 1 MW solar PV system. Thus, the costs and benefits of solar PV system which will generate electricity into are analyzed and assessed based on the government incentive in Turkey. In this study, it is seen from Figure 2 that measured annual average solar radiation to the horizontal surface is 1671 kWh/m². As shown in Figure 3, PV and module loss, inverter loss, and external transform loss are

TABLE 7: Economic indicators for 1 MW solar PV system for NPV.

Economic indicators for 1 MW solar PV system for NPV (USD)										
					Years					
0	1	2	3	4	5	6	7	8	9	10
Income	204961	203692	202473	201308	200201	199156	198179	197275	196449	195709
Expenses	111204	111204	111204	111204	111204	151188	150992	150810	150643	150494
Depreciation expenses	80924	80924	80924	80924	80924	80924	80924	80924	80924	80924
Cash flow	174681	173412	172193	171028	169921	128892	128111	127389	126730	126139
Present value of cash flows	−1287479	162343	149780	138222	127590	117811	83053	76719	70898	65550	60635
Scrap cost (10 years later)	346819										
Present value	1399420										
Net present value	111941										

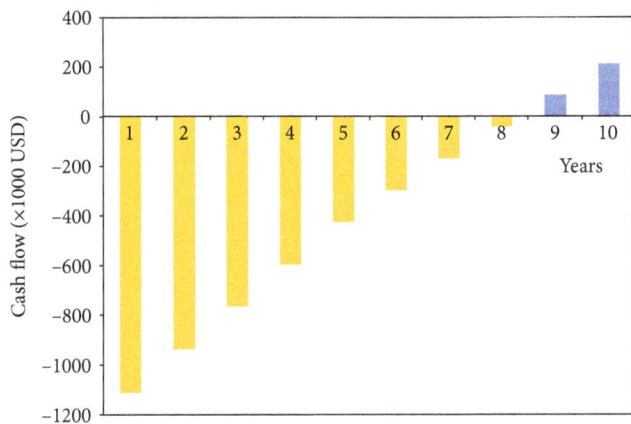

FIGURE 4: Cumulative cash flow, NPV for PV power plant.

TABLE 8: Economic indicators for PBP.

Description	Value (USD)
The total annual benefit	68926
Total initial cost	1287479
Scrap value	346819
Depreciation	80924
Repayment period (year)	6.27

TABLE 9: Economic indicators for IRR.

Discount rate	NPV
$i_1 = 10\%$	+13453
$i_2 = 12\%$	−23277

taken into account while determining total generation electricity of a PV system. Using the study of Verma and Singhal [24], energy injected into grid is calculated as 1544 MWh/year in Osmaniye. The input economical parameters for these three economical methods are summarized in Table 5. The costs of the main parts of the solar PV system are given in Table 6.

PV panels, transportation, and installation are the main equipment for solar PV investment, and its percentage is around 87.4. Balance of plant cost, as the second higher investment cost, accounts for approximately 8% of the total cost. All of these costs and interest rates given in Tables 5 and 6 were used for economic feasibility analysis in this research [1].

NPV is the difference between the present value of investment cash inflow and investment cost. If the NPV is a positive value, the project is potentially feasible. Also, NPV determines whether the project is generally an acceptable investment or not. Table 7 shows that the NPV is greater than zero when the year is approximately 8.3, which is the payback year of this investment. Besides, Figure 4 shows that positive and negative cash flow changes by years.

Table 8 presents the second economic method which is the simple payback method (PBP), and economic indicators for PBP are given in this table. The payback period is 6.27.

The quicker the regaining of the cost of an investment is, the more desirable the investment for the basic assumption of PBP is. Consequently, the result of PBP for Osmaniye seems feasible.

The development of a PV project would be acceptable if the IRR is equal to or greater than the required rate of return. Table 9 shows the IRR indicators and the value of the calculated IRR for Osmaniye. As can be seen from this table, the IRR is 10.36%, greater than the discount rate. So, the considered investment appears profitable.

4. Conclusion

Photovoltaic application can provide clean and reliable energy, without noise, and is an environmentally friendly source of power. In addition, the cost of photovoltaic energy technology is gradually decreasing as the market demand and production of PV systems are increasing. This study makes an economic analysis of a solar PV grid-connected system for 1 MW electricity generation plant in Osmaniye and the technical potential of solar PV electricity generation. This research underlined the obtained results given below.

The total value of measured solar radiation was about 1.7 MWh/m^2/year for one complete year. The highest monthly average daily radiation was found to be 7.3 kWh/

m^2/day, whereas the lowest monthly average solar radiation was 1.6 kWh/m^2/day.

The NPV was found to vary between $100000 and $120000, while the mean value remained as $111941. The payback year of this investment in terms of NPV is approximately 8.3, and the PBP varied between 6 and 7 years. PBP was calculated as 6.27 years, and the mean value of the internal rate of return (IRR) was determined as 10.36%. As a result, based on economic indicators and technical results, Osmaniye was found to be the best site for the development of a PV-based power plant.

Conflicts of Interest

The authors declare that they have no conflicts of interest.

Acknowledgments

The authors acknowledge the financial support of The Office of Scientific Research Projects of Osmaniye Korkut Ata University for the funding under Project no. OKÜ-BAP-2014-PT3-038. It is gratefully acknowledged.

References

[1] S. Rehman, M. A. Bader, and S. A. Al-Moallem, "Cost of solar energy generated using PV panels," *Renewable and Sustainable Energy Reviews*, vol. 11, no. 8, pp. 1843–1857, 2007.

[2] V. V. Tyagi, N. L. Panwar, N. A. Rahim, and R. Kothari, "Review on solar air heating system with and without thermal energy storage system," *Renewable and Sustainable Energy Reviews*, vol. 16, no. 4, pp. 2289–2303, 2012.

[3] C. Ertekin, F. Evrendilek, and R. Kulcu, "Modeling spatiotemporal dynamics of optimum tilt angles for solar collectors in Turkey," *Sensors*, vol. 8, no. 5, pp. 2913–2931, 2008a.

[4] C. Ertekin, R. Kulcu, and F. Evrendilek, "Techno-economic analysis of solar water heating systems in Turkey," *Sensors*, vol. 8, no. 2, pp. 1252–1277, 2008b.

[5] International Energy Agency (IEA), "Report IEA-PVPS T1-24:2014," PVPS Report Snapshot of Global PV 1992–2013 Preliminary Trends Information from the IEA PVPS Programme, 2014.

[6] A. Bianchini, M. Gambuti, M. Pellegrini, and C. Saccani, "Performance analysis and economic assessment of different photovoltaic technologies based on experimental measurements," *Renewable Energy*, vol. 85, pp. 1–11, 2016.

[7] M. S. Cengiz and M. S. Mami, "Price-efficiency relationship for photovoltaic systems on a global basis," *International Journal of Photoenergy*, vol. 2015, Article ID 256101, p. 12, 2015.

[8] N. Caglayan, C. Ertekin, and F. Evrendilek, "Spatial viability analysis of grid-connected photovoltaic power systems for Turkey," *International Journal of Electrical Power & Energy Systems*, vol. 56, pp. 270–278, 2014.

[9] S. Yılmaz, H. R. Ozcalık, S. Kesler, F. Dincer, and B. Yelmen, "The analysis of different PV power systems for the determination of optimal PV panels and system installation-a case study in Kahramanmaras, Turkey," *Renewable and Sustainable Energy Reviews*, vol. 52, pp. 1015–1024, 2015.

[10] A. Batman, F. G. Bagriyanik, Z. E. Aygen, O. Gul, and M. Bagriyanik, "A feasibility study of grid-connected photovoltaic systems in İstanbul, Turkey," *Renewable and Sustainable Energy Reviews*, vol. 16, no. 8, pp. 5678–5686, 2012.

[11] E. S. Hrayshat and M. S. Al-Soud, "Solar energy in Jordan: current state and prospects," *Renewable Energy*, vol. 8, no. 2, pp. 193–200, 2004.

[12] A. H. Al-Badi, M. H. Albadi, A. M. Al-Lawati, and A. S. Malik, "Economic perspective of PV electricity in Oman," *Energy*, vol. 36, no. 1, pp. 226–232, 2011.

[13] A. M. Paudel and H. Sarper, "Economic analysis of a grid-connected commercial photovoltaic system at Colorado State University-Pueblo," *Energy*, vol. 52, pp. 289–296, 2013.

[14] S. Sowe, N. Ketjoy, P. Thanarak, and T. Suriwong, "Technical and economic viability assessment of PV power plants for rural electrification in the Gambia," *Energy Procedia*, vol. 52, pp. 389–398, 2014.

[15] D. A. Quansah, M. S. Adaramola, I. A. Edwin, and E. K. Anto, "An assessment of grid-charged inverter-battery systems for domestic applications in Ghana," *Journal of Solar Energy*, vol. 2016, Article ID 5218704, p. 11, 2016.

[16] E. Toklu, M. S. Güney, M. Işık, O. Comaklı, and K. Kaygusuz, "Energy production, consumption, policies and recent developments in Turkey," *Renewable and Sustainable Energy Reviews*, vol. 14, no. 4, pp. 1172–1186, 2010.

[17] Renewable Energy & Environmental Technologies, *Republic of Turkey Prime Ministry Investment Support and Promotion Agency*, 2013, http://www.invest.gov.tr/enUS/infocenter/publications/Documents/environmental.tech.renewable.industry.pdf.

[18] Yenilenebilir Enerji Genel Müdürlüğü, *Güneş Enerjisi Potansiyel Atlası (GEPA)*, 2016, April 2016, http://www.eie.gov.tr/MyCalculator/Default.aspx.

[19] *Turkish State Meteorological Service*, 2016, April 2016, http://www.mgm.gov.tr/veridegerlendirme/il-ve-ilceler-istatistik.aspx?m=OSMANIYE.

[20] *Photovoltaic Solar Module Properties*, 2016, April 2016, http://www.byd.com/pv/module.html.

[21] M. A. H. Mondal and A. K. M. Sadrul Islam, "Potential and viability of grid-connected solar PV system in Bangladesh," *Renewable Energy*, vol. 36, no. 6, pp. 1869–1874, 2011.

[22] S. Rodrigues, R. Torabikalaki, F. Faria et al., "Economic feasibility analysis of small scale PV systems in different countries," *Solar Energy*, vol. 131, pp. 81–95, 2016.

[23] L. B. Jose and D. Rodolfo, "Economical and environmental analysis of grid connected photovoltaic system in Spain," *Renewable Energy*, vol. 31, no. 8, pp. 1107–1128, 2006.

[24] A. Verma and S. Singhal, "Solar PV performance parameter and recommendation for optimization of performance in large scale grid connected solar PV plant—case study," *Journal of Energy Power Sources*, vol. 2, no. 1, pp. 40–53, 2015.

[25] T. C. M. Bankası, *Enflasyon Raporu 2016-1*, 2016, April 2016, http://www.tcmb.gov.tr/wps/wcm/connect/TCMB+TR/TCMB+TR+Main+Menu/Istatistikler/Piyasa+Verileri/.

4

Influence of the Porosity of the TiO$_2$ Film on the Performance of the Perovskite Solar Cell

Xiaodan Sun,[1] Jia Xu,[2] Li Xiao,[1] Jing Chen,[1,2] Bing Zhang,[1,3] Jianxi Yao,[1,2] and Songyuan Dai[1,3]

[1]State Key Laboratory of Alternate Electrical Power System with Renewable Energy Sources, North China Electric Power University, Beijing 102206, China
[2]Beijing Key Laboratory of Energy Safety and Clean Utilization, North China Electric Power University, Beijing 102206, China
[3]Beijing Key Laboratory of Novel Film Solar Cell, North China Electric Power University, Beijing 102206, China

Correspondence should be addressed to Jianxi Yao; jianxiyao@ncepu.edu.cn and Songyuan Dai; sydai@ipp.ac.cn

Academic Editor: Pushpa Pudasaini

The structure of mesoporous TiO$_2$ (mp-TiO$_2$) films is crucial to the performance of mesoporous perovskite solar cells (PSCs). In this study, we fabricated highly porous mp-TiO$_2$ films by doping polystyrene (PS) spheres in TiO$_2$ paste. The composition of the perovskite films was effectively improved by modifying the mass fraction of the PS spheres in the TiO$_2$ paste. Due to the high porosity of the mp-TiO$_2$ film, PbI$_2$ and CH$_3$NH$_3$I could sufficiently infiltrate into the network of the mp-TiO$_2$ film, which ensured a more complete transformation to CH$_3$NH$_3$PbI$_3$. The surface morphology of the mp-TiO$_2$ film and the photoelectric performance of the perovskite solar cells were investigated. The results showed that an increase in the porosity of the mp-TiO$_2$ film resulted in an improvement in the performance of the PSCs. The best device with the optimized mass fraction of 1.0 wt% PS in TiO$_2$ paste exhibited an efficiency of 12.69%, which is 25% higher than the efficiency of the PSCs without PS spheres.

1. Introduction

Perovskite solar cells based on CH$_3$NH$_3$PbI$_3$ have attracted much attention. Tremendous progress has been made since the seminal work of Kojima et al. in 2009 [1]. In just six years, power conversion efficiencies (PCEs) of PSCs have increased sharply from 3.8% [1] to 22.1% [2], which exceeds the PCEs of polycrystalline silicon solar cells [3–5]. Moreover, their solution processability and low cost endow them with high potential for next generation solar cells [4, 6].

Two typical PSC structures are widely used, including the planar heterojunction architectures [7] and the mesoporous structures [8–10]. Planar perovskite solar cells are advantageous because they have simple and scalable cell configurations [11]. In mesoporous structured PSCs, semiconductors, including TiO$_2$, ZnO, insulating Al$_2$O$_3$ and ZrO$_2$, are employed as the electron transporting layer (ETL) or the scaffold of the perovskite layer [12]. Due to its

excellent physicochemical properties, such as large band gap, chemical stability, photostability, nontoxicity, and low cost [13], mesoporous TiO$_2$ is the most widely used electron transporting material in PSCs. The mp-TiO$_2$ film acts as not only the scaffold of the perovskite layer but also as the pathway for electron transport [14].

It has been reported that the structural properties of the mp-TiO$_2$ layer, such as particle size [15], thickness [16–18] and porosity, have a significant influence on the performance of PSCs [19]. Highly porous mp-TiO$_2$ films promote the easy infiltration of perovskite which subsequently fills the pores. A higher deposition of perovskites in mp-TiO$_2$ film results in increased light absorption and a higher current density [12, 14]. Moreover, the interface between the mp-TiO$_2$ film and perovskite film plays a key role in determining the overall conversion efficiency of PSCs [20]. Increasing the specific surface area and porosity of mp-TiO$_2$ film can promote a deeper infiltration of perovskites in TiO$_2$ films. This superior

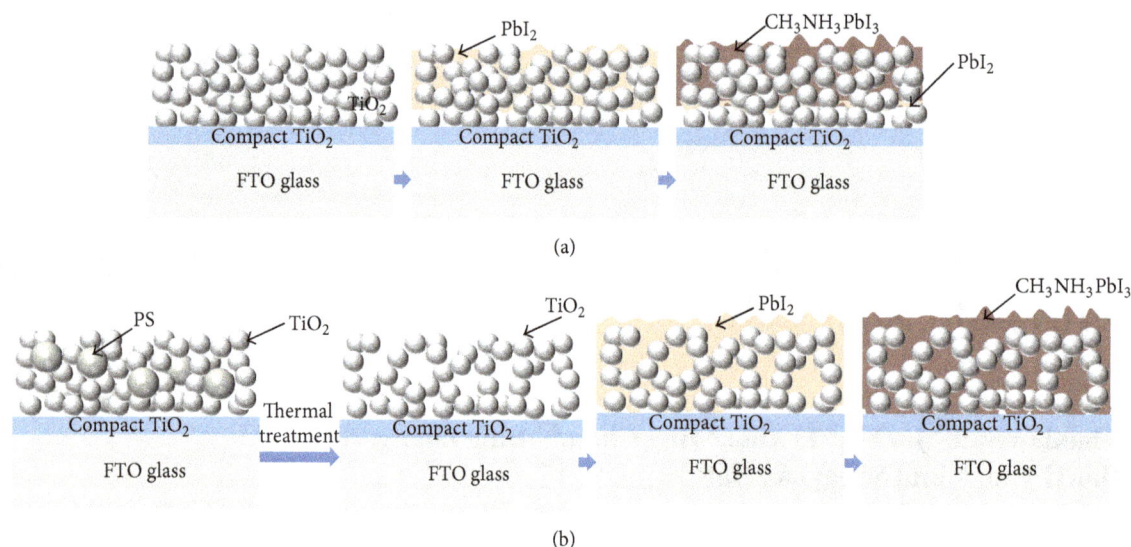

FIGURE 1: Schematic illustration of the distribution of PbI_2 and TiO_2 in mp-TiO_2 film (a) without and (b) with PS spheres.

infiltration is thought as an effective way to decrease the contact barrier between the $TiO_2/CH_3NH_3PbI_3$ interfaces, which could improve the transport of carriers in the PSCs. Dharani et al. [21] used electrospinning to prepare TiO_2 nanofibers as the ETL for PSCs. The TiO_2 nanofiber formed a highly porous structure. The excellent porous network resulted in improved loading of PbI_2. Therefore, the CH_3NH_3I can infiltrate through the pores to completely react with PbI_2. Sarkar et al. [22] prepared well-organized mesoporous TiO_2 photoelectrodes with enlarged pores by block copolymer-induced sol–gel assembly. TiO_2 photoelectrodes with larger pores are favorable for filling of perovskite in the mp-TiO_2 film. Within a certain range, devices based on larger pores showed a higher J_{sc} and superior performance. Rapsomanikis et al. [19] synthesized highly meso-macroporous TiO_2 thin films as ETLs of PSCs using a sol–gel process and Pluronic P-123 block copolymer as the organic template. Their results showed that the high porosity enabled the TiO_2 thin film to act as an ideal host for perovskite. The efficient contact between mp-TiO_2 and perovskite enhanced the electron transport.

Methods of controlling the mesoporous networks of mp-TiO_2 include changing the size of TiO_2 nanoparticles, using amphiphilic block copolymers [14, 22] and using templates. Many researchers reported that polystyrene (PS) spheres can be used as a mesostructured template to fabricate macro and mesoporous TiO_2 films in dye-sensitized solar cells (DSSC) because of its size tunability [23, 24]. By using various preparation method and PS spheres with different sizes, the morphology of the films can be easily controlled. Dionigi et al. [25] used PS spheres as structure-directing agents and coated the PS spheres with titanium dihydroxide to fabricate porous TiO_2 films with ordered pore architectures. Du et al. [24] fabricated hierarchically ordered macro-mesoporous TiO_2 films as the interfacial layer of DSSC using PS spheres

as a template. Because of the periodically ordered structure and large specific surface of the macro-mesoporous TiO_2 films, a higher J_{sc} and a PCE enhanced by 83% were obtained.

However, to our knowledge, there have been no reports on the use of PS spheres to change the mesoporous networks of mp-TiO_2 film in PSCs. In this study, we present a facile method of controlling the porosity of mp-TiO_2 by introducing PS spheres into TiO_2 paste, which can be an effective strategy for the development of mesoporous PSCs. Figure 1 shows the schematic representation of the reaction process studied in this work. As shown in Figure 1(a), TiO_2 nanoparticles are close to each other. There are only few pores between the particles. PbI_2 cannot fully infiltrate into the narrow pores of mp-TiO_2 without the introduction of PS spheres during the preparation process; instead, CH_3NH_3I would react with the superficial PbI_2, which results in a large amount of residual PbI_2 in the TiO_2 film. In Figure 1(b), PS spheres can be observed in the TiO_2 film before heat treatment. Then PS spheres were removed by heat treatment and a large amount of pores were formed in mesoporous TiO_2 film. Therefore, more PbI_2 infiltrated into the film. The subsequently deposited CH_3NH_3I reacted more completely with PbI_2, which reduced the amount of the residual PbI_2. By adjusting the mass fraction of PS spheres in TiO_2 paste, a controllable porous mp-TiO_2 film was obtained. The incorporation of the macropores in mp-TiO_2 films increased the perovskite loading in the film and improved the contact between the TiO_2 and perovskite interface, which effectively suppressed charge recombination in the interface. X-ray diffraction and fluorescence lifetime measurements confirmed that the increased porosity ensured an adequate reaction between PbI_2 and CH_3NH_3I, thus decreasing the amount of residual PbI_2 and enhancing the electron injection from the perovskite to mp-TiO_2 film. The PSCs with the porous TiO_2 films showed an enhanced short-circuit current density and higher efficiency.

2. Materials and Methods

2.1. Materials. Titanium dioxide paste (18-RT) was purchased from Yingkou OPV Tech New Energy Co., Ltd. Ethanol (99.8%) and hydrochloric acid (36%) were purchased from Beijing Chemical Plant (Beijing, China). The 100 nm PS spheres (Mw ~ 100000) were purchased from Janus New-Materials Co., Ltd. PbI_2 (99%) was purchased from Acros. CH_3NH_3I (99.5%) and 2,2′,7,7′-tetrakis-(N,N-dip-methoxyphenylamine)-9,9′-spirobifluorene (Spiro-OMeTAD) (99.7%) were purchased from Borun Chemicals (Ningbo, China). Tris(2-(1H-pyrazol-1-yl)-4-tert-butylpyridine)-cobalt(III)tris(bis(trifluoromethylsulfonyl)imide) (FK209-cobalt(III)-TFSI) was purchased from MaterWin Chemicals (Shanghai, China). N,N-dimethylformamide (DMF) was purchased from Alfa Aesar. Isopropanol was purchased from J&K Scientific Co., Ltd. All chemicals were used as received. Glass substrates with a transparent fluorine-doped tin oxide (FTO, sheet resistance 15 Ω/square) layer were used for the PSCs.

The solution of the compact TiO_2 is prepared by mixing titanium isopropoxide, HCl, and ethanol with a volume ratio of 7 : 22 : 100. In the solution of compact TiO_2, the concentrations of HCl and ethanol are 70.24% and 8.48%, respectively. For the preparation of the TiO_2 paste, we first diluted the TiO_2 paste in ethanol with a ratio of 1 : 3.5. Then, the mixture was added to 100 nm PS spheres with various wt% (0 wt%, 1.0 wt%, and 1.5 wt%) and stirred for 12 h. The spiro-MeOTAD solution was prepared by dissolving 72.3 mg spiro-MeOTAD in 1 mL chlorobenzene, and then 28.8 μL TBP and an 8 μL solution of LiTFSI (520 mg/mL LiTFSI in acetonitrile) were added.

2.2. Device Fabrication. Devices were fabricated on FTO glass substrates with a dimension of 1.5 cm × 1.3 cm. First, FTO was partially etched with Zn powder and HCl. Then, the etched FTO was cleaned using potassium sulfate solution, soap, deionized water, and ethanol, and, finally, it was sintered at 500°C for 30 minutes. A compact TiO_2 layer was prepared by spin coating compact TiO_2 solution, followed by annealing at 500°C for 30 min. The mesoporous TiO_2 film was prepared by spin coating 35 μL TiO_2 paste at 5000 r.p.m. for 30 s. Then, the films were heated to 450°C for 2 hours with a heating rate of 5°C/min. 462 mg of PbI_2 was dissolved in 1 mL DMF under stirring at 70°C for 12 hours. 40 μL PbI_2 solution was spin-coated on the mp-TiO_2 films at 3000 r.p.m. for 30 s. After loading was performed for 4 min, the substrates were dried at 70°C for 30 min. After the films were cooled to room temperature, 90 μL CH_3NH_3I solution in 2-propanol (8 mg/mL) was sprayed on the PbI_2 films, and the films were spun at 4000 r.p.m. for 30 s and then dried at 70°C for 30 min. The hole transporting layer (HTL) was prepared by spinning spiro-MeOTAD on the $TiO_2/CH_3NH_3PbI_3$ film at 3000 r.p.m. for 30 s. Finally, 70 nm gold electrodes were deposited on top of the device by thermally evaporation.

2.3. Characterization and Measurement. The surface morphology of the films was observed with a field emission scanning electron microscope (SEM, SU8010, Hitachi, 20.0 kV, 10.5 μA). The AFM images were obtained by An AC Mode III (Agilent 5500) atomic force microscope (AFM). X-ray diffraction (XRD) patterns were obtained by using a Bruker X-ray diffractometer with a Cu-Kα radiation source (40 kV, 400 mA). The 2θ diffraction angle was scanned from 10° to 80°, with a scanning speed of 1 second per step. The incident-photon-to-electron conversion efficiency (IPCE) curves were measured under ambient atmosphere using a QE-R measurement system (Enli Technology). The current-voltage characteristics (*J-V* curves) were obtained with a Keithley 2400 source meter and a sunlight simulator (XES-300T1, SAN-EI Electric, AM 1.5), which was calibrated using a standard silicon reference cell.

3. Results and Discussion

3.1. Morphology of the Porous TiO2 Film. The size and mass fraction of PS spheres are crucial in determining the porosity and uniformity of the TiO_2 mesoporous layer. The surface morphology of the TiO_2 films was analyzed by scanning electron microscopy. The SEM images of sample PS-1.0 before and after heat treatment are shown in Figure S1 in Supplementary Material available online at https://doi.org/10.1155/2017/4935265. As observed in Figure 2, unlike the TiO_2 layer formed by spinning the paste without PS spheres (PS-0), pores can be observed in the mp-TiO_2 film prepared by TiO_2 paste with PS spheres. For a low mass fraction of 0.5 wt% PS spheres in the paste (PS-0.5), a few pores with an average size of 70 nm (the statistic numbers and histogram of pore size distribution are shown in Table S1 and Figure S2) are formed on the surface of the mp-TiO_2 layer. When the mass fraction of the PS spheres in TiO_2 paste increased to 1.0 wt%, (PS-1.0), lots of pores with an average size of 80 nm are formed. The average pore sizes for both PS-0.5 and PS-1.0 are almost the same, and the pore distribution is uniform. As the mass fraction of PS spheres increases to 1.5 wt% (PS-1.5), pores with relative larger pores (94 nm) were formed due to the high mass fraction of PS spheres and the agglomeration of PS spheres in the TiO_2 paste. The pore distribution is inhomogeneous compared to the PS-1.0 sample. The pore structure could be clearly observed in the cross-sectional SEM images of samples PS-0 and PS-1.0 (Figures 2(e) and 2(f)). As observed in Figures 2(e) and 2(f), the thickness of samples PS-0 and PS-1.0 is 260 nm and 360 nm, respectively. Moreover, the porous structure of sample PS-1.0 is looser than that of sample PS-0. The TiO_2 nanoparticles are densely packed and there are almost no pores among the nanoparticles in sample PS-0. However, pores can be observed from the cross-sectional SEM image in PS-1.0 (Figure 2(f)). The pores emerged not only on the surface of mp-TiO_2 film but also throughout the film. Sample

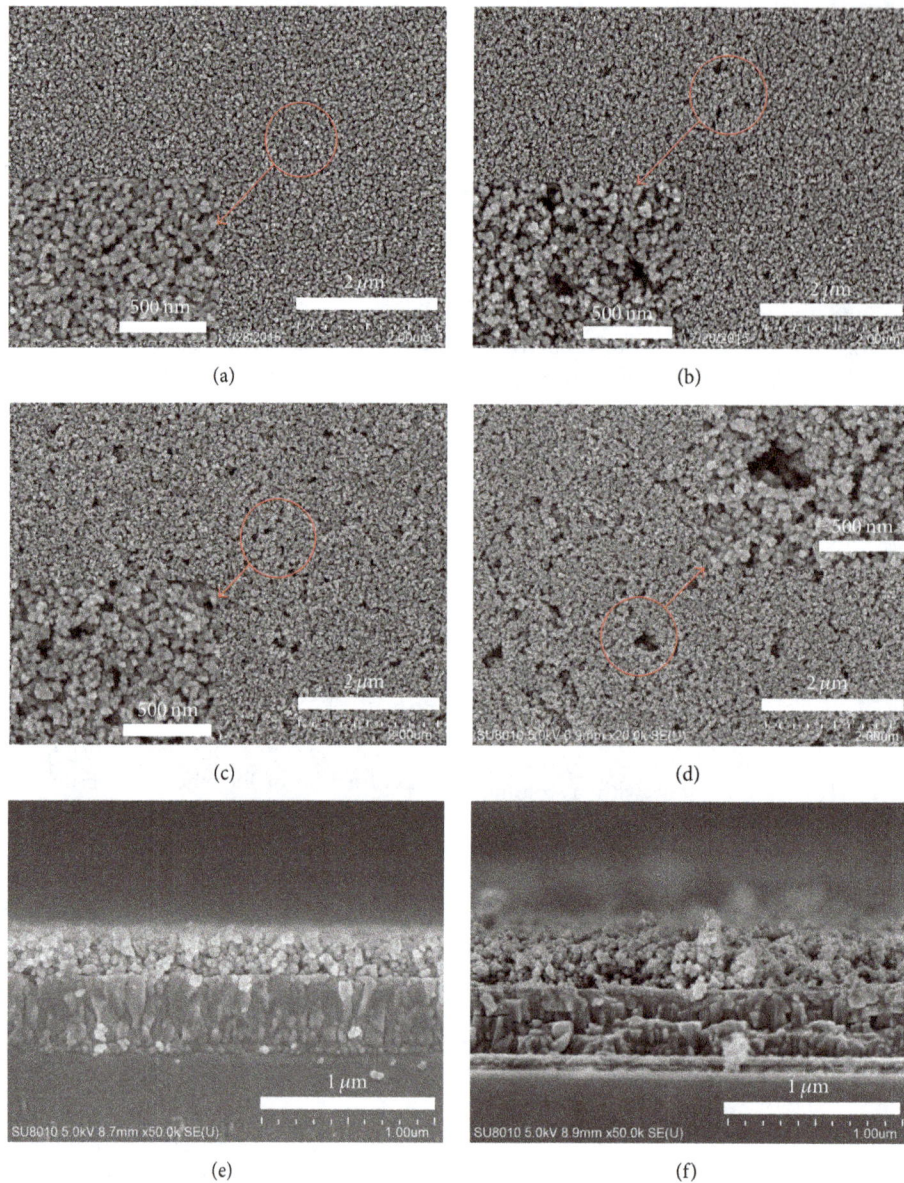

FIGURE 2: Plane-view SEM images of (a) PS-0, (b) PS-0.5, (c) PS-1.0, and (d) PS-1.5. Cross-sectional SEM images under high magnification of (e) PS-0 and (f) PS-1.0.

PS-1.0 is thicker than sample PS-0 due to the presence of more pores in the film.

3.2. Effect of TiO$_2$ Porosity on the Growth of Perovskite.

The mp-TiO$_2$ film plays a key role in determining the structure of the perovskite layer. SEM and AFM images of the perovskite layer were obtained for mp-TiO$_2$ films prepared by TiO$_2$ paste with different mass fraction PS spheres. As shown in Figure 3, the average size of the perovskite is 300 nm regardless of if PS spheres were used in the mp-TiO$_2$ paste or not. The root-mean-square (RMS) roughness values of the CH$_3$NH$_3$PbI$_3$ films based on samples PS-0, PS-0.5, PS-1.0, and PS-1.5 obtained for a 5 μm × 5 μm area were evaluated as 53.3 nm, 55.6 nm, 53.2 nm, and 58.2 nm, respectively. This means that the amount of pores in the mp-TiO$_2$ films does

not change the morphology and roughness of the perovskite film.

To better understand the distribution of perovskite in the mp-TiO$_2$ film, energy dispersive X-ray (EDX) mapping was performed. The EDX mapping of the cross-sectional area shows the distribution of two elements, Ti and Pb, along the mp-TiO$_2$ film thickness. As observed in Figure 4(a), a small amount of Pb is deposited in the sample PS-0. Figure 4(b) shows that more Pb is deposited at deeper levels in the sample PS-1.0 and the distribution of Pb is more uniform along the thickness of the TiO$_2$ film. In addition, CH$_3$NH$_3$PbI$_3$ was uniformly distributed in the mp-TiO$_2$ film owing to the presence of more pores.

Moreover, the perovskite crystallinity and the amount of residual PbI$_2$ are crucial to the performance of perovskite solar cells. It is well known that the residual PbI$_2$ layer exists

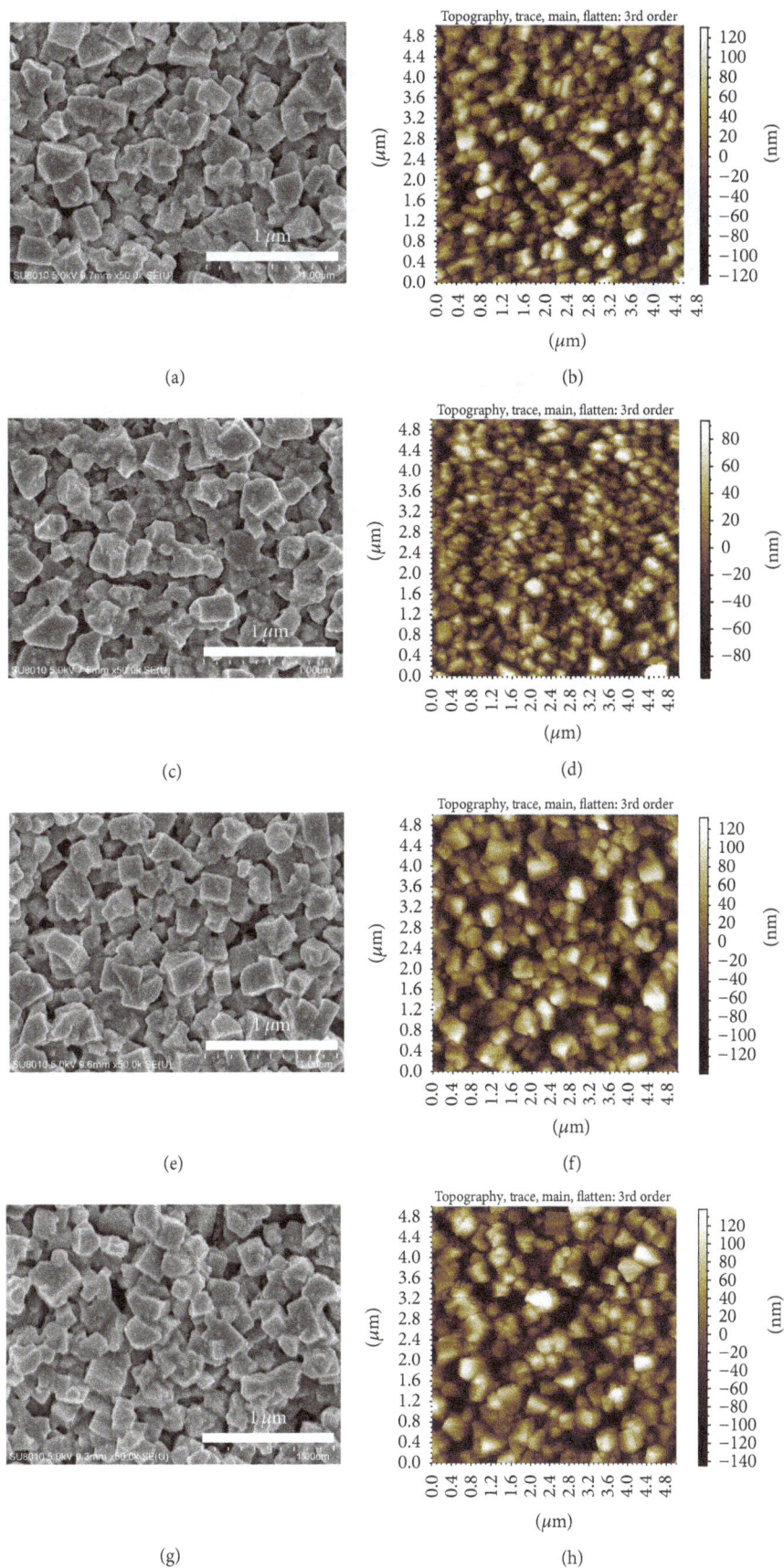

FIGURE 3: SEM and AFM images ($5\,\mu m \times 5\,\mu m$) of the perovskite films based on the mesoporous TiO$_2$: (a), (b) PS-0; (c), (d) PS-0.5; (e), (f) PS-1.0; and (g), (h) PS-1.5.

(a)

(b)

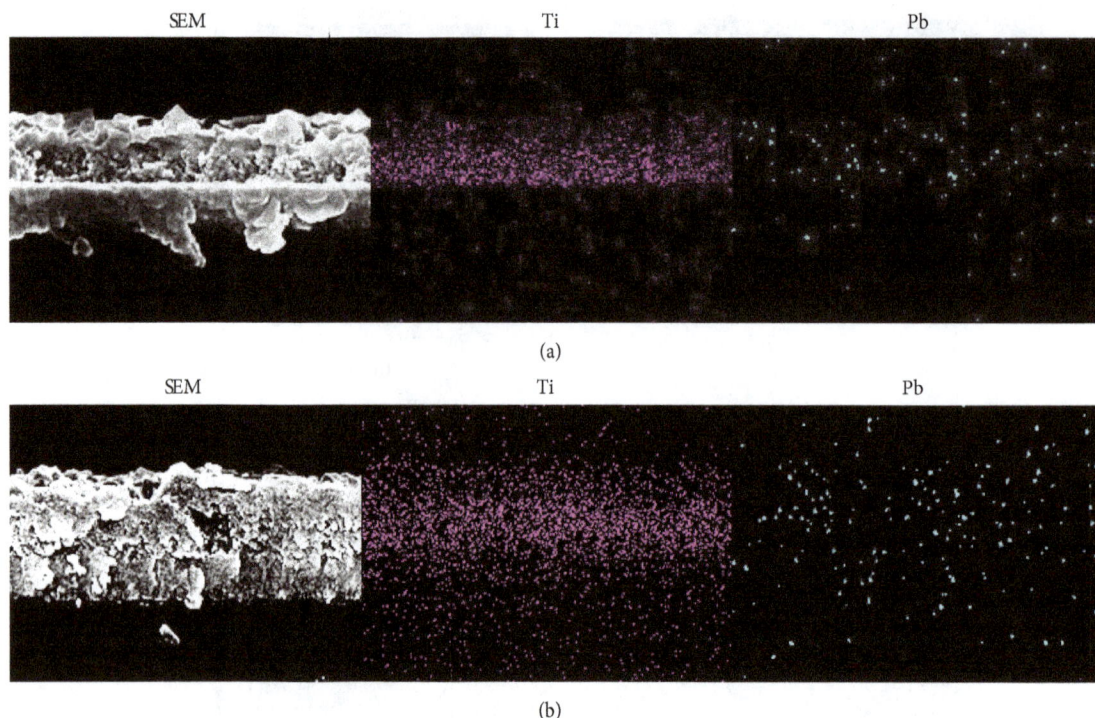

FIGURE 4: SEM-EDX mapping along the mp-TiO$_2$ film thickness to show the change in the distribution of Ti and Pb in CH$_3$NH$_3$PbI$_3$ loaded on (a) PS-0 and (b) PS-1.0.

between the interface of the mp-TiO$_2$ film and perovskite layer due to the incomplete reaction of PbI$_2$ and CH$_3$NH$_3$I [26]. Excessive amounts of residual PbI$_2$ will block the electron injected from the perovskite to mp-TiO$_2$ film, which deteriorates the cell performance of PSCs [27, 28]. However, a trace amount of residual PbI$_2$ acts as a passivation layer and reduces charge recombination at the interface between mp-TiO$_2$ film and the perovskite layer. To investigate the influence of the porosity of TiO$_2$ substrate on the growth of perovskite, X-ray diffraction was performed. In Figure 5, the peaks at 26.8°, 38.1°, and 51.8° correspond to the (110), (200), and (211) planes of FTO. The diffraction peaks marked by stars represent the PbI$_2$ (001) lattice plane, which conforms well with the literature data [29]. The Bragg peaks at 14.08°, 19.92°, 28.40°, 31.85°, 40.46°, and 43.02°, respectively, represent the reflections from the (110), (112), (220), (310), (224), and (314) crystal planes of the tetragonal perovskite structure [30, 31], which means that the change in porosity of the TiO$_2$ mesoporous layer has no influence on perovskite crystallinity. However, as the mass fraction of PS spheres in the TiO$_2$ paste increases, the intensity of the PbI$_2$ peaks reduces. There is only a little residual PbI$_2$ in the sample PS-1.0. By introducing PS spheres in TiO$_2$ paste, more pores were formed in the mp-TiO$_2$ film. The presence of more pores enables a deeper infiltration of PbI$_2$ and CH$_3$NH$_3$I solution, which endows complete reaction to CH$_3$NH$_3$PbI$_3$.

3.3. Photovoltaic Characterization of PSCs.
The current density versus voltage (J-V) characteristics of PCSs based on the mp-TiO$_2$ layer prepared by TiO$_2$ paste with and without PS

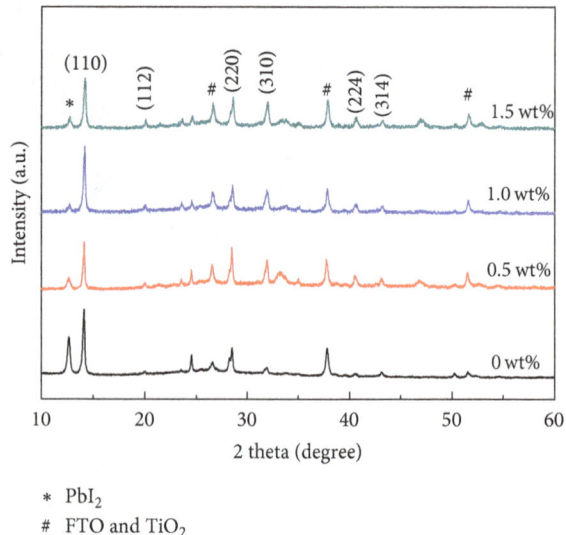

FIGURE 5: X-ray diffraction patterns of CH$_3$NH$_3$PbI$_3$ films based on TiO$_2$ mesoporous layers with different mass fractions of PS spheres.

spheres are shown in Figure 6. The photovoltaic parameters of the devices are summarized in Table 1. PSCs based on the mp-TiO$_2$ film PS-0 showed a reasonable PCE of 10.07% with an open-circuit voltage (V_{oc}) of 0.91 V, a short-circuit current (J_{sc}) of 19.07 mA/cm^2, and a fill factor (FF) of 57.99%. A relatively higher performance was exhibited by the device with PS spheres. After doping 0.5 wt% PS spheres, J_{sc}, V_{oc},

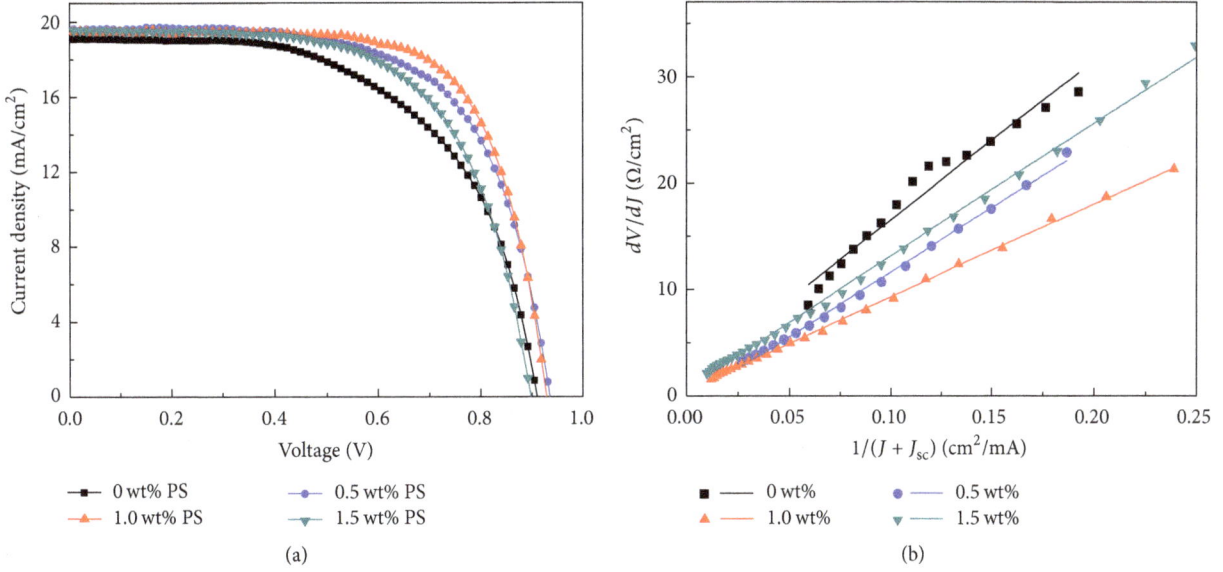

FIGURE 6: (a) Current density-voltage curves and the best-performing solar cells based on mp-TiO$_2$ film with different wt% of PS. (b) Plots of dV/dJ versus $1/(J + J_{sc})$ and the linear fitting curves.

TABLE 1: Photovoltaic performance of CH$_3$NH$_3$PbI$_3$ based devices as a function of different wt% of PS.

wt% of PS	R_{sh} (Ω)	R_s ($\Omega \cdot cm^2$)	J_{sc} (mA/cm^2)	V_{oc} (V)	Fill factor (%)	Efficiency (%)
0	1220	1.57	19.07	0.91	57.99	10.07
0.5	3860	0.43	19.64	0.94	64.91	11.92
1.0	5320	0.56	19.44	0.93	69.91	12.62
1.5	3700	0.71	19.54	0.90	63.47	11.14

FF, and PCE increased to 19.64 mA/cm^2, 0.94 V, 64.91% and 11.92%, respectively. For a 1.0% mass fraction of PS spheres, the best PCE (V_{oc} of 0.93, J_{sc} of 19.44 mA/cm^2, FF of 69.91%, and PCE of 12.62%) was achieved. When the mass fraction of PS spheres increased to 1.5 wt%, the PCS exhibits J_{sc} of 19.54 mA/cm^2, V_{oc} of 0.90 V, FF of 63.47%, and PCE of 11.14%. The decrease in the residual PbI$_2$ contributes to rapid electron injection from the perovskite to TiO$_2$ and higher J_{sc}. The improvements in the performance for the PSCs are mainly due to the increase of FF. Generally, FF depends largely on the series resistance (R_s) and shunt resistance (R_{sh}).

The equivalent circuit of the perovskite solar cell is shown in Figure 7. The output current density J can be expressed by the following equation:

$$J = J_0 \left(e^{q(V - JR_s)/AkT} - 1 \right) + \frac{V - JR_s}{R_{sh}} - J_{sc}, \quad (1)$$

where J_0 represent the reverse saturation current density, A is ideality factor, k is Boltzmann's constant, T represent the temperature, and q represent electron charge. When $R_s \ll R_{sh}$, (1) can be expressed as

$$\frac{dV}{dJ} = R_s + \frac{AkT}{q} \frac{1}{J + J_{sc}}. \quad (2)$$

It is found that, from (2), dV/dJ has a linear relation with $(J_{sc} - J)^{-1}$. The intercept of the linear fitting curve gives

the value of series resistance. Figure 6(b) shows the plot of dV/dJ versus $1/(J + J_{sc})$ and the linear fitting curve. As can be seen in Figure 6(b), the fitting curves are more linear by doping PS spheres in TiO$_2$ paste. Slightly decreased values of R_s from 1.57 $\Omega \cdot cm^2$ to 0.43 $\Omega \cdot cm^2$, 0.56 $\Omega \cdot cm^2$, and 0.71 $\Omega \cdot cm^2$ were evaluated after doping 0.5 wt%, 1.0 wt%, and 1.5 wt% PS spheres in the mp-TiO$_2$ paste. The smaller R_s values are due to the reduction in both the contact resistance and bulk resistance, which means a higher photocurrent will be generated [32]. However, when the mass fraction of PS spheres is 1.5 wt%, the increased R_s is attributed to the higher resistance caused by more pores and the negative contact with TiO$_2$ nanoparticles. The device based on TiO$_2$ mesoporous layer prepared by TiO$_2$ paste with PS spheres showed larger R_{sh} (R_{sh} values of the PSCs are 1220 Ω, 3860 Ω, 5320 Ω, and 3700 Ω for doping with 0 wt%, 0.5 wt%, 1.0 wt%, and 1.5 wt% PS spheres, resp.). A higher R_{sh} can improve FF and electron mobility [33], which is consistent with the results of the J-V test.

R_{sh} is closely related to the charge recombination at interfaces inside solar cells. A lower charge recombination contributes to a higher R_{sh} [34]. To better understand the separation of light-induced charge at the TiO$_2$/CH$_3$NH$_3$PbI$_3$ interface, we performed time-resolved photoluminescence (PL) decay measurements on the CH$_3$NH$_3$PbI$_3$ perovskite-filled mp-TiO$_2$ films prepared by TiO$_2$ paste with different

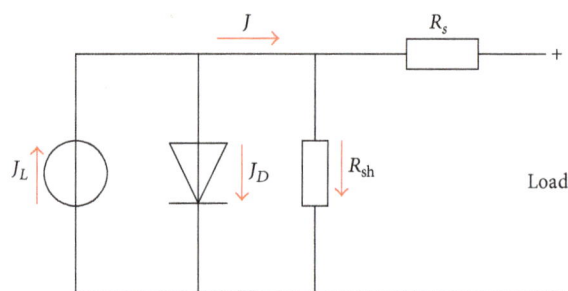

FIGURE 7: Equivalent circuit of the perovskite solar cell.

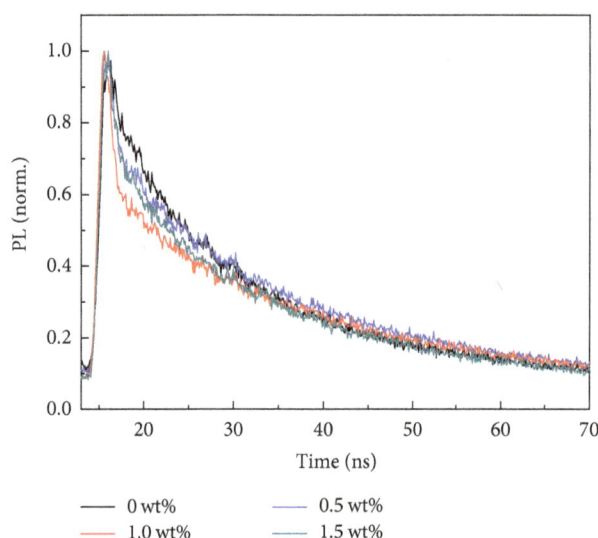

FIGURE 8: Normalized transient PL decay profiles of the perovskites based on TiO_2 with different wt% PS.

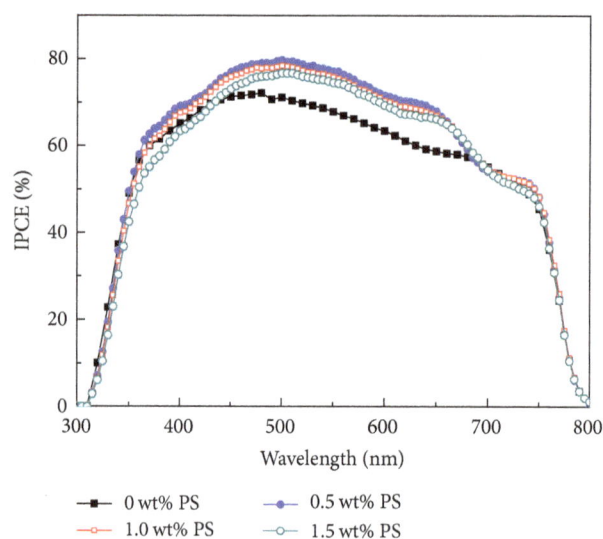

FIGURE 9: IPCE spectra of the best-performing solar cells based on different mass fractions of PS spheres.

mass fractions of PS spheres, which are presented in Figure 8. Using global biexponential fits, the PL decay of the $CH_3NH_3PbI_3$ perovskite in the mp-TiO_2 films without PS spheres and with 0.5 wt%, 1.0 wt% and 1.5 wt% PS spheres exhibits τ_1 values of 22.53 ns, 17.57 ns, 17.71 ns, and 18.93 ns, respectively. By doping PS spheres into the TiO_2 paste, the rate of electron injection from the perovskite into TiO_2 film becomes faster, which results in lower charge recombination at the TiO_2/$CH_3NH_3PbI_3$ interfaces. This could be attributed to the better filling of the $CH_3NH_3PbI_3$ perovskite in the mp-TiO_2 film and the more complete contact between the TiO_2/perovskite as more pores emerge in the mp-TiO_2 film.

The photon-to-electron conversion efficiency (IPCE) spectra with mp-TiO_2 doping for different mass fractions of PS spheres are shown in Figure 9. The convolution of the spectral response with the photon flux of the AM 1.5G spectrum provided the estimated J_{sc} values of 15.537 mA/cm^2, 16.994 mA/cm^2, 16.825 mA/cm^2, and 16.397 mA/cm^2. The calculated J_{sc} values from the IPCE spectrum are well matched with the J_{sc} values obtained from the J-V curves. In addition, the PCSs from mp-TiO_2 films with PS spheres exhibited a higher and broader spectrum from 450 nm to 700 nm. Here, an IPCE of ~80% was obtained at the

maximum peak, while the device based on mp-TiO_2 films without PS spheres exhibited a lower IPCE of ~70%.

To further determine the influence of the porosity of mp-TiO_2 films on the fabricated solar cells, we showed the statistic results of the cells based on the mp-TiO_2 film prepared by TiO_2 paste with different mass fractions of PS spheres in Figure 10. The deviations of J_{sc}, V_{oc}, FF, and PCE have been shown in Table S2. As observed in Figure 10, J_{sc} and FF increase as the mass fraction of the PS spheres increases and they attain their highest values at the mass fraction of 1.0 wt%, which contributes to the increase in the PCE.

4. Conclusions

In conclusion, mp-TiO_2 films with tunable porosities were fabricated by doping PS spheres in TiO_2 paste and applied as the ETL of perovskite solar cells. The results indicate that the porosity of mp-TiO_2 films not only affects the infiltration and residual amounts of PbI_2 but also significantly influences the contact between the mp-TiO_2 film and perovskite layer. By adjusting the mass fraction of PS spheres, the perovskite solar cell based on mp-TiO_2 film prepared by TiO_2 paste with 1.0 wt% PS spheres exhibits the highest power conversion efficiency of 12.62% under a simulated standard AM 1.5 condition.

Competing Interests

The authors declare that they have no competing interests.

Acknowledgments

This work was supported by the National High Technology Research and Development Program of China (863 Program) (no. 2015AA050602), the National Natural Science Foundation of China (no. 51372083), Jiangsu Province Science and

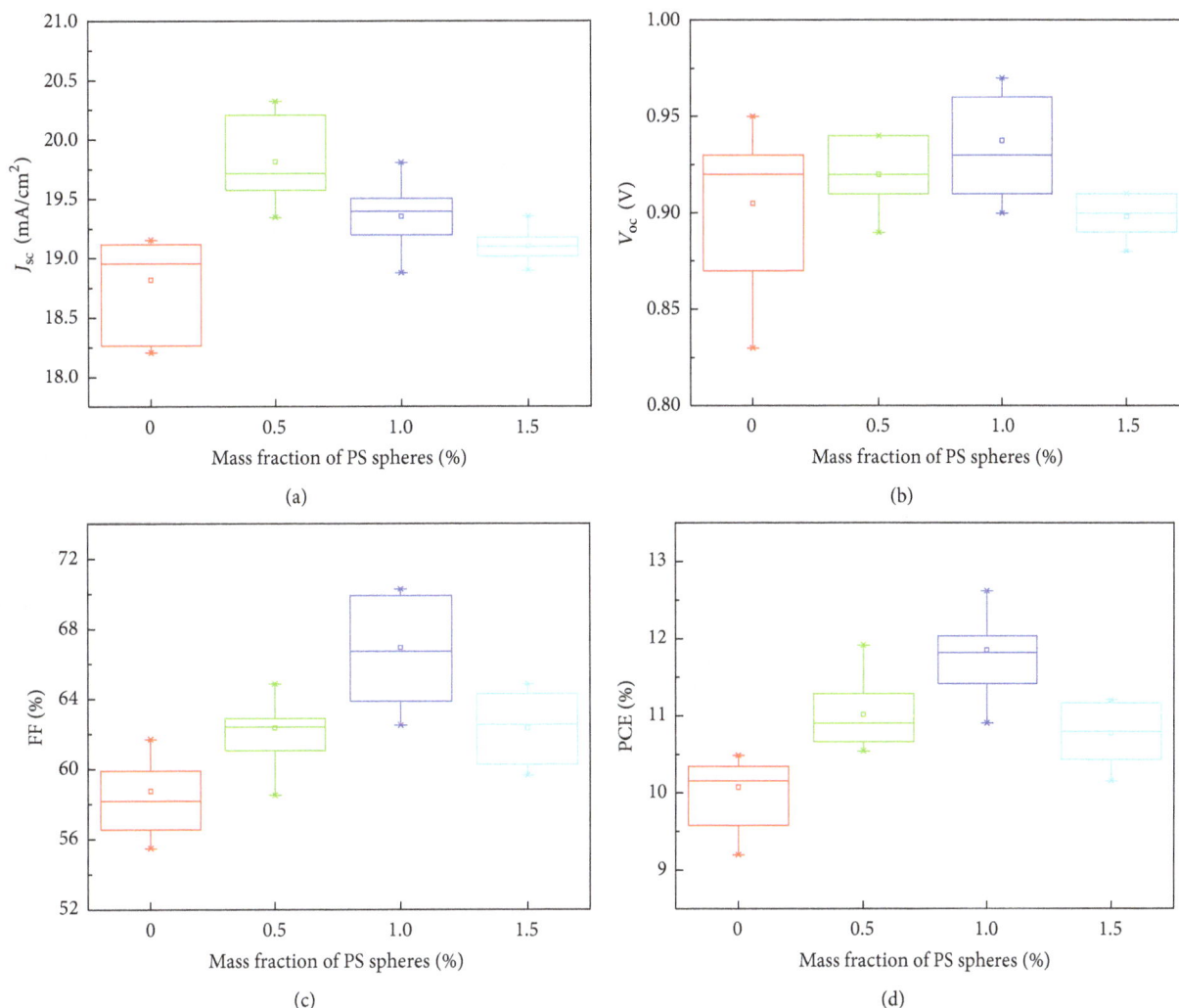

FIGURE 10: Statistic results of the cells based on the mp-TiO$_2$ film with different mass fractions of PS spheres.

Technology Support Program, China (BE2014147-4), and the Fundamental Research Funds for the Central Universities (nos. 2014ZZD07 and 2015ZD11).

References

[1] A. Kojima, K. Teshima, Y. Shirai, and T. Miyasaka, "Organometal halide perovskites as visible-light sensitizers for photovoltaic cells," *Journal of the American Chemical Society*, vol. 131, no. 17, pp. 6050–6051, 2009.

[2] "Best Research-Cell Efficiencie," national renewable energy laboratory, http://www.nrel.gov/pv/assets/images/efficiency_chart .jpg.

[3] W. S. Yang, J. H. Noh, N. J. Jeon et al., "High-performance photovoltaic perovskite layers fabricated through intramolecular exchange," *Science*, vol. 348, no. 6240, pp. 1234–1237, 2015.

[4] J. Qing, H.-T. Chandran, H.-T. Xue et al., "Simple fabrication of perovskite solar cells using lead acetate as lead source at low temperature," *Organic Electronics: physics, materials, applications*, vol. 27, article no. 3243, pp. 12–17, 2015.

[5] M. I. Ahmed, A. Habib, and S. S. Javaid, "Perovskite solar cells: potentials, challenges, and opportunities," *International Journal of Photoenergy*, vol. 2015, Article ID 592308, 13 pages, 2015.

[6] Q. Chen, H. Zhou, Z. Hong et al., "Planar heterojunction perovskite solar cells via vapor-assisted solution process," *Journal of the American Chemical Society*, vol. 136, no. 2, pp. 622–625, 2014.

[7] M. Liu, M. B. Johnston, and H. J. Snaith, "Efficient planar heterojunction perovskite solar cells by vapour deposition," *Nature*, vol. 501, no. 7467, pp. 395–398, 2013.

[8] L. Etgar, P. Gao, Z. Xue et al., "Mesoscopic CH$_3$NH$_3$PbI$_3$/TiO$_2$ heterojunction solar cells," *Journal of the American Chemical Society*, vol. 134, no. 42, pp. 17396–17399, 2012.

[9] H.-S. Kim, C.-R. Lee, J.-H. Im et al., "Lead iodide perovskite sensitized all-solid-state submicron thin film mesoscopic solar cell with efficiency exceeding 9%," *Scientific Reports*, vol. 2, article 591, 2012.

[10] M. M. Lee, J. Teuscher, T. Miyasaka, T. N. Murakami, and H. J. Snaith, "Efficient hybrid solar cells based on mesosuperstructured organometal halide perovskites," *Science*, vol. 338, no. 6107, pp. 643–647, 2012.

[11] Y. Li, J. K. Cooper, R. Buonsanti et al., "Fabrication of planar heterojunction perovskite solar cells by controlled low-pressure vapor annealing," *Journal of Physical Chemistry Letters*, vol. 6, no. 3, pp. 493–499, 2015.

[12] H. Liu, Z. Huang, S. Wei, L. Zheng, L. Xiao, and Q. Gong, "Nano-structured electron transporting materials for perovskite solar cells," *Nanoscale*, vol. 8, no. 12, pp. 6209–6221, 2016.

[13] M. He, D. Zheng, M. Wang, C. Lin, and Z. Lin, "High efficiency perovskite solar cells: from complex nanostructure to planar heterojunction," *Journal of Materials Chemistry A*, vol. 2, no. 17, pp. 5994–6003, 2014.

[14] H. Lu, K. Deng, N. Yan et al., "Efficient perovskite solar cells based on novel three-dimensional TiO_2 network architectures," *Science Bulletin*, vol. 61, no. 10, pp. 778–786, 2016.

[15] Y. Yang, K. Ri, A. Mei et al., "The size effect of TiO_2 nanoparticles on a printable mesoscopic perovskite solar cell," *Journal of Materials Chemistry A*, vol. 3, no. 17, pp. 9103–9107, 2015.

[16] S. Aharon, S. Gamliel, B. E. Cohen, and L. Etgar, "Depletion region effect of highly efficient hole conductor free $CH_3NH_3PbI_3$ perovskite solar cells," *Physical Chemistry Chemical Physics*, vol. 16, no. 22, pp. 10512–10518, 2014.

[17] Y. Zhao, A. M. Nardes, and K. Zhu, "Solid-state mesostructured perovskite $CH_3NH_3PbI_3$ solar cells: charge transport, recombination, and diffusion lengt," *Journal of Physical Chemistry Letters*, vol. 5, no. 3, pp. 490–494, 2014.

[18] H.-S. Kim and N.-G. Park, "Parameters affecting I–V hysteresis of $CH_3NH_3PbI_3$ perovskite solar cells: effects of perovskite crystal size and mesoporous TiO2 layer," *Journal of Physical Chemistry Letters*, vol. 5, no. 17, pp. 2927–2934, 2014.

[19] A. Rapsomanikis, D. Karageorgopoulos, P. Lianos, and E. Stathatos, "High performance perovskite solar cells with functional highly porous TiO_2 thin films constructed in ambient air," *Solar Energy Materials and Solar Cells*, vol. 151, pp. 36–43, 2016.

[20] W. Wang, Z. Zhang, Y. Cai et al., "Enhanced performance of $CH_3NH_3PbI_{3-x}Cl_x$ perovskite solar cells by CH_3NH_3I modification of TiO_2-perovskite layer interface," *Nanoscale Research Letters*, vol. 11, no. 1, article 316, 2016.

[21] S. Dharani, H. K. Mulmudi, N. Yantara et al., "High efficiency electrospun TiO_2 nanofiber based hybrid organic–inorganic perovskite solar cell," *Nanoscale*, vol. 6, no. 3, pp. 1675–1679, 2014.

[22] A. Sarkar, N. J. Jeon, J. H. Noh, and S. I. Seok, "Well-organized mesoporous TiO_2 photoelectrodes by block copolymer-induced sol–Gel assembly for inorganic–organic hybrid perovskite solar cells," *Journal of Physical Chemistry C*, vol. 118, no. 30, pp. 16688–16693, 2014.

[23] Y. J. Kim, Y. H. Lee, M. H. Lee et al., "Formation of efficient dye-sensitized solar cells by introducing an interfacial layer of long-range ordered mesoporous TiO_2 thin film," *Langmuir*, vol. 24, no. 22, pp. 13225–13230, 2008.

[24] J. Du, X. Lai, N. Yang et al., "Hierarchically ordered macro–mesoporous TiO_2–graphene composite films: improved mass transfer, reduced charge recombination, and their enhanced photocatalytic activities," *ACS Nano*, vol. 5, no. 1, pp. 590–596, 2011.

[25] C. Dionigi, P. Greco, G. Ruani, M. Cavallini, F. Borgatti, and F. Biscarini, "3D hierarchical porous TiO_2 films from colloidal composite fluidic deposition," *Chemistry of Materials*, vol. 20, no. 22, pp. 7130–7135, 2008.

[26] S. Wang, W. Dong, X. Fang et al., "Credible evidence for the passivation effect of remnant PbI_2 in $CH_3NH_3PbI_3$ films in improving the performance of perovskite solar cells," *Nanoscale*, vol. 8, no. 12, pp. 6600–6608, 2016.

[27] Y. H. Lee, J. Luo, R. Humphry-Baker, P. Gao, M. Grätzel, and M. K. Nazeeruddin, "Unraveling the reasons for efficiency loss in perovskite solar cells," *Advanced Functional Materials*, vol. 25, no. 25, pp. 3925–3933, 2015.

[28] J. Jiang, H. J. Tao, S. Chen et al., "Efficiency enhancement of perovskite solar cells by fabricating as-prepared film before sequential spin-coating procedure," *Applied Surface Science*, vol. 371, pp. 289–295, 2016.

[29] D. H. Cao, C. C. Stoumpos, C. D. Malliakas et al., "Remnant Pbi2, an unforeseen necessity in high-efficiency hybrid perovskite-based solar cells? A)," *APL Materials*, vol. 2, no. 9, Article ID 091101, 2014.

[30] Y. Zhao and K. Zhu, "Charge transport and recombination in perovskite $(CH_3NH_3)PbI_3$ sensitized TiO_2 solar cells," *The Journal of Physical Chemistry Letters*, vol. 4, no. 17, pp. 2880–2884, 2013.

[31] T. Baikie, Y. Fang, J. M. Kadro et al., "Synthesis and crystal chemistry of the hybrid perovskite (CH3NH3)PbI3 for solid-state sensitised solar cell applications," *Journal of Materials Chemistry A*, vol. 1, no. 18, pp. 5628–5641, 2013.

[32] Y. Tu, J. Wu, M. Zheng et al., "TiO_2 quantum dots as superb compact block layers for high-performance $CH_3NH_3PbI_3$ perovskite solar cells with an efficiency of 16.97%," *Nanoscale*, vol. 7, no. 48, pp. 20539–20546, 2015.

[33] W. Ke, G. Fang, J. Wan et al., "Efficient hole-blocking layer-free planar halide perovskite thin-film solar cells," *Nature Communications*, vol. 6, article 6700, 2015.

[34] W. Ke, G. Fang, Q. Liu et al., "Low-temperature solution-processed tin oxide as an alternative electron transporting layer for efficient perovskite solar cells," *Journal of the American Chemical Society*, vol. 137, no. 21, pp. 6730–6733, 2015.

5

Copper Sulfide Catalyzed Porous Fluorine-Doped Tin Oxide Counter Electrode for Quantum Dot-Sensitized Solar Cells with High Fill Factor

Satoshi Koyasu, Daiki Atarashi, Etsuo Sakai, and Masahiro Miyauchi

School of Materials and Chemical Technology, Tokyo Institute of Technology, 2-12-1 Ookayama, Meguro-ku, Tokyo 152-8552, Japan

Correspondence should be addressed to Masahiro Miyauchi; mmiyauchi@ceram.titech.ac.jp

Academic Editor: Nini R. Mathews

The performance of quantum dot-sensitized solar cell (QDSSC) is mainly limited by chemical reactions at the interface of the counter electrode. Generally, the fill factor (FF) of QDSSCs is very low because of large charge transfer resistance at the interface between the counter electrode and electrolyte solution containing redox couples. In the present research, we demonstrate the improvement of the resistance by optimization of surface area and amount of catalyst of the counter electrode. A facile chemical synthesis was used to fabricate a composite counter electrode consisting of fluorine-doped tin oxide (FTO) powder and CuS nanoparticles. The introduction of a sputtered gold layer at the interface of the porous-FTO layer and underlying glass substrate also markedly reduced the resistance of the counter electrode. As a result, we could reduce the charge transfer resistance and the series resistance, which were 2.5 [Ω] and 6.0 [Ω], respectively. This solar cell device, which was fabricated with the presently designed porous-FTO counter electrode as the cathode and a PbS-modified electrode as the photoanode, exhibited a FF of 58%, which is the highest among PbS-based QDSSCs reported to date.

1. Introduction

The first quantum dot-sensitized solar cell (QDSSC) was reported over two decades ago [1]. Although QDSSCs have a similar structure and working principle to those of dye-sensitized solar cells (DSSCs), they have attracted considerable attention because of several advantageous characteristics, particularly a tunable bandgap in response to particle size, high visible-light absorbance, and possibility of multiple exciton generation [1, 2]. Recently, the high energy conversion efficiencies of QDSSC have been reported for PbS [3, 4] quantum dots (QD) and $CdSe_xTe_{1-x}$ QD [5] with ZnS shell. These QDSSCs exhibited higher short circuit current (J_{sc}) than a DSSC due to their absorption of a wider range of visible light and higher quantum yield. In contrast, the open circuit voltage (V_{oc}) of QDSSCs is only approximately 0.5 V. Therefore, the QDSSCs generally have high J_{sc} but have low V_{oc}. Under the condition with high J_{sc} and low V_{oc} properties, the fill factor (FF) of QDSSCs is highly sensitive to series resistance, as compared to solar cells with low J_{sc} and high V_{oc}. The FF of QDSSCs is reportedly limited by series resistance at the interface of the counter electrode [6, 7]. Consequently, the low V_{oc} and insufficient FF of QDSSCs limit their energy conversion efficiency of solar cells [8].

QDSSCs mainly consist of three components: a quantum dot-modified mesoporous TiO_2 electrode, which functions as a photoanode; a counter electrode, which functions as a cathode; and electrolyte, which is composed of sulfur redox couples and fills the gap between the photoanode and cathode. QDSSC involves several types of resistance, those in bulk or at interfaces. In particular, high resistance at the counter electrode limits the FF and efficiency of QDSSCs [9–11].

Two types of series resistance occur at counter electrodes: normal series resistance (R_s) and charge transfer resistance (R_{ct}). R_{ct} values are strongly dependent on the reduction reaction of the redox couple. Iodine redox couple is usually used in DSSCs system, while S^{2-}/S_x^{2-} redox couple is used in QDSSCs system. R_{ct} of S^{2-}/S_x^{2-} redox couple in QDSSCs is markedly higher than that of the iodine redox couple in DSSCs because of lower reactivity of S^{2-}/S_x^{2-} redox couple. Many researchers have attempted to develop efficient counter electrodes with low resistance for the construction of high FF solar cells. For example, Hodes et al. reported a low-resistance Cu_2S counter electrode, which was prepared by the sulfurization of a metallic copper sheet in polysulfide solution [12]. Also, some recent studies reported the counter electrodes with high surface area and low R_{ct} [13–23]. The origin of R_{ct} is the limitation of the chemical reaction for redox couple on the surface of electrode. Chemical reaction rate strongly depends on catalytic activity and surface area of a counter electrode. Therefore, porous structure with large surface area would have advantage to reduce R_{ct}. In particular, we have focused on the catalyst deposited porous oxide conductor as a counter electrode, which possesses high chemical stability, mechanical strength, catalytic activity, and conductivity. Although recent previous work reported the CuS deposited porous tin-doped indium oxide (ITO) counter electrode [24], optimizations of materials and structural parameters on theses electrodes, such as materials of porous layer, thickness of porous layer, loading amount of CuS catalyst, and the reduction of sheet resistance in a conductive substrate, are still challenging issues for the development of efficient QDSSCs.

In the present study, we aimed to develop the efficient counter electrode of QDSSCs with low R_{ct} as well as low R_s. To achieve low R_{ct}, we designed CuS deposited porous fluorine-doped tin oxide (FTO) electrodes by optimizing the thickness of porous-FTO layer and amount of CuS catalyst. We choose FTO but not ITO as porous layer because conductivity of ITO is extremely reduced by heating. To reduce R_s of electrodes, on the other hand, the reduction of sheet resistance in conductive substrate is very effective, since that of commercial FTO glass is relatively high (about $10\,\Omega/\square$). Generally, the sheet resistance of metal is lower than that of a commercial FTO glass, but only noble metals like gold (Au) or platinum (Pt) are available as a counter electrode of QDSSCs to avoid their corrosion in polysulfide electrolyte solution. However, use of these noble metals as a bulk form is not economical; thus, we introduced a thin Au layer between a commercial FTO glass and porous-FTO layer by sputtering to reduce R_s value. We developed CuS deposited porous-FTO film with underlying thin Au layer as an efficient counter electrode of QDSSC, which yielded quite high fill factor (FF = 58%).

2. Materials and Methods

2.1. Materials.
Commercial FTO-coated glass substrates (AGC, sheet resistance: $10.2\,\Omega/\square$) were used for the fabrication of counter and working electrodes. Pt and Au discs were used as sputtering target materials and were purchased from Kojundo Chemical Laboratory. TiO_2 particles (P25, Degussa) were used for screen printing in the preparation of photoanodes. $Pb(CH_3COO)_2 \cdot 3H_2O$, $Zn(CH_3COO)_2 \cdot 2H_2O$, $Cu(CH_3COO)_2 \cdot H_2O$, and $Na_2S \cdot 9H_2O$ were purchased from Kanto Chemical. Tin chloride ($SnCl_4 \cdot 5H_2O$) and ammonia

aqueous solution (NH_3 aq (25%)) were also purchased from Kanto Chemical, and acetylene carbon black was obtained from Sigma-Aldrich.

2.2. Fabrication of Counter Electrode.
FTO nanopowder was synthesized according to a previously described procedure [25]. Briefly, 14.02 g $SnCl_4 \cdot 5H_2O$, 0.704 g NH_4F, and 0.8 g acetylene carbon black were mixed with 40 mL distilled water in a beaker. Ammonia aqueous solution (25%) was then added until the pH of the solution reached 7. The resulting gel was dried at 100°C for 1 day and was then heated at 650°C for 30 min. After the heating treatment, the obtained powder was further annealed at 750°C for 30 min and was then immersed in pure water. The dispersed FTO particles were collected by centrifugation (6000 rpm for 5 min) and were mixed with ethyl cellulose and α-terpineol to prepare a paste for screen printing. For preparation of the FTO paste, 2 g FTO powder was dispersed in 200 mL ethanol, to which 6 mL ethyl cellulose solution was added. The ethyl cellulose solution was prepared by mixing two different molecular weights of ethyl cellulose (30–60 mPa s: 0.333 g + 5–15 mPa s: 0.666 g) in 10 mL ethanol. Following the addition of ethyl cellulose, 4 mL α-terpineol was added to FTO dispersed solution. After stirring for 10 min, ethanol was evacuated using an evaporator at 50°C to obtain FTO paste. Porous-FTO electrodes were constructed by screen-printing the FTO paste on a commercial dense FTO-coated glass, which was subsequently sintered at 450°C for 30 min. The thickness of the porous-FTO layers was controlled by the mesh size of the screen mask and the number of screen coating times. Using this method, porous-FTO electrodes with thicknesses of 5, 10, 15, and 20 μm were fabricated. We have optimized the thickness to obtain the low resistance as mentioned later. Subsequently, CuS catalyst nanoparticles were loaded onto the constructed porous-FTO electrodes by a successive ionic layer adsorption and reaction (SILAR) method [26, 27]. Two types of aqueous solutions, 0.1 M copper acetate ($Cu(CH_3COO)_2$) and 0.1 M sodium sulfide (Na_2S), were prepared for the SILAR method. Porous-FTO electrodes were immersed in $Cu(CH_3COO)_2$ solution for 30 sec and were then washed with distilled water. The washed substrates were immersed in Na_2S solution for 30 sec, followed by washing with distilled water. The dipping cycles in the two solutions were repeated 5 to 40 times, and the substrates were then dried at room temperature for 1 h. According to the SILAR cycles, we could optimize the loading amount of CuS. In addition to the fabrication of CuS-modified counter electrodes, a powder form of CuS was prepared using the same solutions with the SILAR method for XRD analysis, as the amount and crystallite size of CuS on porous-FTO were too small to determine the crystal phase of CuS.

To investigate the effect of a metal-conducting layer between the dense FTO film and porous-FTO layer, Au thin film was coated on the surface of a commercial FTO glass substrate by DC magnetron sputtering (MSP-30T; Vacuum Device) before the screen printing of the porous-FTO layer. Au was deposited for 3 min at a current of 200 mA under an argon atmosphere pressure of 8.1×10^{-4} Pa. Under these conditions, the thickness of the resulting Au layer was

350 nm, as estimated by the deposition rate and sputtering time. The expected resistance of our Au layer was calculated by considering the electrode structure. The distance from porous-FTO to the position for bonding attaching clip was 15 mm. The width of porous-FTO was 5 mm, and resistivity of gold was about 2×10^{-8} [Ωm]. If we use gold thin film of 350 nm, calculated expected resistance of gold layer by using these values is calculated as 2×10^{-8} [Ωm] $\times 15 \times 10^{-3}$ [m]/(5 $\times 10^{-3}$ [m] $\times 350 \times 10^{-9}$ [m]) = 0.17 [Ω]. For comparison with a conventional counter electrode that is typically used in QDSSCs and DSSC, platinum (Pt) counter electrodes were prepared by sputtering of Pt (MSP-30T; Vacuum Device) onto a commercial FTO-coated glass substrate at 200 mA for 45 sec at a pressure of 8.1×10^{-4} Pa.

2.3. Fabrication of PbS Photoanode. A photoanode was fabricated using a previously reported method [4]. Briefly, TiO_2 paste was prepared from commercial TiO_2 powder (P-25, Degussa) by mixing with ethyl cellulose and α-terpineol. Typically, 1 g TiO_2 powder was dispersed in 200 mL ethanol containing two types of ethyl cellulose with different viscosities, as was described for the FTO paste fabrication. After stirring the solution for 10 min, ethanol was evacuated using an evaporator at 50°C. Porous TiO_2 electrodes were then constructed on commercial dense FTO-coated glass by the screen printing of the TiO_2 paste and subsequent sintering at 450°C for 30 min.

Core-shell-type PbS/ZnS QDs were deposited on the porous TiO_2 electrode using the SILAR method. Previous studies have reported that the ZnS shell prevents the back electron transfer and recombination of electron on the surface of PbS light absorber [4, 28]. QD-modified photoanodes were also prepared by the SILAR method. Two types of solutions, 0.02 M of lead acetate ($Pb(CH_3COO)_2$) dissolved in methanol and 0.02 M of sodium sulfide (Na_2S) in methanol, were used for PbS QD deposition onto porous TiO_2 electrodes. First, porous TiO_2 electrodes were immersed in $Pb(CH_3COO)_2$ solution for 30 sec and washed once with methanol, and the electrodes were then immersed in Na_2S solution for 30 sec, followed by a further wash with methanol. The dipping cycles in the two solutions were repeated twice. In addition to PbS QD deposition, a ZnS thin layer was formed on the surface of the PbS QDs to form a core-shell structure. To form the ZnS thin layer, the PbS-modified TiO_2 electrodes were immersed in 0.02 M of zinc acetate ($Zn(CH_3COO)_2$) in methanol solution for 30 sec, washed once with methanol, and then dipped in 0.02 M of Na_2S methanol solution for 30 sec, followed by a further wash with methanol. The dipping cycles were repeated for 10 times, resulting in the formation of ZnS shells on the PbS QDs. After completion of the SILAR cycles, the substrates were dried at room temperature for 1 h.

2.4. Characterization and Electrochemical Properties of Counter Electrode. The crystal phase of synthesized FTO powder was determined using a X-ray diffraction meter (XRD; Rigaku SmartLab SPI/TISM). Morphologies of the surface and cross section of the porous-FTO electrodes

were recorded using a field emission scanning electron microscope (FE-SEM; Hitachi S4500). Microstructures of CuS/FTO composites removed from the electrodes by scraping were observed using a JEM-2010F transmission electron microscope (TEM; JEOL, Ltd.). The specific surface area (BET) of the porous-FTO electrode was also measured by determination of the nitrogen adsorption-desorption isotherm using a surface area analyzer (Gemini V 2380; Shimadzu Corp.).

Catalytic performance of counter electrodes was studied by conducting electrochemical impedance analysis. Impedance analysis is one of the most useful methods to separately discuss the charge transfer resistance and normal series resistance. For the analysis, the CuS-modified FTO electrode was connected as a working electrode, and the Pt and Ag/AgCl electrodes were used as counter and reference electrodes, respectively. Catalytic activities of CuS-modified FTO electrodes were investigated by measuring series resistance (R_s) and charge transfer resistance (R_{ct}), which were determined from fitting results of Nyquist plots. A potentiostat equipped with a frequency response analyzer (FRA) was used for impedance measurements (HZ-5000; Hokuto Denko). In the electrochemical measurements, the area of the electrode exposed to the electrolyte solution was 0.25 cm^2, and the bias was set at the rest potential condition (0.67 V versus Ag/AgCl), amplitude voltage was 10 mV, and the AC frequency ranged from 20 kHz to 0.01 Hz. The electrolyte was composed of 1.0 M Na_2S and 1.0 M sulfur aqueous solution.

2.5. Fabrication and Performance Evaluation of Solar Cell Device. An aqueous solution composed of 1.0 M sodium sulfide and 1.0 M sulfur was used as an electrolyte in the QDSSC device, which contained a 50 μm gap between the CuS counter electrode and PbS photoanode due to the insertion of an ethylene polymer film spacer (Surlyn; Du Pont, Ltd.).

The electrolyte solution was introduced into the gap between the counter electrode and photoanode through two pinholes that were made in the FTO glass counter electrode using an ultrasonic pinhole maker. After introducing the electrolyte solution into the gap, the pinholes in the counter electrode were covered with masking tape. A solar simulator (Peccell Ltd.) was used to measure the efficiency and FF of the solar cell. Incident photon to current conversion efficiency (IPCE) was measured using a SM-250 DAM (Bunko Keiki Ltd.). A schematic illustration of the solar cell configuration is shown in Figure 1.

3. Results and Discussion

The crystal phase of the synthesized FTO powder was investigated by XRD (Figure 2). All XRD diffraction peaks of the FTO powder were assigned to rutile SnO_2 (ICDD number 01-076-7837, space group: P42/mnm).

UV-Vis diffuse reflectance spectra of the FTO powder revealed that fluorine doping resulted in broad visible-light absorption, indicating that free electrons were formed

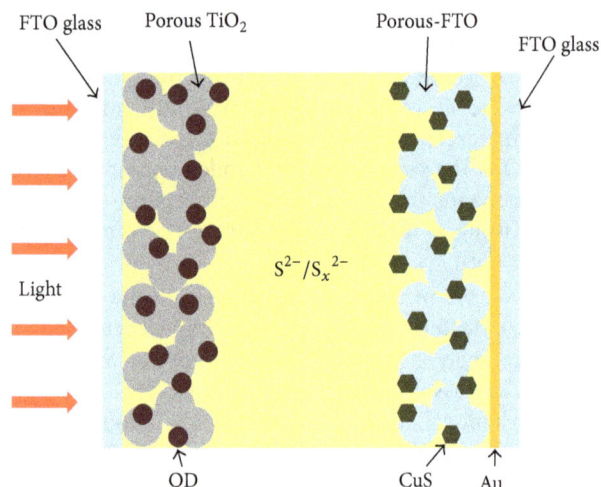

FIGURE 1: Schematic illustration of a QDSSC equipped with a porous-FTO/CuS counter electrode.

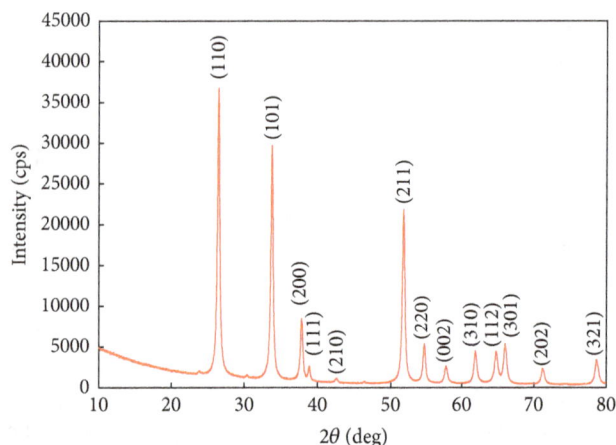

FIGURE 2: XRD pattern of FTO powder.

in the SnO_2 crystal (see the supporting information, Figure S1, in Supplementary Material available online at https://doi.org/10.1155/2017/5461030). Surface area of porous-FTO (thickness 5 μm) was determined by nitrogen adsorption-desorption isotherm. This value was 286 cm^2/cm^2 (real surface area/apparent surface area).

Photographs of the fabricated porous-FTO/CuS counter electrodes were shown in our supporting information (Figure S3), in which one electrode was constructed with an Au film between the porous-FTO and commercial FTO glass substrate, and another was constructed without an Au layer. A cross-sectional SEM image of the counter electrode is shown in Figure 3(a). From the image, it was determined that the thickness of the porous-FTO layer loaded with CuS catalyst was 5 μm, and this layer was uniformly coated on the commercial FTO glass substrate. CuS was deposited on the substrate using 20 SILAR cycles, which was determined to be optimum for this solar cell device, as described in detail below. As the CuS nanoparticles could not be clearly observed by SEM, the CuS/FTO composite was analyzed by TEM after

being scraped from the surface of the counter electrode. Two sizes of particles were observed in the composite: 10 to 20 nm particles (i) with anisotropic shape and 50 nm particles (ii) with spherical shape (Figure 3(b)). To determine the microstructure of the CuS/FTO composite, EDX point analysis was performed (Figure 3(c)). The analysis indicated that the 10 to 20 nm particles (i) were CuS catalyst, whereas the 50 nm spherical particles (ii) were FTO. The CuS particles were highly dispersed and attached onto the FTO surface of the electrode. We also investigated the XRD pattern for CuS powder synthesized by the same starting solution of the SILAR method, and all diffraction peaks were assigned to covellite CuS crystal phase (see our supporting information, Figure S4).

We optimized various parameters, including loading amount of CuS, porous-FTO thickness, and Au layer insertion on the substrate. R_s, R_{ct}, and internal resistance (R_{in}) values were determined from fitting results using the equivalent circuit (Figure S5 in supporting information). We used an inductance in the equivalent circuit, because we used thin conducting wire which have small inductance for impedance measurement. Constant phase element (CPE) contains two parameters T and P. Impedance (Z) of CPE was defined by the equation $Z = 1/(T \times (i\omega)^P)$. In our electrode, CPE1-T value (like a capacitance) was smaller than CPE2-T value. Therefore, R_{in} was attributed to the high-frequency semicircle, and R_{ct} was attributed to the low-frequency semicircle. Figures 4(a)(1), 4(a)(2), 4(a)(3), and 4(a)(4) are Nyquist plots of the samples fabricated by SILAR cycle 5, 10, 20, and 40 times with a thickness of 5 μm for porous-FTO. Two semicircles were observed in the plots of all samples. Figure 4(b) shows the influence of the number of SILAR cycles used for CuS deposition on the resistance derived from the semicircles in Figure 4(a). Although the internal resistance (R_{in}) and R_s values were not markedly affected by the number of SILAR cycles, R_{ct} was decreased with increasing number of SILAR cycles. The reduction of R_{ct} was saturated at approximately 15 Ω when 20 SILAR cycles were used for CuS deposition. Warburg Impedance (W_o) was decreased with increasing number of SILAR cycles. The origin of W_o is attributed to diffusion of S_x^{2-} redox, and the decrease of W_o would be owing to the efficient reduction reaction of S_x^{2-} at CuS sites. The electrode of 40 SILAR cycles exhibited the lowest W_o resistance, but the electrode of 20 SILAR cycles also exhibited sufficiently low resistance. Additionally, the film of 40 SILAR cycles was unstable and easily peeled off, since inner pressure was increased by excessive crystal growth of CuS by SILAR (Figure S2).

Based on the optimization of loading amount of CuS, we used the electrode with 20 SILAR cycles for achieving the low resistance of the porous-FTO/CuS electrodes. The origin of R_{ct} was attributed to the resistance of charge transfer between CuS and electrolyte, because R_{ct} was decreased with increasing number of SILAR cycles. In contrast, R_{in} was not so affected by SILAR cycle, since R_{in} was attributed to resistance of CuS/FTO interface or grain boundary of FTO particles.

The influence of the porous-FTO film thickness on resistance was next examined using porous-FTO/CuS electrodes

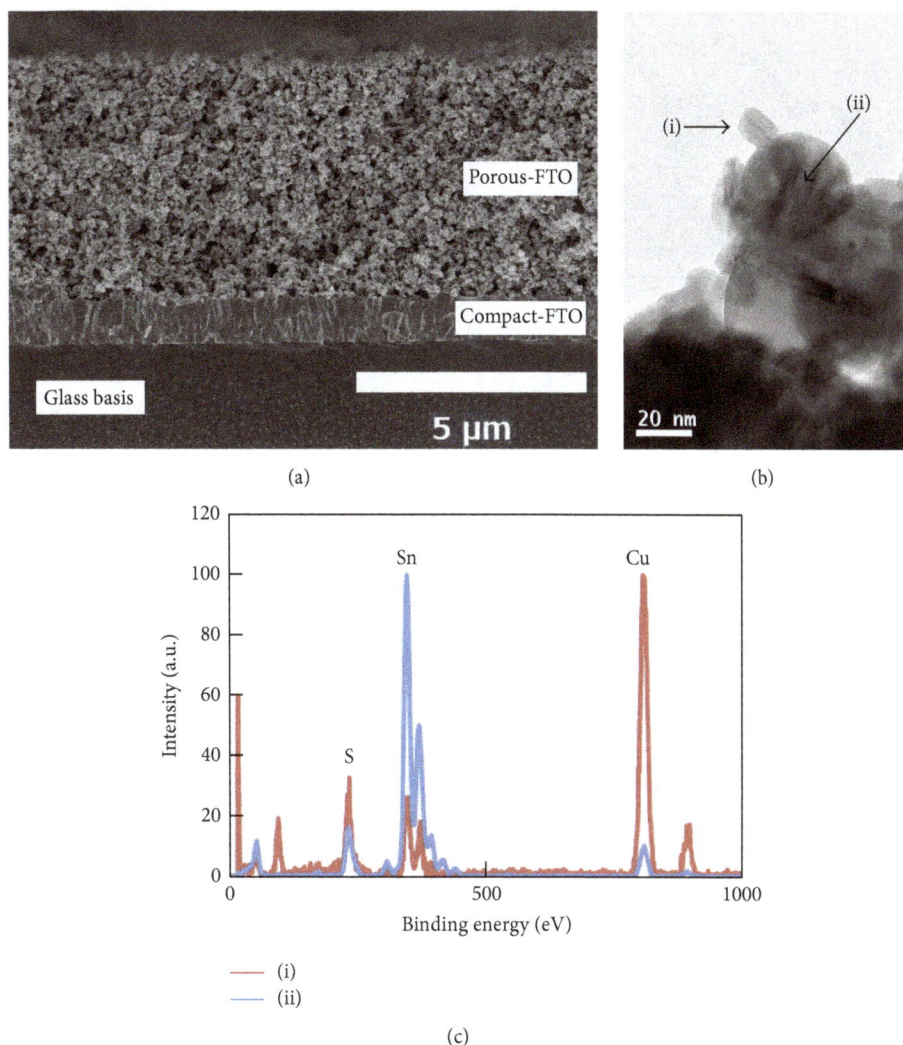

FIGURE 3: Cross-sectional SEM image of a porous-FTO electrode (a), TEM image of the porous-FTO (ii)/CuS (i) composite removed from the counter electrode surface (b), and EDX point analysis at points (i) and (ii) in the TEM image (c).

constructed using 20 CuS SILAR cycles. Nyquist plots of electrodes with porous-FTO films ranging from 5, 10, 15, and 20 μm are shown in Figures 5(a)(1), 5(a)(2), 5(a)(3), and 5(a)(4) and the thickness dependence on the resistance is shown in Figure 5(b). In contrast to R_{in} and W_o, R_{ct} was significantly decreased when the thickness of porous-FTO layer was increased. The decrease of R_{ct} is attributable to the increase of the electrode surface area that is available to drive catalytic reactions. R_{in} and W_o were slightly increased with increasing thickness of porous-FTO. We speculate that the increase of R_{in} was caused by grain boundaries in porous-FTO as increase of FTO thickness, and W_o was also increased by blocking effect of S_x^{2-} ion diffusion in thick porous structure. As the reduction of resistance was saturated around 20 μm, we concluded that the optimum structural condition for the QDSSC counter electrode is a porous-FTO layer of 20 μm thickness and 20 SILAR cycles for CuS deposition.

We speculate that the CuS exhibited high catalytic activity for the reduction of S^{2-}/S_x^{2-} redox couples due to the efficient adsorption of S_2^{2-} ion onto CuS crystals. R_{ct} of the counter electrode is lower than that previously reported for counter electrodes containing CuS catalyst [13, 24], owing to our optimization of the CuS synthesis process. Specifically, we used copper acetate as a starting precursor for CuS synthesis, as the nucleation and crystal growth of metal sulfide particles are limited in strong acid conditions, such as nitrate and chloride salts. CuS has a unique crystal configuration, in which there are two types of sulfur sites, namely, S^{2-} and S_2^{2-} ions. Crystal structure of CuS (covellite) is shown in our supporting information (Figure S6). Notably, the solid-state S_2^{2-} site has structural similarity with the polysulfide ion in the electrolyte used in our system and plays an important role in the catalytic reactivity with sulfur redox couples in the electrolyte. We investigated the relationship between the crystal structures of various metal sulfide catalysts and their catalytic properties, and the data were shown in our supporting information (Figure S7). Among CuS, FeS_2, and PbS catalysts, CuS exhibited the highest redox efficiency.

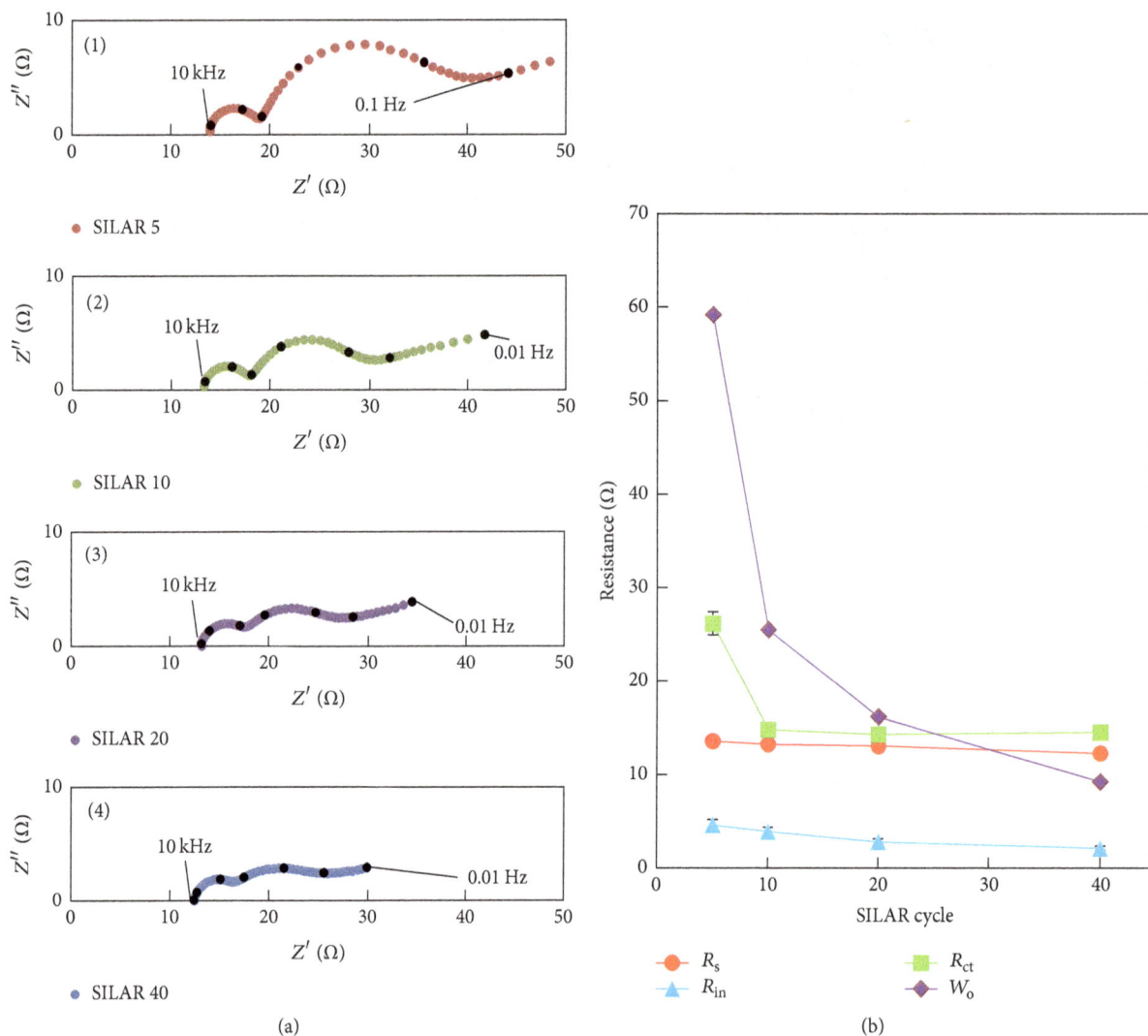

FIGURE 4: Nyquist plots of porous-FTO (5 μm)/CuS electrodes for various SILAR cycles. Panels (a)(1), (a)(2), (a)(3), and (a)(4) are the results for SILAR cycles with 5, 10, 20, and 40 times, respectively. SILAR cycles dependence on the resistance is shown in panel (b). Black dots in panels (a)(1) to (a)(4) indicate the data recorded under each decade of the frequency.

Notably, cathodic current densities of CuS and FeS$_2$ were higher than that of PbS. Both CuS and FeS$_2$ have S$_2^{2-}$ sites in their crystal, while PbS does not. Further, we also compared our CuS loaded porous-FTO electrode with a conventional Pt counter electrode, and the conductivity of our electrode was much superior than that of a Pt electrode (see our supporting information, Figure S8). Electrodes' stability is very important for a practical application; thus, we studied the stability of the CuS-modified porous-FTO electrode under the repeated redox reaction for 20 cycles (supporting information, Figure S9). And, as a result, the performance of our electrode remained stable, suggesting its high stability under the repeated redox reactions.

We also investigated the effect of inserting a layer of Au between the dense FTO layer of commercial FTO-coated glass and the porous-FTO layer. Figures 6(a) and 6(b) show Nyquist plots for counter electrodes (20 SILAR cycles, 20 μm FTO layer) with and without an Au layer. The

structural features of electrode, other than the Au layer, were fixed at using the optimal conditions described above. The introduction of the Au layer at the interface between the porous and dense FTO layers led to a drastic reduction in R_s. This result suggests that the sheet resistance of the dense FTO layer of commercial substrate is largely decreased by the sputtered Au layer. We have comprehensively designed and analyzed the nanostructure and interfaces of the counter electrode and concluded that a 20 μm porous-FTO layer, 20 SILAR cycles for CuS deposition, and the inclusion of an Au layer at the interface between the porous and dense FTO layers are optimal for decreasing the resistance of the counter electrode. Under these conditions, the lowest R_s, R_{in}, and R_{ct} values achieved were 2.52, 2.86, and 6.04 Ω, respectively.

We fabricated a sandwich-type solar cell device using the optimized counter electrode and PbS-modified photoanode. As a comparison, we also prepared a QDSSC with a Pt counter electrode. The I-V curves of the two QDSSCs under a 1.5

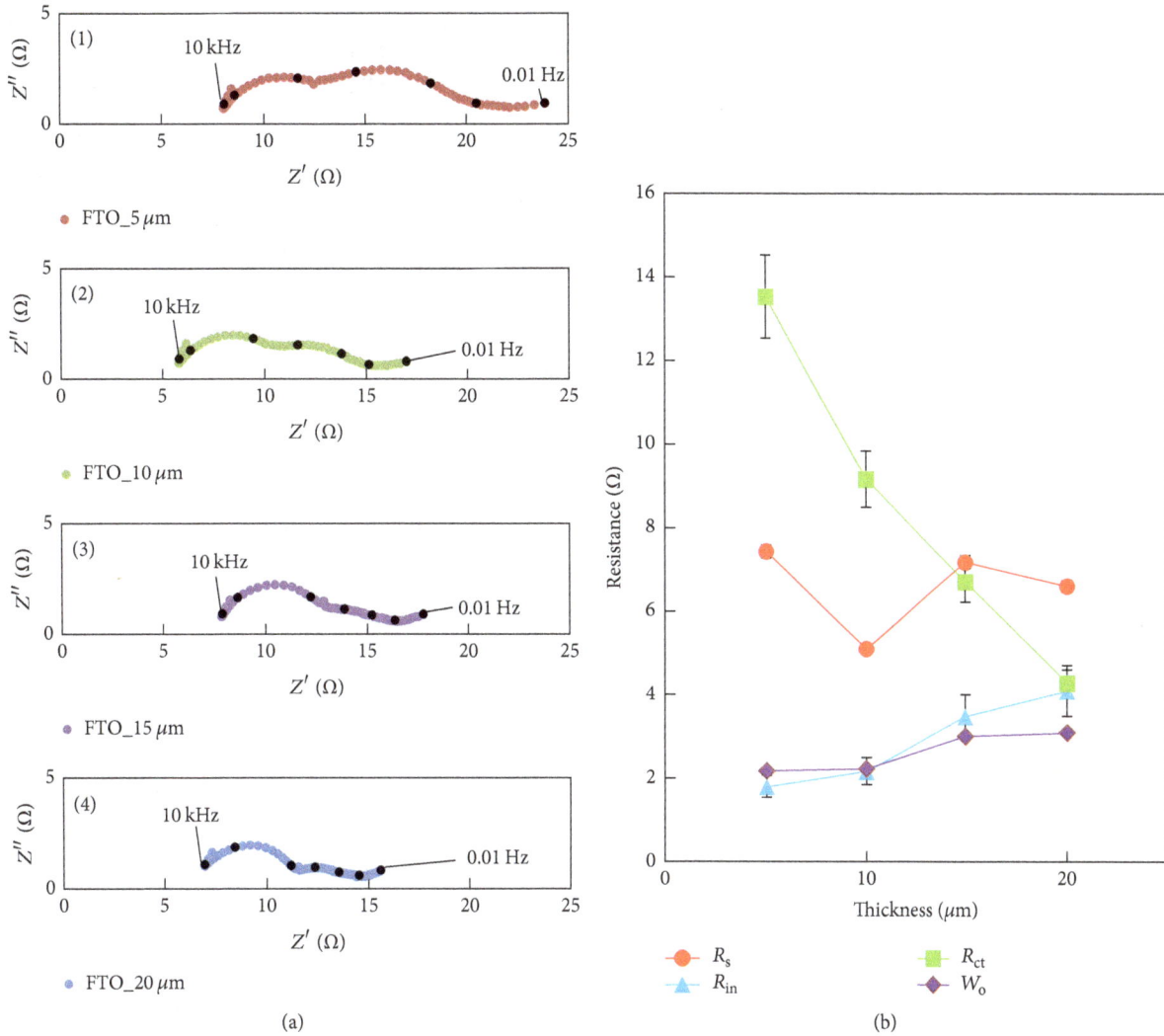

FIGURE 5: Nyquist plots for porous-FTO/CuS electrodes for various FTO thicknesses. Panels (a)(1), (a)(2), (a)(3), and (a)(4) are the results for FTO thicknesses of 5, 10, 15, and 20 μm, respectively. FTO thickness dependence on the resistance is shown in panel (b). CuS SILAR cycle for all samples was fixed at 20 times. Black dots in panels (a)(1) to (a)(4) indicate the data recorded under each decade of the frequency.

AM solar simulator are shown in Figure 7. The QDSSC with a Pt counter electrode exhibited very low FF (16%) and limited energy conversion efficiency (0.61%). In contrast, the QDSSC with the porous-FTO/CuS (thickness 5 μm, SILAR cycle 5) counter electrode exhibited relatively high FF (51.6%) and energy conversion efficiency (1.12%). The QDSSC with the optimized Au/porous-FTO/CuS (thickness 20 μm, SILAR cycle 20) counter electrode exhibited quite high FF (58%) and energy conversion efficiency (1.59%). Resistance values of counter electrodes and solar cells properties were summarized in Table 1. Notably, J_{sc} and V_{oc} were not influenced by the type of counter electrode but were affected by the photoanode, indicating that our optimized counter electrode could be used in devices constructed with other types of photoanodes, such as CdS/CdSe. J_{sc} and V_{oc} of the QDSSC with the Pt counter electrode were larger than those of the QDSSC constructed with the porous-FTO/CuS counter electrode. These results are due to differences in

the reflectivity of the counter electrode, as the Pt electrode reflects more visible light back to the photoanode. However, if the amount of visible-light absorption by the photoanode is sufficient, the reflection from the counter electrode becomes negligible. Thus, the FF of QDSSC in the present study is the most efficient reported to date and is generally applicable for use with various types of photoanodes for the development of efficient QDSSCs with high FF.

TABLE 1: Resistance values and solar cell properties.

	R_s [Ω]	R_{in} [Ω]	R_{ct} [Ω]	J_{sc} [mA/cm^2]	V_{oc} [V]	FF [%]	η [%]
Pt	12.7		7519	8.80	0.43	16	0.61
CuS5/FTO5 μm	13.6	4.7	26.2	5.81	0.37	52	1.12
CuS20/FTO20 μm/Au	2.5	2.9	6.0	7.14	0.38	58	1.59

(a)

(b)

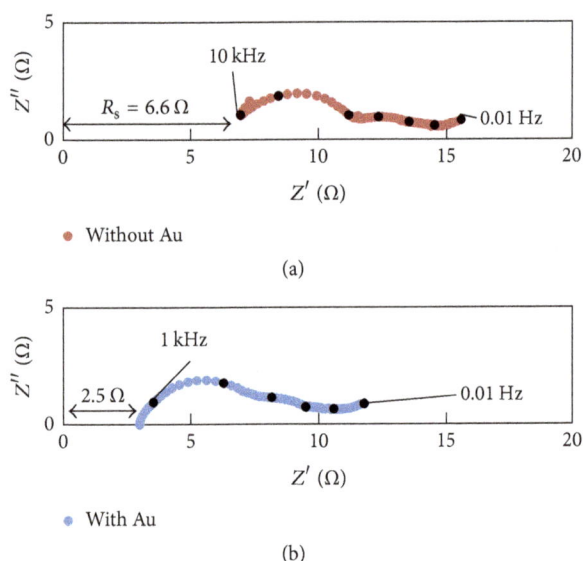

FIGURE 6: Nyquist plots of counter electrodes constructed without Au (a) and with Au layer (b). Black dots indicate the data recorded under each one decade of the frequency.

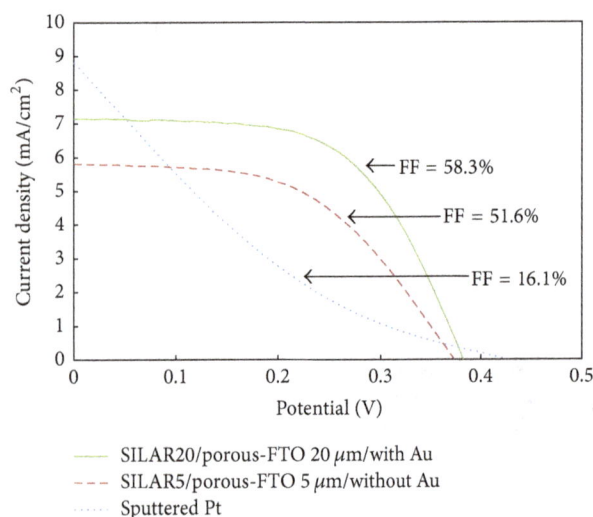

FIGURE 7: I-V curves of QDSSCs constructed with various counter electrodes under solar simulator. Blue dotted line: Pt counter electrode, red dashed line: porous-FTO/CuS (porous-FTO thickness: $5\,\mu m$, SILAR cycles: 5 times), and solid green line: Au/porous-FTO/CuS (porous-FTO thickness: $20\,\mu m$, SILAR cycles: 20 times), respectively.

4. Conclusions

We have fabricated the efficient counter electrode for constructing QDSSCs, on the basis of the CuS deposited porous-FTO electrode. According to the optimization on the thickness of porous-FTO layer and loading amount of CuS nanoparticles, R_{ct} value was drastically decreased. We also found that introduction of an Au layer between the porous-FTO layer and dense FTO led to a marked reduction in R_s. Finally, very low R_s ($2.5\,\Omega$) and R_{ct} ($6.0\,\Omega$)

were achieved. When we combined the optimized counter electrode with a PbS-based photoanode, the highest FF (58%) was achieved among previously reported PbS and polysulfide electrolyte based QDSSCs. Our CuS-modified porous-FTO counter electrode can be combined with various types of photoanodes, in addition to PbS-based photoanodes. The strategy to design the counter electrode is expected to be applicable for other types of QDs for the development of QDSSCs with even higher efficiency.

Competing Interests

The authors declare that they have no competing interests.

Acknowledgments

This work is financially supported by JST PRESTO, JST ACT-C programs, and JSPS KAKENHI Grant no. 26410234. The authors also thank Mr. J. Koki and Mr. A. Genseki for electron microscope observation. The manuscript was carefully read by Mr. G. Newton.

References

[1] R. Vogel, K. Pohl, and H. Weller, "Sensitization of highly porous, polycrystalline TiO₂ electrodes by quantum sized CdS," Chemical Physics Letters, vol. 174, no. 3-4, pp. 241–246, 1990.

[2] Y. Wang, A. Suna, W. Mahler, and R. Kasowski, "PbS in polymers. From molecules to bulk solids," The Journal of Chemical Physics, vol. 87, no. 12, pp. 7315–7322, 1987.

[3] S. D. Sung, I. Lim, P. Kang, C. Lee, and W. I. Lee, "Design and development of highly efficient PbS quantum dot-sensitized solar cells working in an aqueous polysulfide electrolyte," Chemical Communications, vol. 49, no. 54, pp. 6054–6056, 2013.

[4] S. Hachiya, Q. Shen, and T. Toyoda, "Effect of ZnS coatings on the enhancement of the photovoltaic properties of PbS quantum dot-sensitized solar cells," Journal of Applied Physics, vol. 111, no. 10, Article ID 104315, 2012.

[5] Z. Pan, K. Zhao, J. Wang, H. Zhang, Y. Feng, and X. Zhong, "Near infrared absorption of CdSexTe1-x alloyed quantum dot sensitized solar cells with more than 6% efficiency and high stability," ACS Nano, vol. 7, no. 6, pp. 5215–5222, 2013.

[6] J. G. Radich, R. Dwyer, and P. V. Kamat, "Cu₂S reduced graphene oxide composite for high-efficiency quantum dot solar cells. Overcoming the redox limitations of S_2^-/S_n^{2-} at the counter electrode," The Journal of Physical Chemistry Letters, vol. 2, no. 19, pp. 2453–2460, 2011.

[7] Y. Jiang, X. Zhang, Q. Q. Ge et al., "ITO@Cu2S tunnel junction nanowire arrays as efficient counter electrode for quantum-dot-sensitized solar cells," Nano Letters, vol. 14, no. 1, pp. 365–372, 2014.

[8] M. Wolf and H. Rauschenbach, "Series resistance effects on solar cell measurements," Advanced Energy Conversion, vol. 3, no. 2, pp. 455–479, 1963.

[9] J. Jiao, Z.-J. Zhou, W.-H. Zhou, and S.-X. Wu, "CdS and PbS quantum dots co-sensitized TiO2 nanorod arrays with improved performance for solar cells application," Materials Science in Semiconductor Processing, vol. 16, no. 2, pp. 435–440, 2013.

[10] L. Meng, F. Zhao, J. Zhang et al., "Preparation of monodispersed PbS quantum dots on nanoporous semiconductor thin film by

two-phase method," *Journal of Alloys and Compounds*, vol. 595, pp. 51–54, 2014.

[11] A. N. Jumabekov, F. Deschler, D. Böhm, L. M. Peter, J. Feldmann, and T. Bein, "Quantum-dot-sensitized solar cells with water-soluble and air-stable PbS quantum dots," *The Journal of Physical Chemistry C*, vol. 118, no. 10, pp. 5142–5149, 2014.

[12] G. Hodes, J. Manassen, and D. Cahen, "Electrocatalytic electrodes for the polysulfide redox system," *Journal of the Electrochemical Society*, vol. 127, no. 3, pp. 544–549, 1980.

[13] W. Ke, G. Fang, H. Lei et al., "An efficient and transparent copper sulfide nanosheet film counter electrode for bifacial quantum dot-sensitized solar cells," *Journal of Power Sources*, vol. 248, pp. 809–815, 2014.

[14] K. Zhao, H. Yu, H. Zhang, and X. Zhong, "Electroplating cuprous sulfide counter electrode for high-efficiency long-term stability quantum dot sensitized solar cells," *The Journal of Physical Chemistry C*, vol. 118, no. 11, pp. 5683–5690, 2014.

[15] S.-Q. Fan, B. Fang, J. H. Kim, J.-J. Kim, J.-S. Yu, and J. Ko, "Hierarchical nanostructured spherical carbon with hollow core/mesoporous shell as a highly efficient counter electrode in CdSe quantum-dot-sensitized solar cells," *Applied Physics Letters*, vol. 96, no. 6, Article ID 063501, 2010.

[16] M. Seol, D. H. Youn, J. Y. Kim et al., "Mo-compound/CNT-graphene composites as efficient catalytic electrodes for quantum-dot-sensitized solar cells," *Advanced Energy Materials*, vol. 4, no. 4, Article ID 1300775, 2014.

[17] S.-Q. Fan, B. Fang, J. H. Kim et al., "Ordered multimodal porous carbon as highly efficient counter electrodes in dye-sensitized and quantum-dot solar cells," *Langmuir*, vol. 26, no. 16, pp. 13644–13649, 2010.

[18] Z. Tachan, M. Shalom, I. Hod, S. Rühle, S. Tirosh, and A. Zaban, "PbS as a highly catalytic counter electrode for polysulfide-based quantum dot solar cells," *Journal of Physical Chemistry C*, vol. 115, no. 13, pp. 6162–6166, 2011.

[19] E. Ramasamy, W. J. Lee, D. Y. Lee, and J. S. Song, "Spray coated multi-wall carbon nanotube counter electrode for tri-iodide (I3-) reduction in dye-sensitized solar cells," *Electrochemistry Communications*, vol. 10, no. 7, pp. 1087–1089, 2008.

[20] H. Yu, H. Bao, K. Zhao, Z. Du, H. Zhang, and X. Zhong, "Topotactically grown bismuth sulfide network film on substrate as low-cost counter electrodes for quantum dot-sensitized solar cells," *The Journal of Physical Chemistry C*, vol. 118, no. 30, pp. 16602–16610, 2014.

[21] M.-H. Yeh, C.-P. Lee, C.-Y. Chou et al., "Conducting polymer-based counter electrode for a quantum-dot-sensitized solar cell (QDSSC) with a polysulfide electrolyte," *Electrochimica Acta*, vol. 57, no. 1, pp. 277–284, 2011.

[22] Q. Zhang, Y. Zhang, S. Huang et al., "Application of carbon counterelectrode on CdS quantum dot-sensitized solar cells (QDSSCs)," *Electrochemistry Communications*, vol. 12, no. 2, pp. 327–330, 2010.

[23] Y. Jiang, X. Zhang, Q.-Q. Ge et al., "Engineering the interfaces of ITO@Cu2S nanowire arrays toward efficient and stable counter electrodes for quantum-dot-sensitized solar cells," *ACS Applied Materials & Interfaces*, vol. 6, no. 17, pp. 15448–15455, 2014.

[24] H. Chen, L. Zhu, H. Liu, and W. Li, "ITO porous film-supported metal sulfide counter electrodes for high-performance quantum-dot-sensitized solar cells," *The Journal of Physical Chemistry C*, vol. 117, no. 8, pp. 3739–3746, 2013.

[25] C.-H. Han, S.-D. Han, J. Gwak, and S. P. Khatkar, "Synthesis of indium tin oxide (ITO) and fluorine-doped tin oxide (FTO) nano-powder by sol–gel combustion hybrid method," *Materials Letters*, vol. 61, no. 8-9, pp. 1701–1703, 2007.

[26] Y. F. Nicolau, "Solution deposition of thin solid compound films by a successive ionic-layer adsorption and reaction process," *Applications of Surface Science*, vol. 22-23, no. 2, pp. 1061–1074, 1985.

[27] Y. F. Nicolau, M. Dupuy, and M. Brunel, "ZnS, CdS, and $Zn_{1-x}Cd_xS$ thin films deposited by the successive ionic layer adsorption and reaction process," *Journal of the Electrochemical Society*, vol. 137, no. 9, pp. 2915–2924, 1990.

[28] B. O. Dabbousi, J. Rodriguez-Viejo, F. V. Mikulec et al., "(CdSe)ZnS core-shell quantum dots: synthesis and characterization of a size series of highly luminescent nanocrystallites," *Journal of Physical Chemistry B*, vol. 101, no. 46, pp. 9463–9475, 1997.

6

An Optimal Charging Strategy for PV-Based Battery Swapping Stations in a DC Distribution System

Shengjun Wu,[1] Qingshan Xu,[1] Qun Li,[2] Xiaodong Yuan,[2] and Bing Chen[2]

[1]School of Electrical Engineering, Southeast University, No. 2 Sipailou, Nanjing 210096, China
[2]Jiangsu Electric Power Research Institute, No. 1 Paweier, Nanjing 211103, China

Correspondence should be addressed to Qingshan Xu; xuqingshan@seu.edu.cn

Academic Editor: Santolo Meo

Photovoltaic- (PV-) based battery swapping stations (BSSs) utilize a typical integration of consumable renewable resources to supply power for electric vehicles (EVs). The charging strategy of PV-based BSSs directly influences the availability, cost, and carbon emissions of the swapping service. This paper proposes an optimal charging strategy to improve the self-consumption of PV-generated power and service availability while considering forecast errors. First, we introduce the typical structure and operation model of PV-based BSSs. Second, three indexes are presented to evaluate operational performance. Then, a particle swarm optimization (PSO) algorithm is developed to calculate the optimal charging power and to minimize the charging cost for each time slot. The proposed charging strategy helps decrease the impact of forecast uncertainties on the availability of the battery swapping service. Finally, a day-ahead operation schedule, a real-time decision-making strategy, and the proposed PSO charging strategy for PV-based BSSs are simulated in a case study. The simulation results show that the proposed strategy can effectively improve the self-consumption of PV-generated power and reduce charging cost.

1. Introduction

Renewable energy and electric vehicles (EVs) have emerged as methods for ensuring energy security and for reducing emissions in many countries. However, for photovoltaics (PV) and other intermittent renewable energy sources with output volatility, large-scale renewable energy connected to the power grid still has many barriers to overcome. Moreover, it is difficult to maintain stable operation of the power grid and optimize scheduling for large populations of EVs with uncertain charging characteristics [1–3]. In addition, if the charging power is generated by coal-fired power plants, the emission advantage for EV is not obvious [4, 5]. The direct integration of PV with EV charging devices is an effective way to reduce EV emissions and the impact of EV charging on the power grid [6, 7].

There are three types of EV charging modes: conventional charging, fast charging, and battery swapping [8, 9]. The battery swapping mode replaces depleted batteries with fully charged batteries and charges the depleted batteries on a charging platform. The battery swapping mode is mainly applied in public transport, such as electric buses and electric taxis. As an important charging mode, the State Grid Corporation of China has invested in and constructed dozens of battery swapping stations (BSSs) in large cities in China [10, 11]. The centralized charging strategy in BSSs is constructive for the integration of renewable energy generation.

Most of the scholarships on EV charging and renewable energy generation have focused on design, operation, and optimal charging strategies [12–16]. A conceptual architecture and an assessment framework were proposed to explore integration scenarios of EVs and renewable energy generation in distribution networks [17]. The combination of PV energy and EVs in uncontrolled charging and smart charging strategies has been studied [18], as was a two-stage framework for the economic operation of an EV parking deck with renewable energy generation [19]. Heuristic optimization algorithms, such as particle swarm optimization (PSO) and genetic algorithm (GA), have been used to solve the multiobjective and nonlinear optimization problems associated with the management of EV charging involved in renewable energy generation. A PSO algorithm was utilized to allocate charging stations of plug-in electric vehicles [20] and to provide a coordinated charging/discharging scheme to increase revenues and incentives [21]. An optimization algorithm was developed based on the well-established PSO and interior point method for the optimal dispatch of EVs and wind power [22]. Two approaches based on the fuzzy genetic algorithm and fuzzy discrete PSO were proposed to minimize energy cost and grid losses by coordinating EV charging [23]. A PSO algorithm can randomly search to achieve satisfactory solutions by using a population of particles and is a practical algorithm for solving EV charging optimization problems involved in renewable energy generation.

The aforementioned charging strategies proposed for integrating PV and a charging station are not suitable for battery charging in a BSS. The charging strategy of a BSS is different from the general EV charging mode. In a BSS, a depleted EV battery system is replaced with a fully charged battery system in several minutes, and the depleted batteries are stored on a centralized charging platform. Recently, several studies have investigated charging strategies for BSSs without renewable energy generation. A novel centralized charging strategy for BSSs considered optimal charging priority and charging location [24]. A direct projection method was developed to compute an optimal charging schedule of the BSS efficiently [25]. However, these charging strategies have not properly considered the uncertain features of PV generation and battery swapping requirements.

A charging strategy for operating a PV-based BSS should take into account battery swapping demand, fluctuation of PV generation, charging cost, and forecast errors. The primary mission of a BSS is to ensure service availability for battery swapping. Battery charging of the BSS must be scheduled so that every EV that arrives will be provided with fully charged batteries. However, the battery charging schedule is compromised with forecasting errors for PV generation and battery swapping demand. In this paper, we propose a PSO charging algorithm to decrease the impact of forecast uncertainties on the availability of battery swapping services and to minimize the charging power cost.

The major contributions of this paper are as follows:

(1) A battery swapping service model and a battery charging model are introduced to improve the service availability of BSSs.

(2) A PSO-based charging strategy is proposed to optimize the charging operation and to achieve self-consumption of PV energy for PV-based BSSs.

(3) The effectiveness of the proposed PSO-based charging strategy is evaluated in a real-time swapping operation while minimizing the total charging cost.

(4) The simulation results indicate that the proposed PSO-based charging strategy can achieve satisfactory effects under different battery demand forecast errors.

The content of this paper is organized as follows: Section 2 describes the system structure and operation model. Section 3 proposes three indexes used to evaluate the operation performance. Section 4 introduces the PSO-based charging strategy, including the basic PSO algorithm and its implementation in the battery charging operation. A case study and analysis are presented in Section 5, and conclusions are drawn in Section 6.

2. System Structure and Operation Model

2.1. System Structure and Components. A typical PV-based BSS structure is shown in Figure 1; the structure is mainly composed of a PV system, a battery charging system, a grid-connected system, and an energy management system.

The PV system consists of PV modules and DC/DC converters. The PV modules directly convert solar energy into electric energy and deliver power to the DC bus through the DC/DC converter. The actual output power of the PV system is mainly affected by solar irradiation and environmental temperature [26].

The battery charging system is comprised of DC chargers and batteries. An EV is driven by a battery system consisting of a fixed set of battery packs that is charged by a DC charger. In general, the battery is initially charged with a constant current until the state of charge (SOC) reaches a certain value, after which it is charged with a constant voltage. Recently, batteries have also been charged with a constant power [6].

The grid-connected system consists of a bidirectional AC/DC inverter and a distribution transformer. The grid-connected system balances power between the distribution grid and the DC bus and maintains a stable voltage in the DC bus.

The energy management system monitors and controls the operation of each component in the PV-based BSS. The optimal charging strategy will run on the energy management system to achieve the operational objectives of the BSS.

2.2. Battery Swapping Service Model. The battery swapping service model is shown in Figure 2. The most important objective of the BSS is to maintain the availability of the battery swapping service. In the initial development period of the BSS, the quantity of batteries reserved in the BSS is limited to a relatively small swapping demand and the high cost of purchasing batteries. Therefore, studying the battery swapping service model is necessary to improve service availability and reduce charging cost.

To analyse the battery swapping service model, we divide one day into I periods, and each period has a length of Δt. The total number of power battery systems and chargers in the BSS is defined as N_B and N_{CH}, respectively. The number of EVs that the BSS can serve is defined as N_S.

For any time slot i, the batteries in the BSS can be divided into the following states: (1) available status, in which the battery is fully charged and ready for the swapping service, defined as $N_A(i)$; (2) charging status, in which the battery is in a charging status, defined as $N_C(i)$, and the battery will complete charging in time slot i, defined as $N_{NA}(i)$; and (3) waiting for charging status, in which the battery is replaced in time slot i, defined as $N_{WB}(i)$.

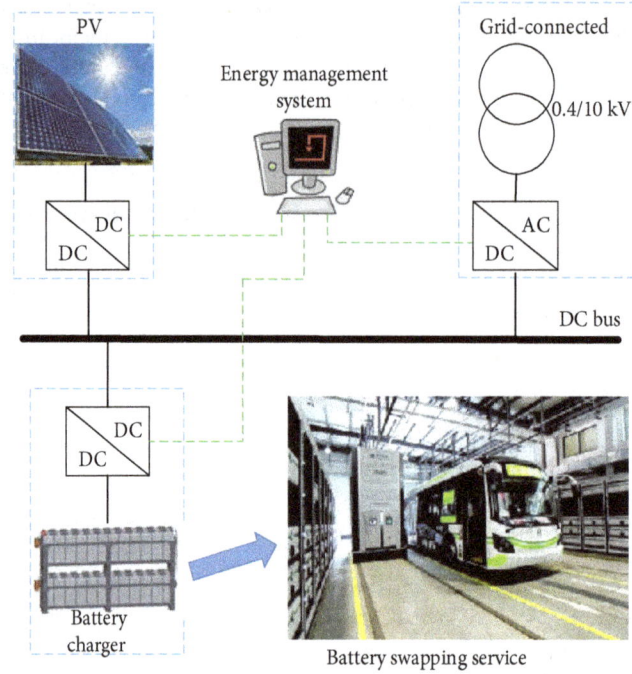

FIGURE 1: Structure of a PV-based BSS.

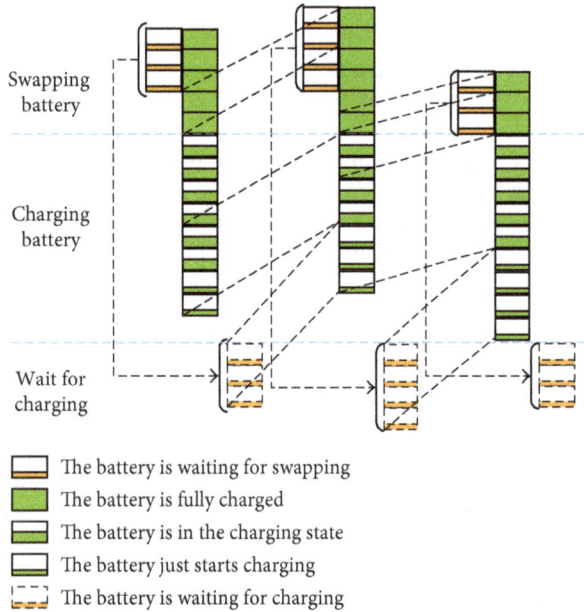

☐ The battery is waiting for swapping
▨ The battery is fully charged
▨ The battery is in the charging state
☐ The battery just starts charging
⊡ The battery is waiting for charging

FIGURE 2: The battery swapping service model.

The total number of EVs waiting to swap a battery in time slot i is defined as $N_{EV}(i)$. The number of EVs coming to the BSS to swap a battery in time slot i is defined as $N_{NEV}(i)$. The number of EVs waiting to swap a battery that has not yet completed battery swapping in time slot i due to limited battery swapping equipment or a lack of fully charged batteries is defined as $N_{WEV}(i)$. The number of EVs that have completed battery swapping in time slot i is defined as $N_{SEV}(i)$. $N_{EV}(i)$ can be calculated as follows:

$$N_{EV}(i) = N_{NEV}(i) + N_{WEV}(i-1) \tag{1}$$

$$N_{EV}(i) = N_{SEV}(i) + N_{WEV}(i). \tag{2}$$

The number of batteries in different states varies dynamically with time. The number of fully charged available battery systems in time slot i can be calculated as follows:

$$N_A(i) = N_A(i-1) - N_{SEV}(i-1) + N_{NA}(i-1). \tag{3}$$

After the battery is fully charged, it should rest for a period of time to reach a steady state before discharge. Thus, the battery becomes available in the time slot after it has completed charging. Similarly, the replaced battery reaches the charging state in the following time slot.

The number of charging state battery systems in time slot i can be calculated as follows:

$$N_C(i) = N_C(i-1) - N_{NA}(i-1) + N_{SEV}(i-1). \tag{4}$$

Since the BSS configures a charger for every replaced battery system on the charging platform, the replaced battery systems can be charged in the next time slot.

2.3. Battery Charging Model. The replaced batteries begin charging in the following time slot. The replaced batteries are charged in a constant power charging mode. In this mode, the charging time for a replaced battery is related to the initial SOC and the final SOC of the charging battery.

$$T_{B\_Cha} = \frac{(SOC_{end} - SOC_{init}) \times W_B}{P_{Cha}}, \tag{5}$$

where T_{B_Cha} represents the battery charging time; SOC_{init} and SOC_{end} represent the initial SOC and finished SOC of a charging battery, respectively; W_B represents the rated capacity of a battery; and P_{Cha} represents the constant charging power.

To maintain the availability of the battery swapping service, there should be enough fully charged batteries to meet the EV swapping demand in each time slot.

$$N_A(i) > N_{EV}(i). \tag{6}$$

In the next time slot $i + 1$, the new fully charged batteries transition to an available state, and new EVs come to the battery swapping service:

$$N_A(i) + N_{NA}(i) > N_{EV}(i) + N_{NEV}(i + 1). \tag{7}$$

Based on (3) and (7), the following equation can be derived:

$$
\begin{aligned}
N_{NA}(i) > &\ N_{NEV}(i) + N_{NEV}(i + 1) + N_{WEV}(i - 1) - N_A(i - 1) \\
&+ N_{SEV}(i - 1) - N_{NA}(i - 1),
\end{aligned} \tag{8}
$$

where $N_{NA}(i)$ represents the number of battery systems that are fully charged in the time slot i and $N_{NEV}(i)$ and $N_{NEV}(i + 1)$ are the numbers of EVs coming to the BSS for battery swapping in the time slots i and $i + 1$, respectively, which are the forecasted values. The other parameters in (8) are known in time slot i, so $N_{NA}(i)$ can be determined.

However, the batteries may not be fully charged, as it may take several time slots to complete battery charging. The charging time for a battery can be calculated using (5). Thus, (8) needs to be applied to time slot $i + n$.

$$
\begin{aligned}
&N_{NA}(i) + N_{NA}(i + 1) + \cdots + N_{NA}(i + n) \\
&> N_{NEV}(i) + N_{NEV}(i + 1) + N_{NEV}(i + 2) + \cdots \\
&\quad + N_{NEV}(i + 1 + n) + N_{WEV}(i - 1) - N_A(i - 1) \\
&\quad + N_{SEV}(i - 1) - N_{NA}(i - 1).
\end{aligned} \tag{9}
$$

The lower limit of the charging quantity of battery systems in time slot i is decided by $N_{NA}(i + n)$. Thus, the minimum charging power can be calculated as follows:

$$P_{Cha\_min}(i) = P_{Cha} \times N_{NA}(i + n). \tag{10}$$

In the actual battery swapping service, the number of EVs coming to the BSS is a typical stochastic value. As a result, the forecast value $N_{NEV}(i)$ always includes uncertainty errors. Having more fully charged batteries available in reserve should be considered to respond to forecast errors; however, keeping many fully charged batteries in reserve is not an economical solution. Therefore, an optimal charging strategy is needed to account for both forecast errors and charging economy.

3. Evaluation Indexes for Operation Performance

Considering the purpose and characteristics of the battery swapping service, the evaluation indexes for the operation performance are given based on three points [27, 28]:

(1) self-consumption of PV energy, (2) availability of the battery swapping service, and (3) economic cost of charging.

3.1. The Self-Consumed PV in Total EV Charging Energy. There are two power sources for EV charging, namely, PV energy and distribution grid energy. This index evaluates the PV self-consumption rate in total EV charging energy, which is represented as the percentage of self-consumed PV energy in total EV charging energy (PPTC):

$$\text{PPTC} = \frac{\sum_{i=1}^{I}\left(E_{EV}(i) - E_{grid}(i)\right)}{\sum_{i=1}^{I}E_{EV}(i)} \times 100\%, \tag{11}$$

where $E_{EV}(i)$ represents the charging energy of the battery charging platform in time slot i and $E_{grid}(i)$ represents the energy supplied by the distribution grid in time slot i.

3.2. Waiting Time for Battery Swapping Service. EVs have to wait for battery swapping primarily due to a lack of fully charged batteries or limited swapping equipment. The battery swapping time for one EV is approximately 6–8 minutes and is not included in the waiting time. This index evaluates the availability of the battery swapping service, which is represented as the waiting time of the battery swapping service (WTBSS):

$$\text{WTBSS} = \frac{\sum_{i=1}^{I}\sum_{j=1}^{N_{SEV}(i)}\text{WT}(i, j)}{\sum_{i=1}^{I}N_{SEV}(i)}, \tag{12}$$

where $\text{WT}(i, j)$ represents the time that EVs spend waiting for battery swapping.

3.3. The Price of Charging Energy. The charging energy price differs based on the energy sources. The energy from the distribution grid is more expensive than the energy from PV generation. The charging energy price is calculated as follows:

$$\text{PR}_{EV} = \frac{\text{PR}_{PV}\sum_{i=1}^{I}\left(E_{EV}(i) - E_{grid}(i)\right) + \text{PR}_{grid}\sum_{i=1}^{I}E_{grid}(i)}{\sum_{i=1}^{I}E_{EV}(i)}, \tag{13}$$

where PR_{EV} represents the average electricity price of the charging energy, $E_{grid}(i)$ represents the charging energy value from the distribution grid in time slot i, PR_{PV} represents the electricity price from PV energy, and PR_{grid} represents the electricity price from the distribution grid.

4. Charging Strategy

4.1. PSO. The optimal charging strategy for PV-based BSSs is a stochastic program, as traditional linear optimization methods cannot handle the random calculations. The probabilistic transition rules used in a PSO make it an appropriate solution for the stochastic problem of the battery charging and swapping operations. A PSO simulates the group behaviours of flocking birds to find a particular optimal objective.

Initially, the PSO algorithm chooses candidate positions randomly within the search space. It then searches for the best positions by updating the generations. During each iteration of the algorithm, the objective function evaluates the fitness of each candidate position. The best fitness and positions are updated by comparing the newly evaluated fitness against the previous best fitness. Finally, the PSO algorithm maintains the best fitness and positions achieved among all particles in the swarm. The velocity and position of each particle are updated as follows [29]:

$$
\begin{aligned}
&v_{id}^{k+1} = wv_{id}^k + c_1 r_1 \left(p_{id}^k - x_{id}^k \right) + c_2 r_2 \left(p_{gd}^k - x_{id}^k \right) \\
&\text{if } v_{id}^{k+1} > v_{\max}, v_{id}^{k+1} = v_{\max}; \\
&\text{if } v_{id}^{k+1} < v_{\min}, v_{id}^{k+1} = v_{\min}; \\
&x_{id}^{k+1} = x_{id}^k + v_{id}^{k+1},
\end{aligned} \tag{14}
$$

where k represents the iteration number; w is the inertia weight; c_1 and c_2 are the learning factors; r_1 and r_2 are two random numbers in the range $[0, 1]$; and v_{id}^k, x_{id}^k, p_{id}^k, and p_{gd}^k are the d-dimensional vectors of velocity, position, best position, and global best position for the ith particle at the kth iteration, respectively.

4.2. PSO-Based Strategy. Forecast data are provided in time slots, which includes PV generation and EV battery swapping demands. Then, the BSS operation plans are created for each time slot based on the battery swapping service model and the battery charging model. The BSS operation plans should guarantee the availability of battery swapping and maintain a low charging cost for each time slot.

However, the forecast values of PV generation and battery swapping demands always include uncertainty errors, and operational plans cannot be updated according to the real-time data, leading to suboptimal solutions. The PSO algorithm can search for the best charging solutions with the data from the previous time slot. A group of positions and velocities is randomly selected from the feasible solution space to initiate the PSO. The charging cost is the objective function with constraints on the charging power. Then, the positions and velocities are updated to search for the best solutions by evaluating the fitness. The best solutions of the PSO are used as the BSS operation plans for the next time slot. There are I time slots in one day, and the PSO algorithm is run to optimize the operation plans. The flowchart for the proposed PSO-based charging strategy is shown in Figure 3.

The charging cost is evaluated based on the electricity price of the charging energy, considering that the electricity quantities vary in the solutions. The charging power constraints are the minimum and maximum charging power.

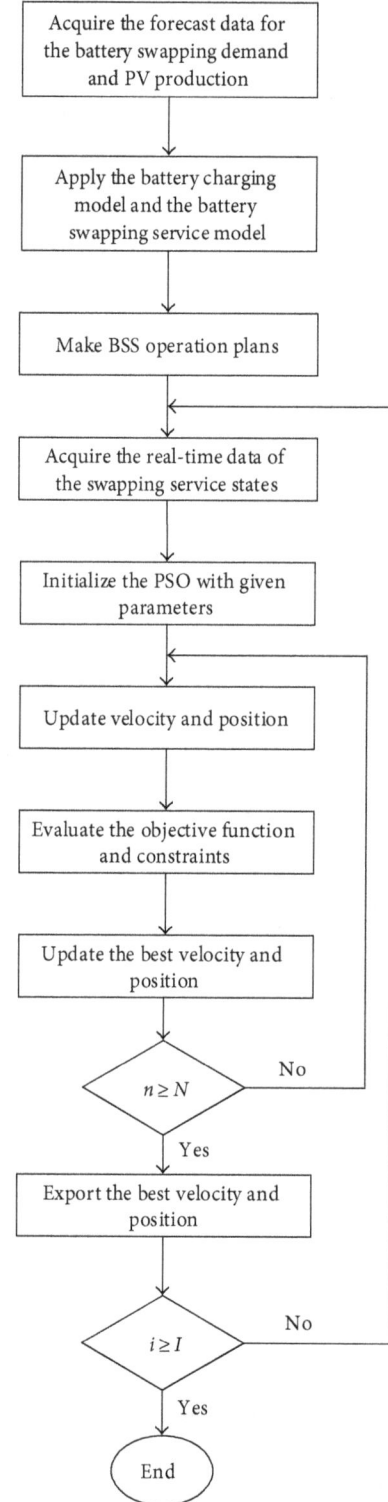

FIGURE 3: Flowchart for the proposed PSO strategy.

The minimum charging power is calculated using (10). The maximum charging power is decided by the quantity of chargers in the constant power charging mode.

The number of batteries being charged is the decision variable for the optimization. The PSO stops when it reaches

TABLE 1: BSS case study configuration parameters.

Parameters	Values
PV capacity	500 kW
Capacity of an EV battery system	160 kWh
Rated capacity of a charger	60 kW
Battery charging rate	1/3 C
Total number of EV battery systems	34
Total number of EV buses	20
PV electricity price	$0.0567/kWh
Distribution grid electricity price	$0.1243/kWh

the number of iterations that ensures convergence to the global best position.

5. Case Study

5.1. Simulation Environment. A simulation was conducted based on the PV-based BSS located in Nanjing, China; the configuration parameters are shown in Table 1. The BSS in the case study provides a battery swapping service for twenty EV buses running along three bus lines. The capacity of the battery system is 160 kWh, with five big battery packs and four small battery packs for each EV bus. The charging mode is constant power charging at a rate of 1/3 C.

According to the actual operation of the BSS, the working time of the battery swapping service runs from 6:00 to 23:00. A robot automatically swaps the batteries, and the battery swapping time of one EV bus is set to 7 minutes. The battery swapping times and the SOC of the replaced battery are acquired from the BSS operation database. The parameters of the PSO in the simulation are set as follows: the number of particles N is 20, the learning factors c_1 and c_2 are both 2, the inertia weight w is 0.8, and the iteration number k is 50.

In the simulation, the duration of a single time slot is set to 30 minutes, and there are 34 time slots for the battery swapping service in one day. The simulation analyses and compares a day-ahead operation schedule [30, 31], a real-time decision-making (RTDM) strategy [28], and the PSO charging strategy in a typical day.

5.2. Simulation Results and Analysis. The day-ahead operation schedule provides the BSS operators with a set of plans for the entire day's operations according to the forecast data. Uncertain forecast errors may affect the availability of battery swapping service and the cost of charging. The RTDM strategy updates the forecast parameters in real time as more operational data become available and repeatedly adjusts the schedule to reduce the impact of forecast errors. We compare the BSS using the day-ahead operation schedule and the PSO charging strategy and analyse the evaluation indexes of these three strategies under different forecast errors.

The same PV production and battery swapping demand data are used to simulate the real-time operations of the BSS. We perform these two approaches to battery charging for 34 time slots in one day. The forecast and actual PV production are shown in Figure 4, and the forecast and actual values of the battery swapping demand are shown in Figure 5. The simulation results show the BSS operation under two charging approaches, as depicted in Figures 6–8.

Figure 6 shows the number of batteries charged by the day-ahead operation schedule and the PSO charging strategy. The day-ahead operation schedule charges batteries 158 times in total, and the PSO strategy charges 156 times. During the real-time operation, the PSO charging strategy takes advantage of PV production to charge more batteries, while the operation schedule uses less PV energy due to the battery demand forecast errors. Figures 7 and 8 show the numbers of EVs swapping batteries and the number of EVs waiting for battery swapping using these two approaches in all 34 time slots. In the schedule operation, eight EVs must wait for battery swapping mainly due to the battery demand forecast errors.

The power distribution results from the two charging approaches can be obtained from the simulation, as shown in Figure 9. Both charging approaches achieve a good PV power consumption performance; only a little of the PV energy is supplied to the grid at noon. Charging the batteries in the BSS mitigates the fluctuation of PV generation, and the charging load of the BSS on the distribution grid decreases. In addition, the PSO strategy performs better than the schedule operation in terms of the self-consumption of PV energy, particularly at noon.

To analyse the performance of the proposed charging strategy in real-time operations, the evaluation results of the indexes proposed in Section 3 are listed in Table 2. The results show that the performances of all the charging strategies are influenced by the battery demand forecast errors.

The evaluation index PPTC is influenced by both the PV production and battery demand forecast errors. When PV output power is greater than the battery charging demand, PV self-consumption is related to the charging strategy. The charging strategies of the day-ahead operation schedule and RTDM track PV generation power, and the charging strategy of PSO is to minimize the charging cost. The global search ability of PSO enables the charging strategy to perform a little better than the other two strategies in the boundary between PV production and battery charging demand under the influence of forecast errors, as shown in Figure 9. The results of the evaluation index PPTC show that all the strategies achieve a good performance on the self-consumption of PV energy. The PPTC of the PSO strategy is greater than that of the other strategies by 3-4 points on average. The battery demand forecast errors have little impact on PV energy self-consumption.

For charging decisions that are made based on battery swapping demand, the availability of the swapping service is seriously influenced by battery demand forecast errors, as indicated on the evaluation index WTBSS. The WTBSS of the schedule strategy increases from 1.3 min to 3.8 min, with battery demand forecast errors growing from 10% to 40%. The increased forecast errors have relatively less impact on the other charging strategies. The RTDM and PSO strategies update the operation data in real time and adjust charging plans to reduce the impact of the accumulated forecast errors.

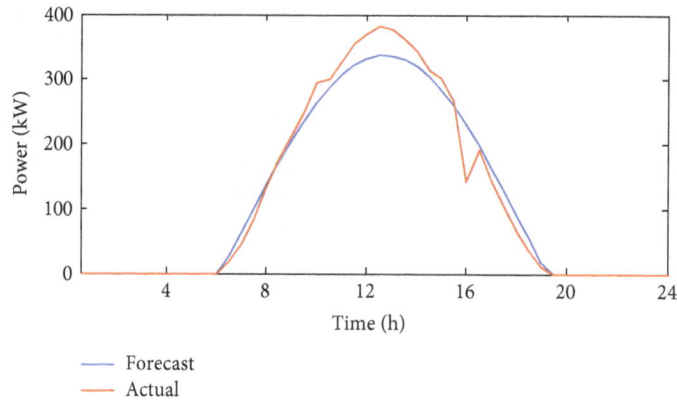

FIGURE 4: Forecast and actual PV power generation.

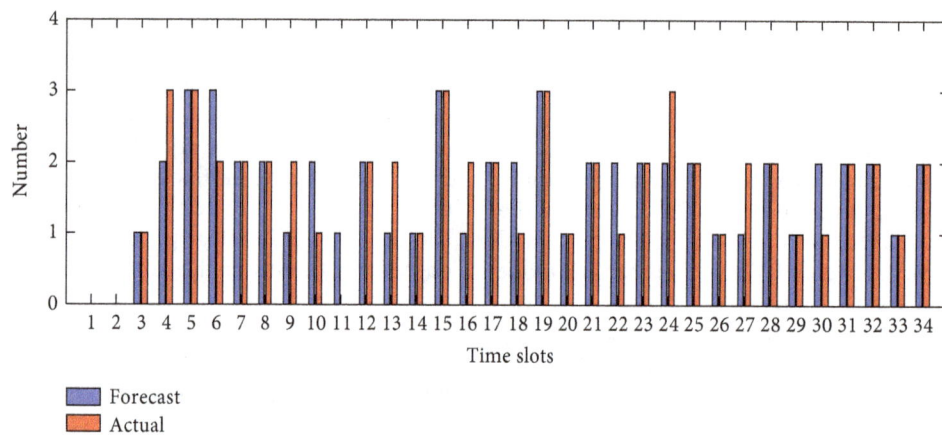

FIGURE 5: Forecast and actual values of battery swapping demand.

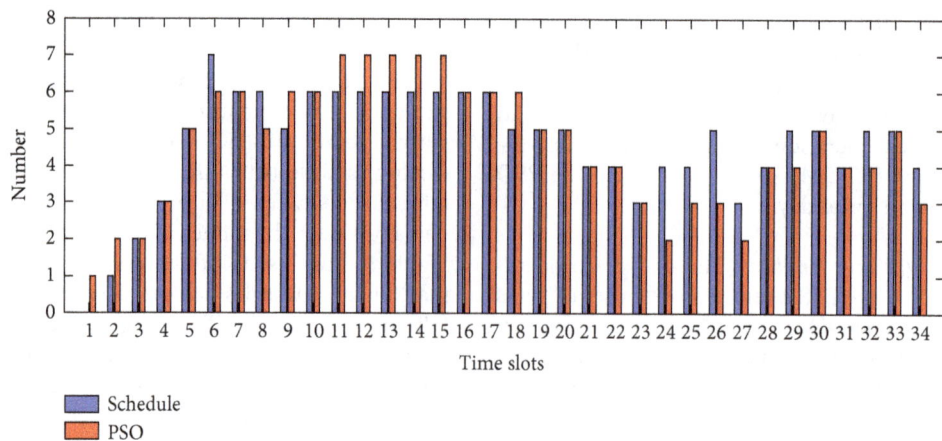

FIGURE 6: The number of batteries charged using the schedule and PSO strategies.

The PSO strategy achieves satisfactory results under 30% forecast errors. When the forecast errors increase to 40%, the WTBSS of the PSO strategy reaches 1.2 min, which is not as good as the RTDM strategy. However, the forecast errors cannot be eliminated. The availability of the battery swapping service can be improved by reserving more fully charged batteries, though this strategy increases charging cost, and purchasing batteries is expensive.

The charging cost evaluation index PR_{EV} is related to the charging energy source. The electricity price of PV generation is lower than that of the distribution grid. As forecast errors increase, PV self-consumption reduces, and the

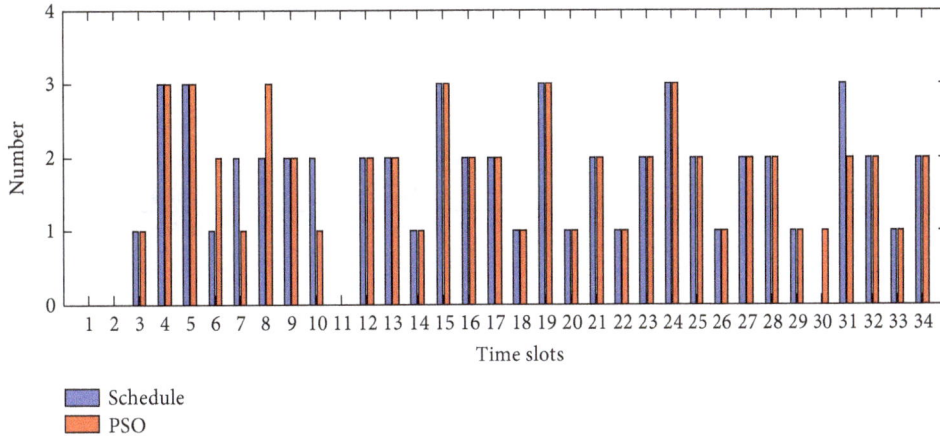

FIGURE 7: The number of batteries swapped using the schedule and PSO strategies.

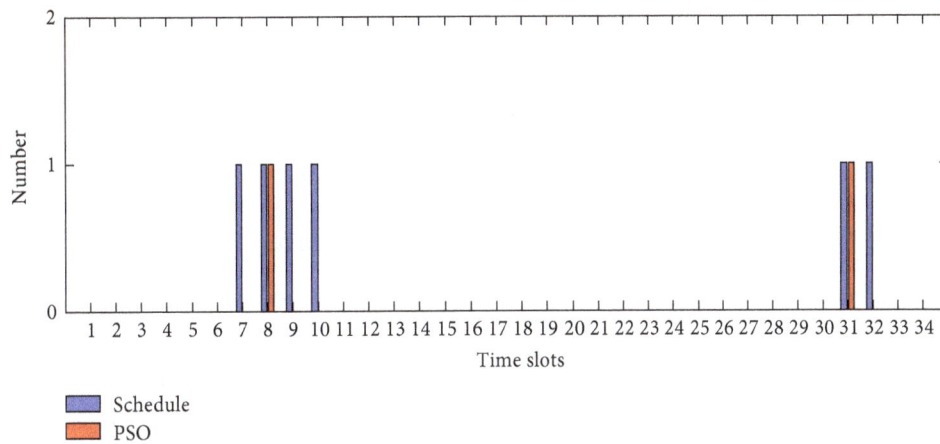

FIGURE 8: The number of EVs waiting using the schedule and PSO strategies.

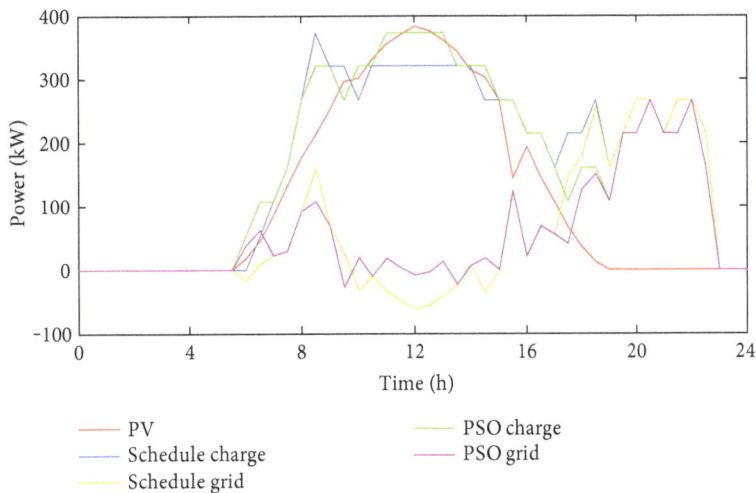

FIGURE 9: Power distribution results using the schedule and PSO strategies.

TABLE 2: Evaluation results for the different strategies.

Indexes	Battery demand forecast errors	Schedule	RTDM	PSO
PPTC	10%	63.1%	63.9%	67.1%
	20%	62.3%	63.2%	66.9%
	30%	62.0%	62.8%	66.5%
	40%	61.2%	62.4%	65.7%
WTBSS	10%	1.3 min	0.5 min	0.3 min
	20%	1.6 min	0.7 min	0.5 min
	30%	2.7 min	0.8 min	0.7 min
	40%	3.8 min	1.1 min	1.2 min
PR_{EV}	10%	\$0.0816/kWh	\$0.0811/kWh	\$0.0790/kWh
	20%	\$0.0822/kWh	\$0.0816/kWh	\$0.0791/kWh
	30%	\$0.0824/kWh	\$0.0819/kWh	\$0.0793/kWh
	40%	\$0.0829/kWh	\$0.0825/kWh	\$0.0799/kWh

charging cost increases. PR_{EV} is negatively correlated with PV self-consumption. The PSO strategy achieves higher PV self-consumption than the other two charging strategies, so PR_{EV} is lower in the same forecast errors. The PR_{EV} of the PSO strategy is lower than that of the other two strategies in all cases. Even when forecast errors expand to 40%, PR_{EV} still reaches \$0.0799/kWh.

The results clearly show that the evaluation indexes of the proposed PSO charging strategy are better than those of the day-ahead operation schedule and RTDM strategy. The proposed strategy is sufficiently sophisticated to handle the forecast errors during real-time operations. The proposed PSO charging strategy can provide a charging reference to optimize the manual operation in the BSS at present, and it may be applied in automatic charging operation in the future.

6. Conclusions

This paper proposes an optimal charging strategy that considers the impact of forecast errors to improve service availability and reduces the operating cost of PV-based BSSs. A battery swapping service model is used to describe the regular battery swapping operation of a BSS, and a battery charging model is developed to determine the charging time and quantity. Then, a PSO-based strategy is proposed to optimize the charging time and the charging number of batteries based on PV production and battery demand forecast errors. The simulation results show that the PSO charging strategy can effectively improve the self-consumption of PV energy and the availability of the swapping service. In addition, the charging cost of the PSO strategy is lower than that of the day-ahead operation schedule and the RTDM strategy. The PSO charging strategy updates the charging plans in each time slot according to the actual data to avoid the influence of cumulative errors. Therefore, the proposed PSO charging strategy can perform well for the real-time operation of PV-based BSSs.

It is difficult to improve the service availability and PV self-consumption of PV-based BSSs without increasing the charging cost. A common practice to improve service

availability is to reserve more fully charged batteries, but this increases the charging cost. Deploying energy storage systems with battery second use may be a good solution to the problem. The authors will continue the optimal charging strategy of PV-based BSS considering battery second use in the future work.

Nomenclature

c_1, c_2:	Learning factors
$E_{EV}(i)$:	Charging energy of the BSS in time slot i
$E_{grid}(i)$:	Charging energy supplied by the distribution grid in time slot i
I:	Total number of time slots in a day
k:	Iteration number
N_B:	Number of battery systems in the BSS
N_{CH}:	Number of chargers in the BSS
N_S:	Number of EVs that the BSS serves
$N_A(i)$:	Number of battery systems available for swapping in time slot i
$N_C(i)$:	Number of battery systems charging in time slot i
$N_{NA}(i)$:	Number of battery systems that complete charging in time slot i
$N_{WB}(i)$:	Number of battery systems waiting for charging status in time slot i
$N_{EV}(i)$:	Number of EVs waiting to swap a battery in time slot i
$N_{NEV}(i)$:	Number of EVs that arrive to swap a battery in time slot i
$N_{SEV}(i)$:	Number of EVs that complete battery swapping in time slot i
$N_{WEV}(i)$:	Number of EVs that have not yet completed battery swapping in time slot i
p_{id}^k:	d-dimensional vectors of best position
p_{gd}^k:	d-dimensional vectors of global best position
P_{Cha}:	Constant charging power of a battery system
$P_{Cha\_min}(i)$:	Minimum charging power of the BSS in time slot i
PR_{EV}:	Average electricity price of the charging energy

PR_{grid}:	Electricity price of the distribution grid
PR_{PV}:	Electricity price of PV energy
r_1, r_2:	Random numbers in the range $[0, 1]$
SOC_{init}:	Initial SOC of a battery system when charging begins
SOC_{end}:	Final SOC of a battery system when charging ends
Δt:	Length of every time slot
T_{B_Cha}:	Charging time of a battery system when charging ends
cv_{id}^k:	d-dimensional vectors of velocity
w:	Inertia weight
W_B:	Rated capacity of a battery system
$WT(i, j)$:	The time that EVs wait for battery swapping
x_{id}^k:	d-dimensional vectors of position.

Conflicts of Interest

The authors declare no conflicts of interest.

Acknowledgments

This work was supported by the Fundamental Research Funds for the Central Universities of China (Grant no. 2242016K41064).

References

[1] P. J. Tulpule, V. Marano, S. Yurkovich, and G. Rizzoni, "Economic and environmental impacts of a PV powered workplace parking garage charging station," *Applied Energy*, vol. 108, pp. 323–332, 2013.

[2] A. Rabiee, M. Sadeghi, J. Aghaeic, and A. Heidari, "Optimal operation of microgrids through simultaneous scheduling of electrical vehicles and responsive loads considering wind and PV units uncertainties," *Renewable and Sustainable Energy Reviews*, vol. 57, pp. 721–739, 2016.

[3] C. Shao, X. Wang, X. Wang, C. Du, and B. Wang, "Hierarchical charge control of large populations of EVs," *IEEE Transactions on Smart Grid*, vol. 7, no. 2, pp. 1147–1155, 2016.

[4] X. Xiao, J. Wen, S. Tao, and Q. Li, "Study and recommendations of the key issues in planning of electric vehicles' charging facilities," *Transactions of China Electrotechnical Society*, vol. 29, no. 8, pp. 1–10, 2014.

[5] W. P. Schill and C. Gerbaulet, "Power system impacts of electric vehicles in Germany: Charging with coal or renewables?" *Applied Energy*, vol. 156, pp. 185–196, 2015.

[6] A. R. Bhatti, Z. Salam, M. J. B. A. Aziz, K. P. Yee, and R. H. Ashique, "Electric vehicles charging using photovoltaic: Status and technological review," *Renewable and Sustainable Energy Reviews*, vol. 54, pp. 34–47, 2016.

[7] A. R. Bhatti, Z. Salam, M. J. B. A. Aziz, and K. P. Yee, "A critical review of electric vehicle charging using solar photovoltaic," *International Journal of Energy Research*, vol. 40, no. 4, pp. 439–461, 2016.

[8] R. Philipsen, T. Schmidt, J. van Heek, and M. Ziefle, "Fast-charging station here, please! User criteria for electric vehicle fast-charging locations," *Transportation Research Part F: Traffic Psychology and Behaviour*, vol. 40, pp. 119–129, 2016.

[9] I. Rahman, P. M. Vasant, B. S. Singh, M. Abdullah-Al-Wadud, and N. Adnan, "Review of recent trends in optimization techniques for plug-in hybrid, and electric vehicle charging infrastructures," *Renewable and Sustainable Energy Reviews*, vol. 58, pp. 1039–1047, 2016.

[10] Y. Li, C. Davis, Z. Lukszo, and M. Weijnen, "Electric vehicle charging in China's power system: Energy, economic and environmental trade-offs and policy implications," *Applied Energy*, vol. 173, pp. 535–554, 2016.

[11] X. Xu, L. Yao, P. Zeng, Y. Liu, and T. Cai, "Architecture and performance analysis of a smart battery charging and swapping operation service network for electric vehicles in China," *Journal of Modern Power Systems and Clean Energy*, vol. 3, no. 2, pp. 259–268, 2015.

[12] A. Schuller, C. M. Flath, and S. Gottwalt, "Quantifying load flexibility of electric vehicles for renewable energy integration," *Applied Energy*, vol. 151, pp. 335–344, 2015.

[13] M. van der Kam and W. van Sark, "Smart charging of electric vehicles with photovoltaic power and vehicle-to-grid technology in a microgrid; a case study," *Applied Energy*, vol. 152, pp. 20–30, 2015.

[14] G. R. Chandra Mouli, P. Bauer, and M. Zeman, "System design for a solar powered electric vehicle charging station for workplaces," *Applied Energy*, vol. 168, pp. 434–443, 2016.

[15] R. Rao, X. Zhang, J. Xie, and L. Ju, "Optimizing electric vehicle users' charging behavior in battery swapping mode," *Applied Energy*, vol. 155, pp. 547–559, 2015.

[16] J. Van Roy, N. Leemput, F. Geth, J. Buscher, R. Salenbien, and J. Driesen, "Electric vehicle charging in an office building microgrid with distributed energy resources," *IEEE Transactions on Sustainable Energy*, vol. 5, no. 4, pp. 1389–1396, 2014.

[17] A. Chaouachi, E. Bompard, G. Fulli, M. Masera, M. De Gennaro, and E. Paffumi, "Assessment framework for EV and PV synergies in emerging distribution systems," *Renewable and Sustainable Energy Reviews*, vol. 55, pp. 719–728, 2016.

[18] F. Fattori, N. Anglani, and G. Muliere, "Combining photovoltaic energy with electric vehicles, smart charging and vehicle-to-grid," *Solar Energy*, vol. 110, pp. 438–451, 2014.

[19] Y. Guo, J. Xiong, S. Xu, and W. Su, "Two-stage economic operation of microgrid-like electric vehicle parking deck," *IEEE Transactions on Smart Grid*, vol. 7, no. 3, pp. 1703–1712, 2016.

[20] E. Pashajavid and M. A. Golkar, "Optimal placement and sizing of plug in electric vehicles charging stations within distribution networks with high penetration of photovoltaic panels," *Journal of Renewable and Sustainable Energy*, vol. 5, no. 5, Article ID 053126, 2013.

[21] M. Ghofrani, A. Arabali, and M. Ghayekhloo, "Optimal charging/discharging of grid-enabled electric vehicles for predictability enhancement of PV generation," *Electric Power Systems Research*, vol. 117, pp. 134–142, 2014.

[22] J. Zhao, F. Wen, Z. Y. Dong, Y. Xue, and K. P. Wong, "Optimal dispatch of electric vehicles and wind power using enhanced particle swarm optimization," *IEEE Transactions on Industrial Informatics*, vol. 8, no. 4, pp. 889–899, 2012.

[23] S. Hajforoosh, M. A. S. Masoum, and S. M. Islam, "Real-time charging coordination of plug-in electric vehicles based on hybrid fuzzy discrete particle swarm optimization," *Electric Power Systems Research*, vol. 128, pp. 19–29, 2015.

[24] Q. Kang, J. Wang, M. Zhou, and A. C. Ammari, "Centralized charging strategy and scheduling algorithm for electric vehicles under a battery swapping scenario," *IEEE*

Transactions on Intelligent Transportation Systems, vol. 17, no. 3, pp. 659–669, 2016.

[25] P. You, Z. Yang, Y. Zhang, S. H. Low, and Y. Sun, "Optimal charging schedule for a battery switching station serving electric buses," *IEEE Transactions on Power Systems*, vol. 31, no. 5, pp. 3473–3483, 2016.

[26] R. Carbone, "PV plants with distributed MPPT founded on batteries," *Solar Energy*, vol. 122, pp. 910–923, 2015.

[27] X. Lu, N. Liu, Q. Tang, and J. Zhang, "Optimal capacity configuration of electric vehicle battery swapping station considering service availability," *Automation of Electric Power System*, vol. 38, no. 14, pp. 77–83, 2014.

[28] N. Liu, Q. Chen, X. Lu, J. Liu, and J. Zhang, "A charging strategy for PV-based battery switch stations considering service availability and self-consumption of PV energy," *IEEE Transactions on Industrial Electronics*, vol. 62, no. 8, pp. 4878–4889, 2015.

[29] R. Poli, J. Kennedy, and T. Blackwell, "Particle swarm optimization," *Swarm Intelligence*, vol. 1, no. 1, pp. 33–57, 2007.

[30] N. G. Paterakis, O. Erdinç, A. G. Bakirtzis, and J. P. S. Catalao, "Optimal household appliances scheduling under day-ahead pricing and load-shaping demand response strategies," *IEEE Transactions on Industrial Informatics*, vol. 11, no. 6, pp. 1509–1519, 2015.

[31] M. R. Sarker, H. Pandzic, and M. A. Ortega-Vazquez, "Optimal operation and services scheduling for an electric vehicle battery swapping station," *IEEE Transactions on Power Systems*, vol. 30, no. 2, pp. 901–910, 2015.

7

Improving the Hybrid Photovoltaic/Thermal System Performance Using Water-Cooling Technique and Zn-H$_2$O Nanofluid

Hashim A. Hussein, Ali H. Numan, and Ruaa A. Abdulrahman

Electromechanical Engineering Department, University of Technology, Baghdad, Iraq

Correspondence should be addressed to Hashim A. Hussein; hashim171967@gmail.com

Academic Editor: Md. Rabiul Islam

This paper presented the improvement of the performance of the photovoltaic panels under Iraqi weather conditions. The biggest problem is the heat stored inside the PV cells during operation in summer season. A new design of an active cooling technique which consists of a small heat exchanger and water circulating pipes placed at the PV rear surface is implemented. Nanofluids (Zn-H2O) with five concentration ratios (0.1, 0.2, 0.3, 0.4, and 0.5%) are prepared and optimized. The experimental results showed that the increase in output power is achieved. It was found that, without any cooling, the measuring of the PV temperature was 76°C in 12 June 2016; therefore, the conversion efficiency does not exceed more than 5.5%. The photovoltaic/thermal system was operated under active water cooling technique. The temperature dropped from 76 to 70°C. This led to increase in the electrical efficiency of 6.5% at an optimum flow rate of 2 L/min, and the thermal efficiency was 60%. While using a nanofluid (Zn-H2O) optimum concentration ratio of 0.3% and a flow rate of 2 L/min, the temperature dropped more significantly to 58°C. This led to the increase in the electrical efficiency of 7.8%. The current innovative technique approved that the heat extracted from the PV cells contributed to the increase of the overall energy output.

1. Introduction

Photovoltaic (PV) systems represent a solution for the problem of low carbon, nonfossil fuel used to generate electricity. Solar radiation absorbed and converted by semiconductor devices (solar cells) can provide a supply of electricity to meet energy needs. An energy source with less emissions of carbon, no dependence on fossil fuels, massive potential for developing countries, and well suited to be distributed, PV, is considered as a medium and long range energy prospect as presented by Firth [1]. The photovoltaic system has advantages compared to other systems, such as low maintenance, unattended operation, reliable long life between 20 and 30 years, no fuel and no fumes, easy to install, and low recurrent costs as presented by Oi [2]. Basically, the solar PV/T system can be broadly categorized into two systems: photovoltaic and solar thermal system. The PV/T system refers to a system that uses heat transfer fluid to extract heat from the panel. The fluid is water or air and sometimes both. The photovoltaic thermal system (PV/T) has been developed for several reasons; one of the main reasons is that the PV/T system can give higher efficiency than PV alone and thermal collector system as presented by Teo [3]. The application of nanofluid in solar collectors leads to a homogeneous temperature distribution inside the receiver. In addition, greater light absorption, a high absorption at visible wavelengths, and a low emissivity at infrared wavelengths can be achieved, and sunlight can be directly converted into useful heat as presented by Taylor et al. [4]. Nanoparticles have the following advantages in solar power plants: (1) the extremely small size of the particles ideally allows them to pass through pumps and plumbing without adverse impacts, (2) nanofluids can absorb energy directly skipping intermediate heat transfer steps, (3) the nanofluids can be optically selective (i.e., high absorption in the solar range and low remittance in the infrared), (4) a more uniform receiver temperature can be achieved inside the collector (reducing material constraints), (5) enhanced heat transfer via greater convection and thermal conductivity may enhance receiver performance, and (6) absorption efficiency may be improved by tuning the

(1) Photovoltaic panel (PV), (2) solar meter, (3) oscillatory collector (copper pipes), (4) DC pump, (5) flow meters (number 2), (6) heat exchanger, (7) battery, (8) charge controller, (9) voltage regulator, (10) ammeter, (11) voltmeter, (12) temperature recorder, (13) thermocouple, (14) boost converter, (15) DC pump circulation, (16) water tank (number 2).

(a)

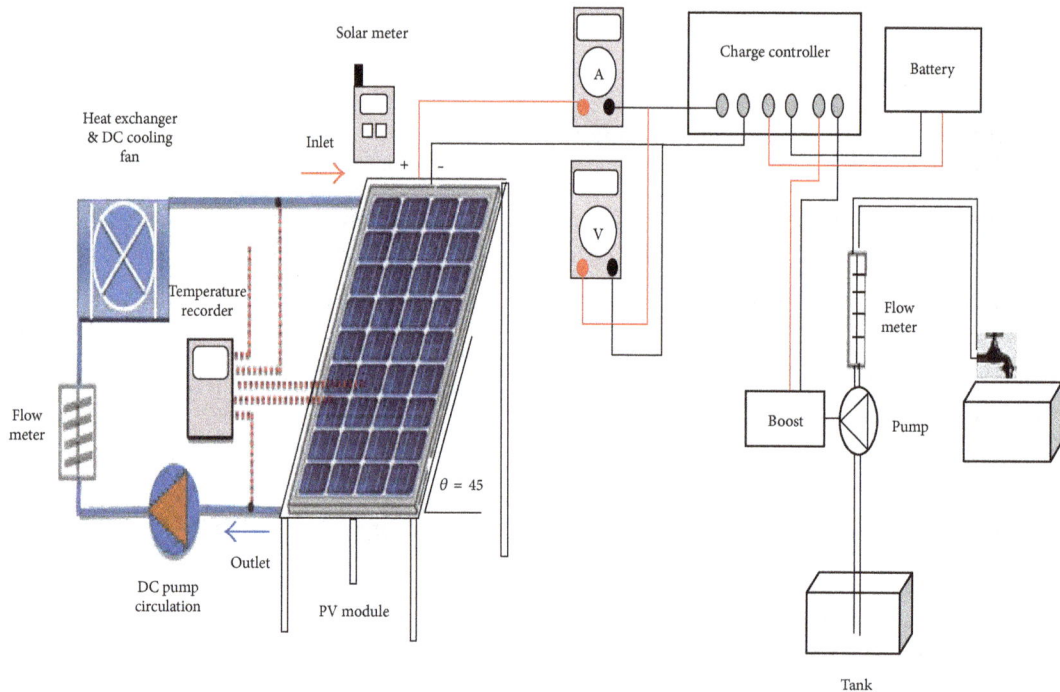

(b)

FIGURE 1: Photograph (a) and schematic diagram (b) of the experimental setup.

nanoparticle size and shape that suit the application as presented by Abdolzadeh and Ameri [5].

EL-Basit et al. [6] investigated a simple one-diode mathematical model, by applying a MATLAB script. The results showed that the variation in irradiation mainly affects the output current, while the variations in temperature mainly affect the output voltage. Odeh and Behnia [7] developed a cooling technique by trickling water on the upper surface of a PV module to improve the performance of a proposed PV pumping system. The results of their experimental rig showed that an increase of about 15% in the output of the system was achieved at peak radiation conditions due to the heat loss by convection between the water and the upper surface of the PV panel. Long-term performance of the system was estimated by integrating the test results in a commercial transient simulation package using the data of site radiation and ambient temperature. The simulation results of annual performance of system showed an increase of 5% in delivered energy from the PV module during dry and warm seasons.

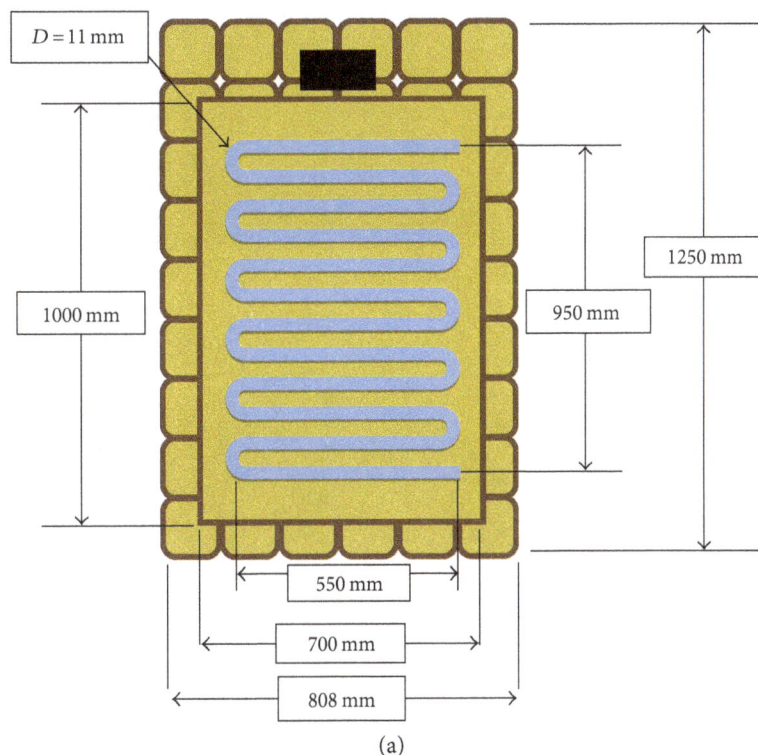

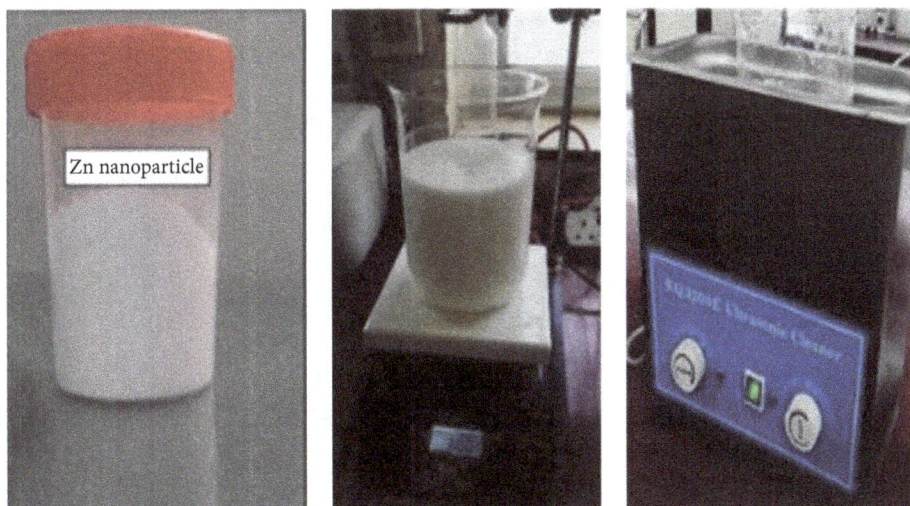

FIGURE 2: (a) Dimensions of thermal pipes mounted on the backside of PV. (b) Steps of nanofluid preparation.

Chaji et al. [8] studied the effect of various concentrations of TiO_2 nanoparticles in water with three values of flow rates, namely, 36, 72, and 108 L/hr. They investigated four particles' concentration ratios (0, 0.1, 0.2, and 0.3% wet). The results showed that adding nanoparticles to water increased the initial efficiency of flat plate solar collector by 3.5 to 10.5% and the index of collector total efficiency by 2.6 to 7% relative to that of the base fluid.

The major problem in PV is the accumulation of heat, which reduces the electrical performance obviously; therefore, heat must be dissipated. In Iraq, the problem becomes much serious, because of a hot weather in most of the year;

this makes the electrical efficiency of PV cells to decrease with the increase of the heat inside the PV cells. The active solution for this problem can be using a water-cooling technique to decrease the heat effects by transferring the heat to the water which can be used in many applications as a hot water. Thermal conductivity enhancement can be achieved by using nanofluid applications such as Zn-H_2O. The originality of the current work is the use of a new design of a cooling technique including copper pipes placed on PV rear surface to absorb the heat accumulated inside the PV cells. This aim was achieved through evaluation of the performance of photovoltaic panels

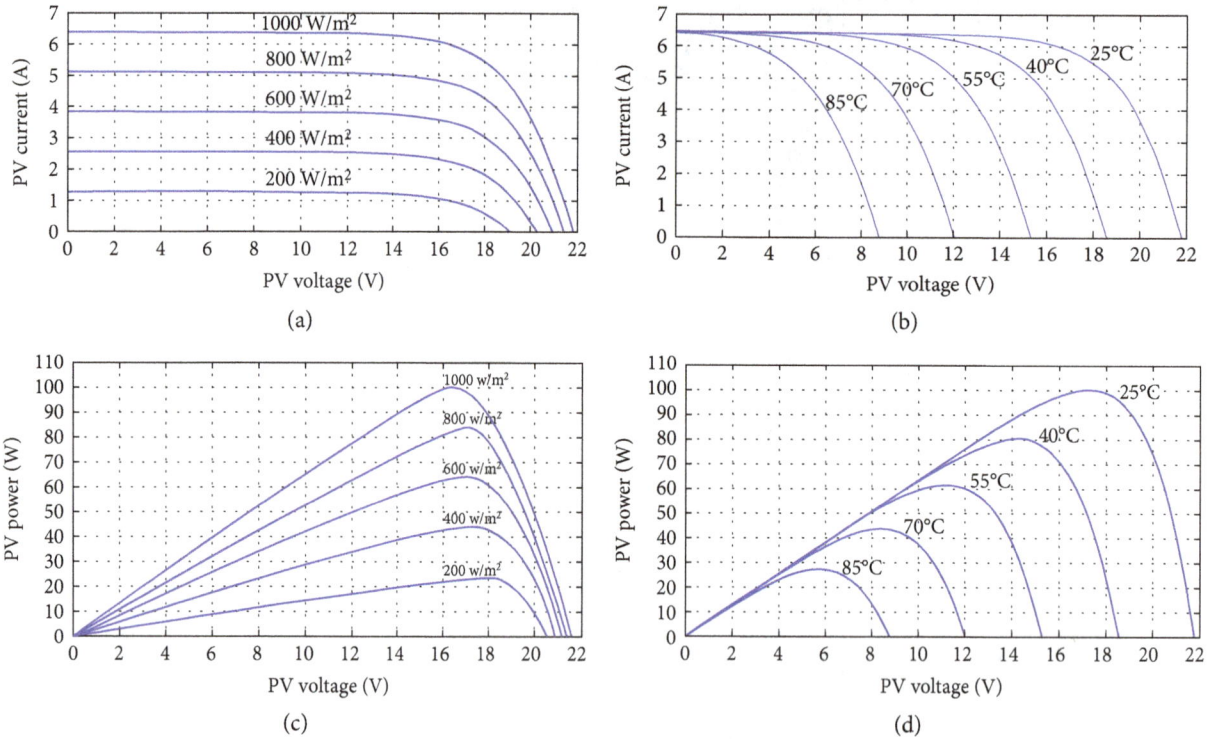

FIGURE 3: (a) Theatrical *I-V* characteristics with radiation at constant temp. (25˚C). (b) *P-V* characteristics with radiation at constant temp. (25˚C). (c) *I-V* characteristics with temperature at constant radiation (1000 W/m^2). (d) *P-V* characteristics with temperature at constant radiation (1000 W/m^2).

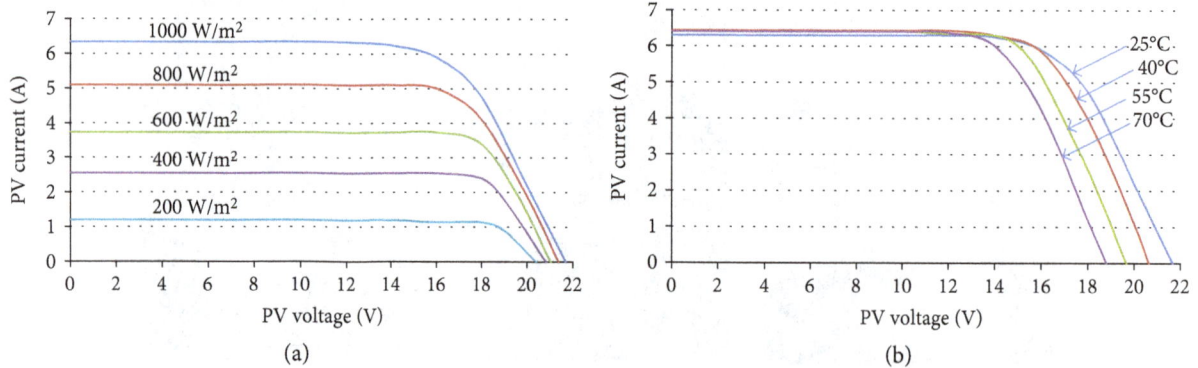

FIGURE 4: (a) Experimental *I-V* characteristics with radiation at constant temp. (25˚C). (b) *I-V* characteristics with temp. at constant radiation (1000 W/m^2).

under different operating conditions, enhancement of the electrical and thermal performance for the photovoltaic/thermal system with water pumping system at different water mass flow rates, and studying the effect of using nanofluid (Zn) as a working fluid in water-circulating pipes at different concentration ratios (0.1, 0.2, 0.3, 0.4, and 0.5).

2. Mathematical Modeling

2.1. Overall Performance of PV/T System. The equations of the nominal electrical efficiency (η_0) presented by Ben [9] are as follows:

$$
\begin{aligned}
\eta_0 &= \frac{V_{mp} I_{mp}}{GA}, \\
\eta_{elec} &= \eta_0 [1 - \beta(T_c - T_0)], \\
Q &= \dot{m} C_P (T_0 - T_i).
\end{aligned}
\tag{1}
$$

The thermal efficiency is evaluated by the following equations, presented by El-Seesy et al. [10]:

$$
\begin{aligned}
\eta_{th} &= \frac{\dot{m} cp (T_0 - T_i)}{A_c G}, \\
\eta_{total} &= \eta_{th} + \eta_{elec} = \frac{\dot{m} cp \int (T_0 - T_i) dt + \int V I dt}{A_c \int G(t) dt}.
\end{aligned}
\tag{2}
$$

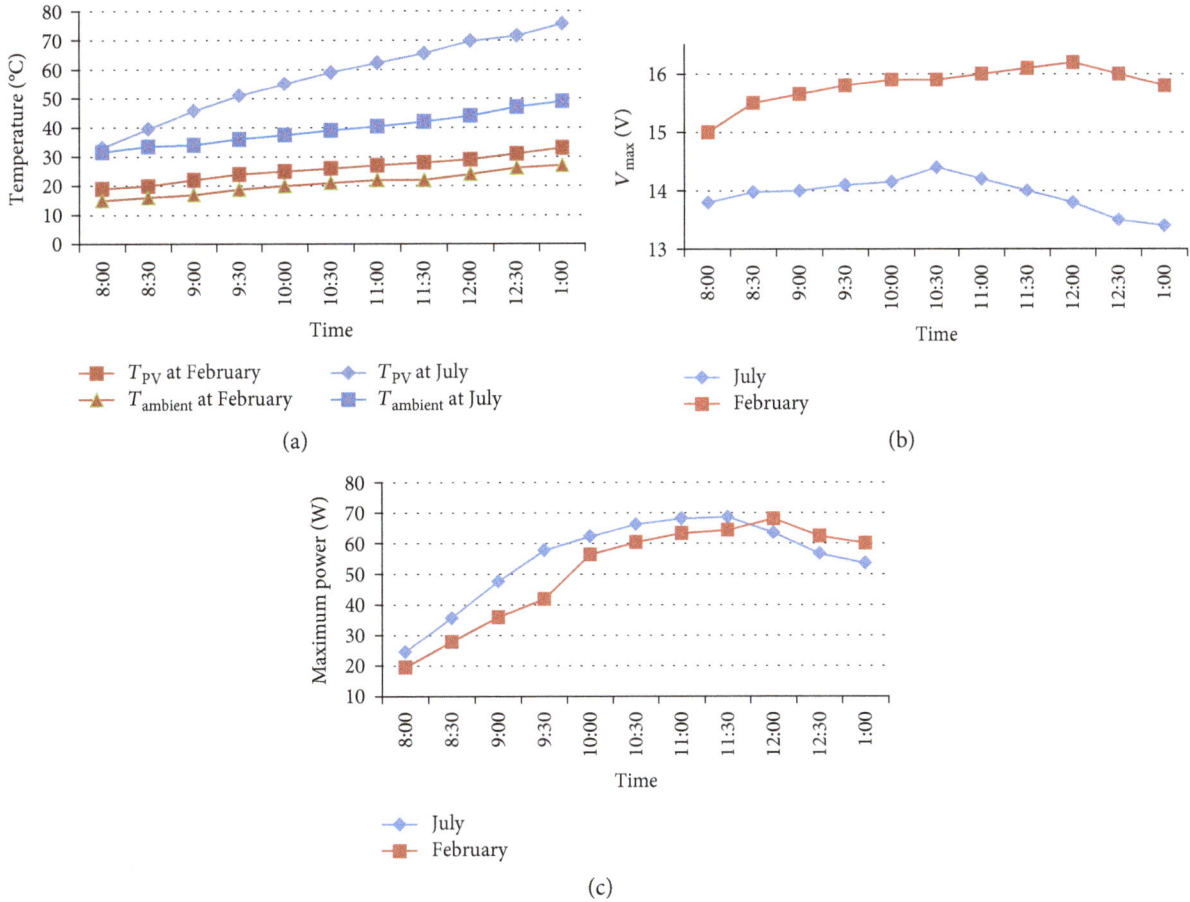

Figure 5: (a) Temperature variations at climatic conditions. (b) Comparison between the voltages at climatic conditions. (c) Comparison between the output powers at climatic conditions.

2.2. Thermophysical Properties of Zn-H₂O Nanofluid.

Thermophysical properties of the working fluid (Zn-H$_2$O nanofluid) are changed due to influence of the nanoparticles. These properties for conventional fluids can be found from standard tables or equations as presented by Darby [11]. The properties of nanofluids can be estimated by using the following equations, as presented by Albadr and Hussein [12, 13]:

$$\rho_{\mathrm{nf}} = (1 - \phi)\rho_{\mathrm{f}} + \phi\rho_{\mathrm{p}},$$
$$(\rho Cp)_{\mathrm{nf}} = (1 - \phi)(\rho Cp)_{\mathrm{f}} + \phi(\rho Cp)_{\mathrm{p}},$$
$$\mu_{\mathrm{nf}} = (1 + 2.5\,\phi)\,\mu_{\mathrm{w}},$$
$$K_{\mathrm{nf}} = \left[\frac{K_{\mathrm{p}} + 2K_{\mathrm{f}} - 2(K_{\mathrm{f}} - K_{\mathrm{p}})\phi}{K_{\mathrm{p}} + 2K_{\mathrm{f}} - (K_{\mathrm{f}} - K_{\mathrm{p}})\phi}\right] K_{\mathrm{f}},$$
$$\phi = \frac{m_{\mathrm{p}}/\rho_{\mathrm{p}}}{(m_{\mathrm{p}}/m_{\mathrm{p}}) + (m_{\mathrm{f}}/\rho_{\mathrm{f}})},$$
$$\propto_{\mathrm{nf}} = \frac{K_{\mathrm{nf}}}{\rho_{\mathrm{nf}}\, Cp_{\mathrm{nf}}},$$
$$\nu = \frac{\mu}{\rho}. \tag{3}$$

Calculation of Reynolds, Peclet, and Prandtl numbers is as follows [13]:

$$\mathrm{Re} = \frac{VD}{\nu},$$
$$\mathrm{Pe} = \frac{VD}{\propto_{\mathrm{nf}}}, \tag{4}$$
$$\mathrm{Pr} = \frac{\nu_{\mathrm{nf}}}{\propto_{\mathrm{nf}}}.$$

Friction factors (f) and Nusselt numbers (Nu) for single-phase flow have been calculated from the following equations:

$$f = [1.58 \ln \mathrm{Re} - 3.82]^{-2},$$
$$\mathrm{Nu} = \frac{(0.125\,f)(\mathrm{Re} - 1000)\mathrm{Pr}}{1 + 12.7\,(0.125\,f)^{0.5}\,(\mathrm{Pr}^{2/3} - 1)}. \tag{5}$$

Friction factor of each flow rate for nanofluid which can be found in single-phase flow cannot be used for calculating friction factor as well as Nusselt number as presented by Hussein [13].

$$f = 0.961\,\mathrm{Re}^{-0.375}\,\phi^{0.052},$$
$$\mathrm{Nu} = 0.074\,\mathrm{Re}_{\mathrm{nf}}^{0.707}\mathrm{Pr}_{\mathrm{nf}}^{0.385}\,\phi^{0.074}. \tag{6}$$

(a)

(b)

(c)

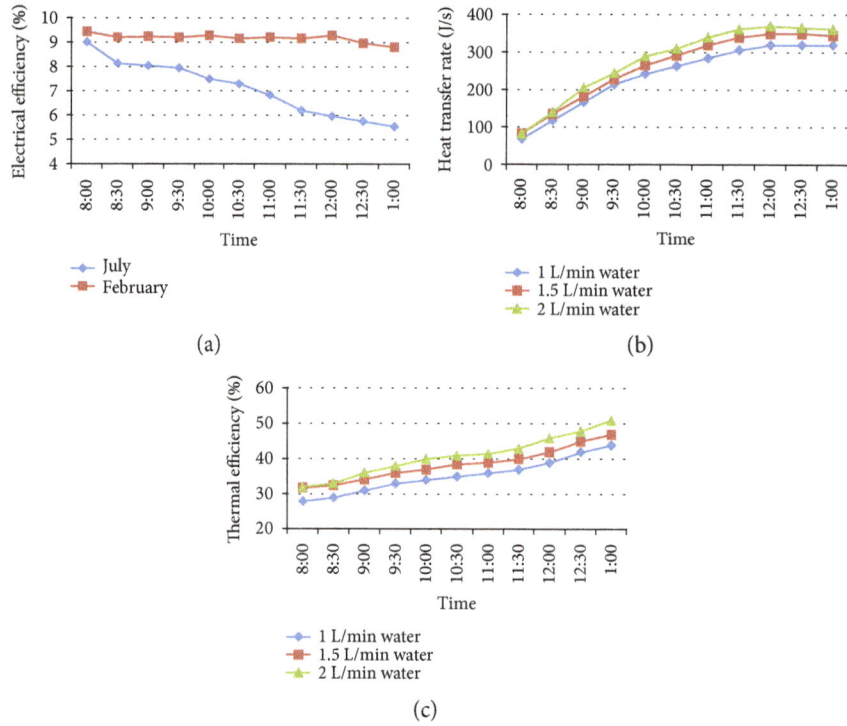

FIGURE 6: (a) Comparison of the electrical efficiency at climatic conditions. (b) Heat transfer rate with different mass flow rates. (c) Effect of the mass flow rates of water on the thermal efficiency.

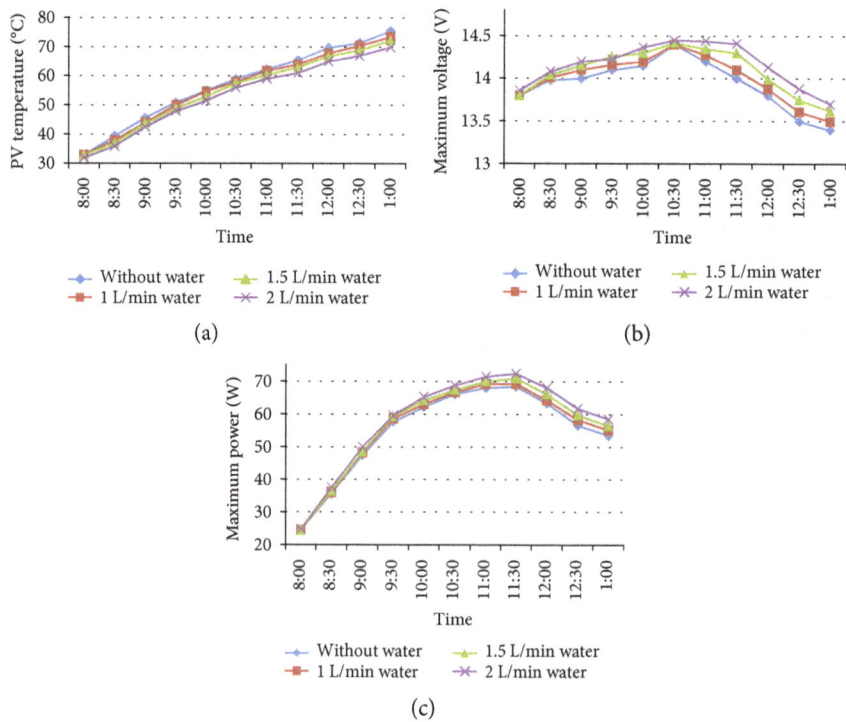

(a)

(b)

(c)

FIGURE 7: (a) Effect of mass flow rates on the PV temperature. (b) Effect of mass flow rates on the voltage. (c) Effect of mass flow rates on the PV power.

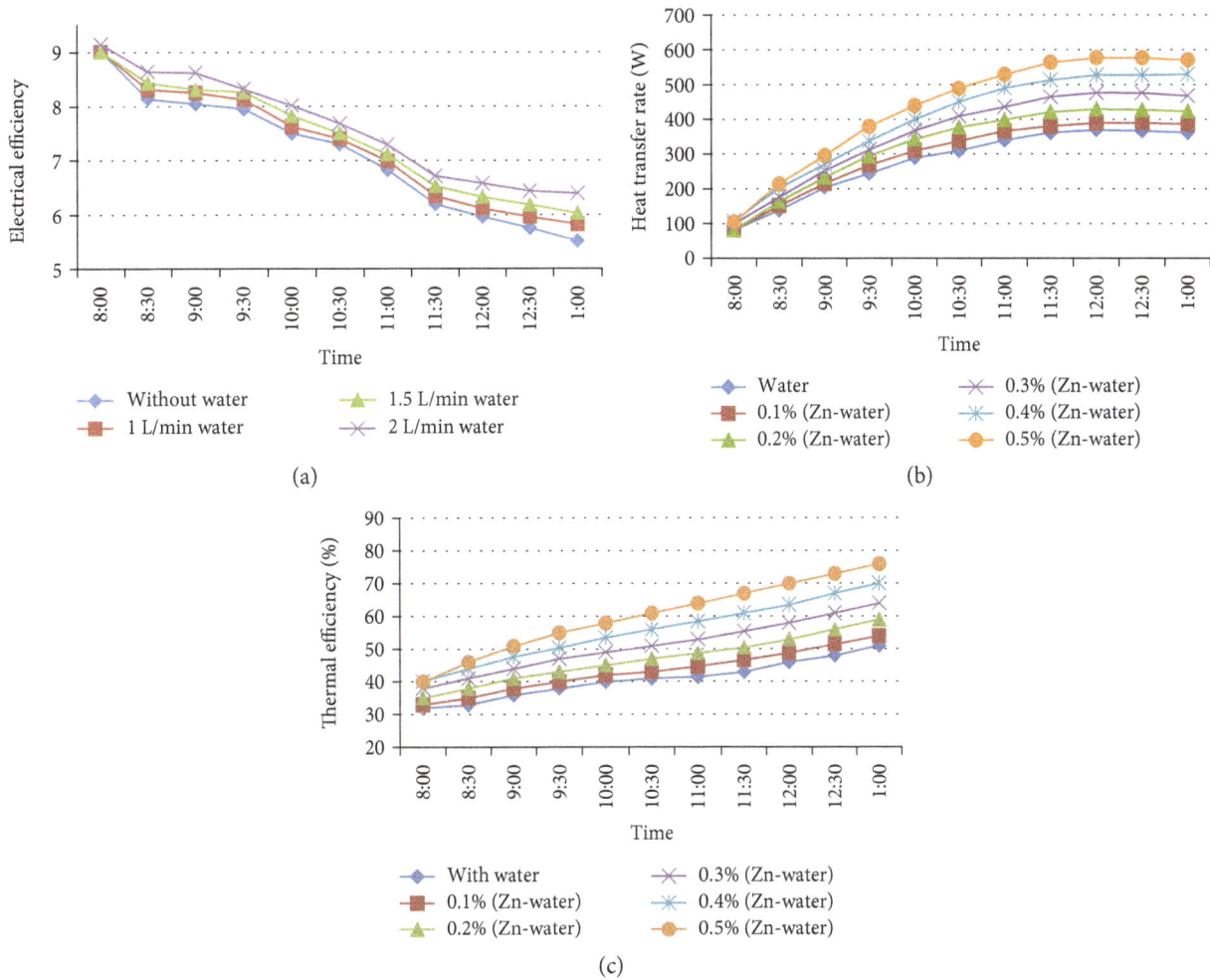

FIGURE 8: (a) Effect of mass flow rates on the electrical efficiency. (b) Heat transfer rate at constant flow rate (2 L/min) at different nanofluid concentrations. (c) Effect of Zn-H$_2$O nanofluid at 2 L/min on thermal efficiency.

3. Material and Methods

3.1. Experimental Setup. The prototype of PV/T system is where the water pumping system was used for all experimental investigations of electrical and thermal effects on system performance and suggested improvements. The setup comprises PV panel, charge controller, battery, DC-DC boost converter, PMDC motor used as a pumping system load, copper pipes fixed on the backside of PV panel, and a radiator with a fan and circulation pump for cooling hot water. The experiment was performed on the site of Electromechanical Engineering Department, University of Technology, in summer and winter seasons.

The photograph of the setup as shown in Figures 1(a) and 1(b) explains the schematic diagram of the complete experimental setup. The major component of the experimental setup is the PV panel that produces direct current (DC) electricity. In this work, the SR-100S PV panel which was made from a monocrystalline semiconductor has been used. The PV panel consists 9×6 cells which have generated 100 Watts maximum power under standard test condition (STC) and typically can generate nearly 5.8 A at maximum solar radiation. The quantities measured during the experiment were as follows: (1) Digital solar meter mounted on the plane of a photovoltaic panel is used to indicate the change of solar irradiance. (2) Five K-type thermocouples connected to the 12-channel digital temperature recorder (type Lutron BTM-4208SD) were used to measure the temperature of PV panel, working fluid, and ambient temperature. (3) The maximum current and voltage, short-circuit current and open-circuit voltage, of the PV panel were recorded manually using multimeters. (4) The mass flow rate of the working fluid (Zn-H$_2$O nanofluid) was measured using flow rate meter.

3.2. PV/T Description. In this work, a specially made serpentine flow collector has been designed. The PV/T collector comprises PV module and thermal collector which are made of copper sheet and pipe. The copper sheet and the piping are paste directly to the back side of PV panel. Copper material has been used due to its high thermal conductivity with the pipe's inner diameter of 11 mm and thickness of 1 mm to transfer the temperature from PV panel to the working fluid. Thermal sink was used between the bottom surface of PV

(a)

(b)

(c)

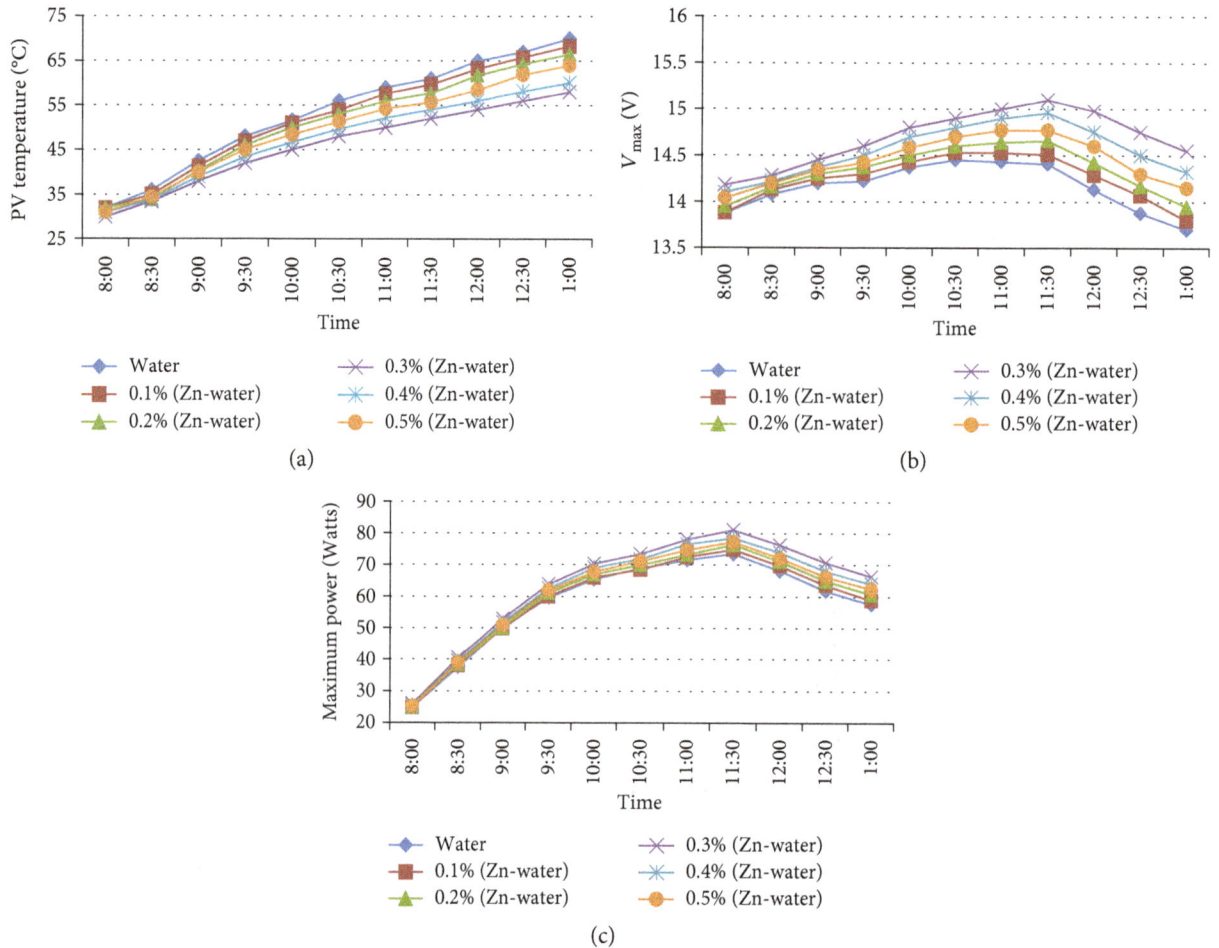

FIGURE 9: (a) Effect of Zn-H_2O nanofluid at 2 L/min on the photovoltaic temperature. (b) Effect of Zn-H_2O nanofluid at 2 L/min on the voltage. (c) Effect of Zn-H_2O nanofluid at 2 L/min on the power.

panel and the surface of 2 mm copper plate to increase thermal conductivity.

The copper pipes are linked using a welding machine. The storage capacity of piping system is 1.5 liters welded on the copper sheet with a height and length and then fixed on the back surface of standard PV panel. The welding method is with 40% tin and 60% silver. The oscillatory flow has at least one inlet and outlet to permit working fluid to enter and to exit from the copper pipes, respectively. Water enters the pipes with low temperature and travel as hot water. The hot water can be consumed or stored for later use. However, this work is dedicated to water pumping system and thus there is no need for hot water output from the proposed PV/T system. In this way, solar radiation energy can be fully used for solar heating applications. The dimension of the thermal collector is shown in Figure 2(a).

3.3. Preparation of Nanofluid. After studying the impact of a water-cooling technique on the performance of PV/T system, Zn-water nanofluid was prepared at five concentration ratios (0.1, 0.2, 0.3, 0.4, and 0.5%) by mixing the particles with 1.5 liters of ionized water. Figure 2(b) shows that Zn-water nanofluid has been prepared in the corrosion laboratory of

the Materials Engineering Department at the University of Technology. Nanopowder was purchased, and a type of Zn nanoparticle was used in this study. The diameter of the nanoparticle is 30 nm.

4. Results and Discussion

4.1. Simulation of the PV Output Characteristics. The PV Matlab model that has been developed is tested to assess the solar radiation effects and PV temperature variations. From the results, it notes that the current increases proportionally with the increase of solar radiation, but the voltage increases nonlinearly with solar radiation and then increases the level of power output as shown in Figure 3. On the other hand, the temperature primarily affects the PV voltage. The rising temperature of PV panel primarily influences the PV voltage more than the PV current. That reason subsequently leads to decrease the power. When the temperature of PV module increases from 25 to 85 at irradiation of 1000 W/m^2, the PV open-circuit voltage is decreased from 21.8 to 18.8 volts and this leads to decrease the PV power generated from 100 to 84 Watts which represents the variation of current-voltage (I-V) and power-voltage (P-V)

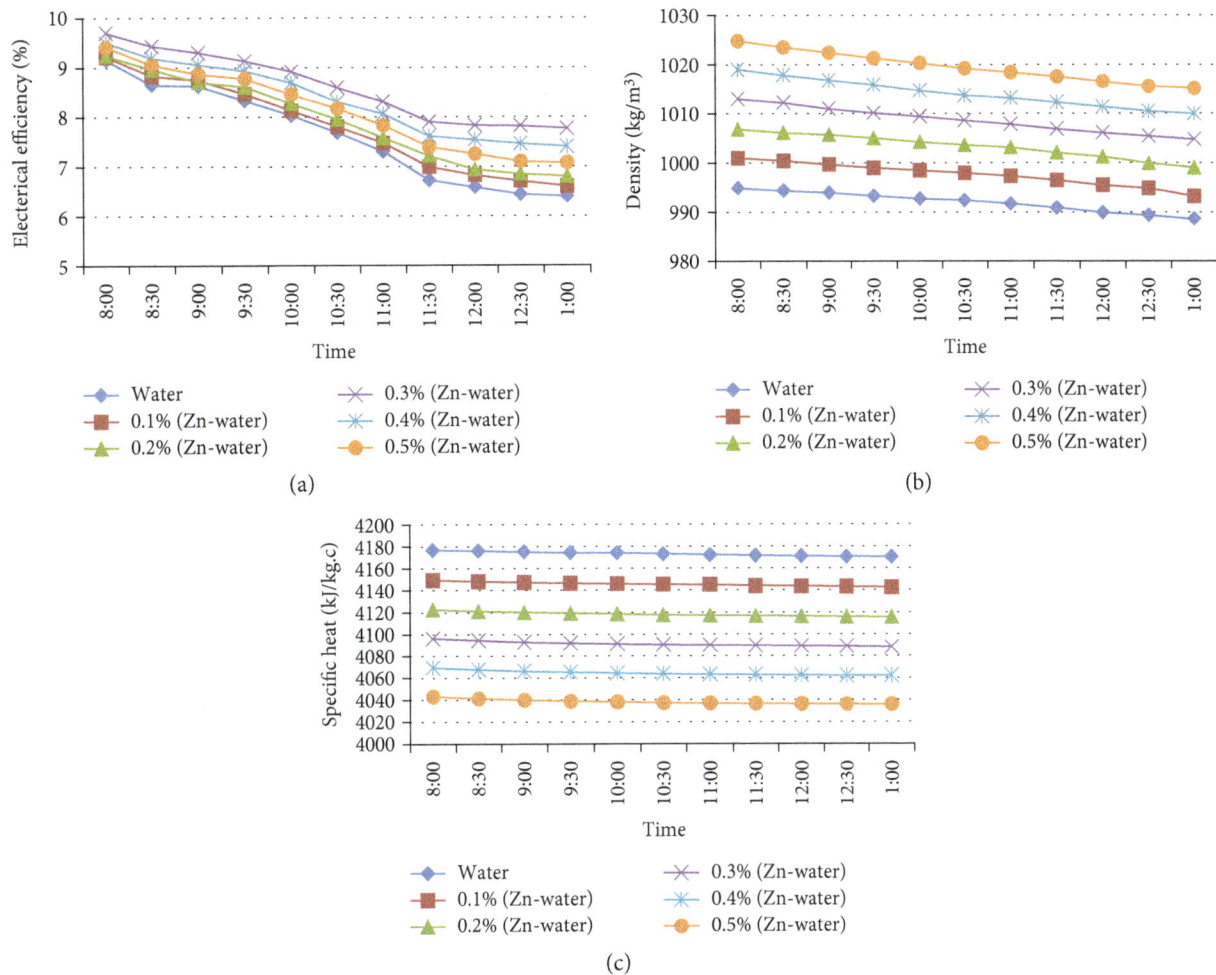

FIGURE 10: Effect of Zn-H$_2$O nanofluid concentrations at 2 L/min on (a) electrical efficiency, (b) working fluid density, and (c) specific heat.

characteristics. After solving the governing equations of the electrical and thermal performance of the system as mentioned in the introduction section which helped to determine and obtained the results, the thermal performance of PV/T system is shown in Figure 4.

4.2. PV Performance with Water-Cooling Technique. Figure 5 shows the temperature difference, voltage, and maximum power variations of the hybrid system PV/T, respectively. These curves represented the changing of temperature difference depending on the solar radiation and ambient temperature with time. The temperature difference between inlet and outlet is almost in linear relationship with the solar radiation at changing value of radiation from 200 W/m^2 to 900 W/m^2 and then falls to 700 W/m^2 nearly. It is observed that when the increase of flow rate causes a decrease in the output temperature and the temperature difference and when the decrease of flow rate leads to increase in the output temperature and the temperature difference and then gets the best thermal gain, this is due to the fluid which takes a long time to absorb heat from the surface of PV module.

Figure 6 shows that the 2 L/min flow rate gives the best performance for the thermal efficiency of the PV/T system due to the increase in heat transfer rate of fluid in pipes which represents that the 2 L/min flow rate of working fluid (water) gives good improvement in current and voltage for photovoltaic/thermal system due to the reduction in photovoltaic temperature at this value of flow rate and the cooling process gives improvement on power generated from photovoltaic, but the better power produced at the 2 L/min flow rate is because more heat dissipated in the radiator with increasing flow rate of working fluid circulated.

It is observed that the electrical efficiency of the PV module increases with increasing the flow rate of fluid. The best electrical efficiency is obtained at optimum flow rate (2 L/min) because all the performance is improved at this rate. The results show that the operation of pumping system depends deeply on the performance of the photovoltaic system and the peak power of the photovoltaic system. The DC voltage influences the speed of running motor. It is observed that low voltage generated from PV module due to high operating temperature leads to a decrease in the output of DC pump, while high voltage leads to an increase

(a)

(b)

(c)

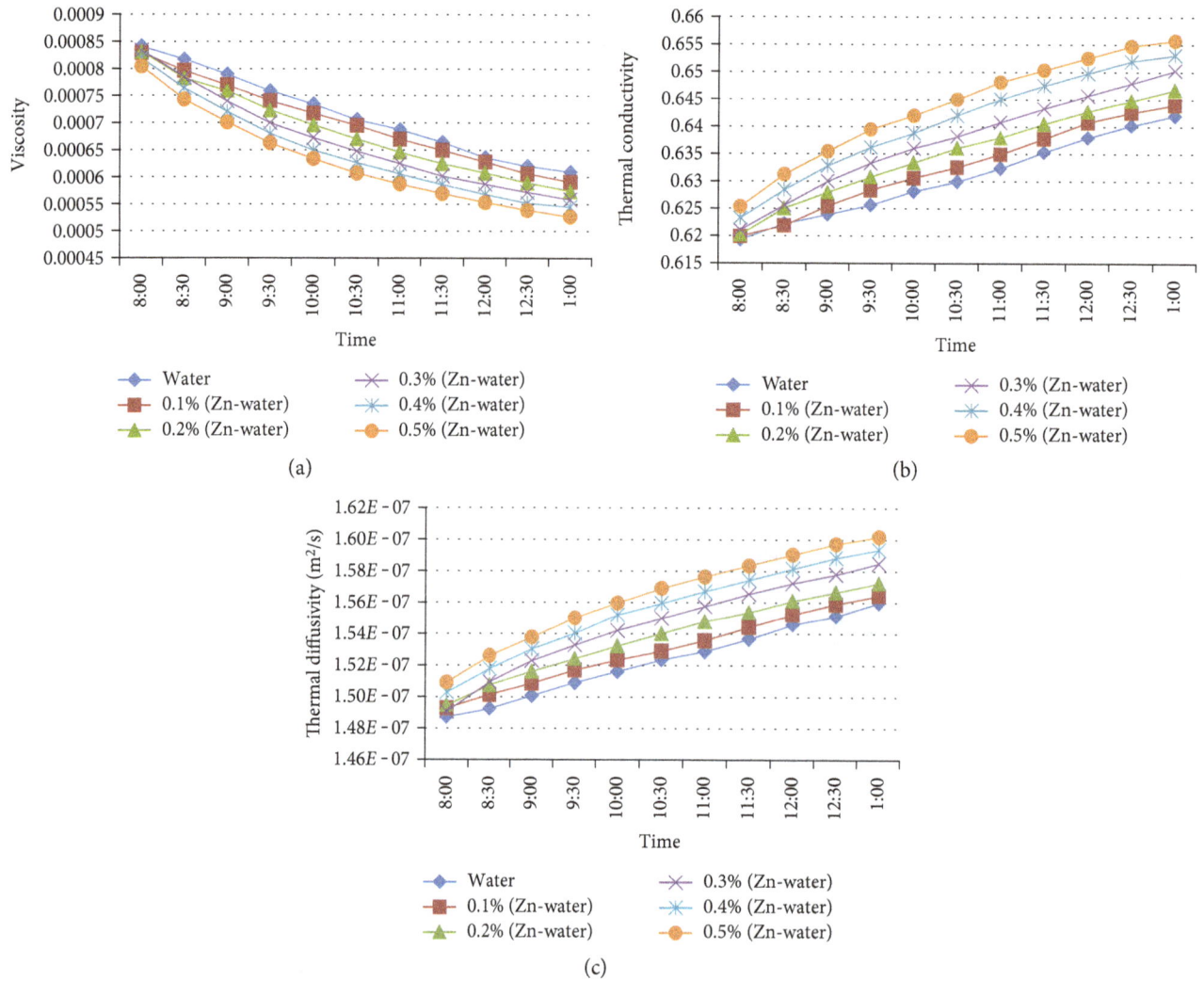

FIGURE 11: Effect of Zn-H_2O nanofluid at 2 L/min on (a) viscosity, (b) thermal conductivity, and (c) thermal diffusivity.

in the output of DC pump. It is observed that circulating the fluid through pipes at photovoltaic cells' rear surface strongly enhances the performance of system and subsystem, since motor pump can receive most of the power of cells by improving the performance of PV module as shown in Figure 7.

4.3. PV Performance with the Use of Zn-H_2O Nanofluid. The thermal conductivity of Zn metal is higher than the water: 112.2 W/m.k for Zn metal while for the water, 0.596 W/m.k. This feature gives an increase in the thermal conductivity of working fluid. Figure 8(b) shows that the heat transfer rate increases with the increase of volume concentration ratio of nanofluid because the nanofluid thermal conductivity increases as the concentration ratios of nanofluid increases and that led to an increase in the thermal performance of photovoltaic/thermal system as shown in Figure 8(c) that is due to the increase of heat transfer rate with the concentration ratio. It is found that 2 L/min of mass flow rate gives the best thermal performance and electrical performance under water test of PV/T.

Figure 9 explains that the value 0.3% gives the best cooling for photovoltaic. This is due to the increase in thermal conductivity of Zn-H_2O nanofluid at this ratio which led to more absorption of heat from photovoltaic surface. If the concentration ratio increases more than 0.3%, the PV temperature will increase because of the increase in density and viscosity of working fluid with the rising of concentration ratio, and this gives reverse impact of improvement. By decreasing PV temperature with the use of nanofluid, the maximum power produced from the PV module will be increased. It was noticed that the better maximum power generated is at 0.3% nanofluid concentration ratio because this volume ratio gives good cooling for PV module; also, it was observed that there is an improvement in I_{max} and V_{max} which leads to enhancement in PV power when using nanofluid and a good case at 0.3% concentration ratio. The electrical efficiency of PV module is improved by using nanofluid at 0.3% volume concentration ratio and reduced when it is greater than 0.3% because of the increase of PV temperature as the volume concentration ratio increases above 0.3%, as shown in Figure 10.

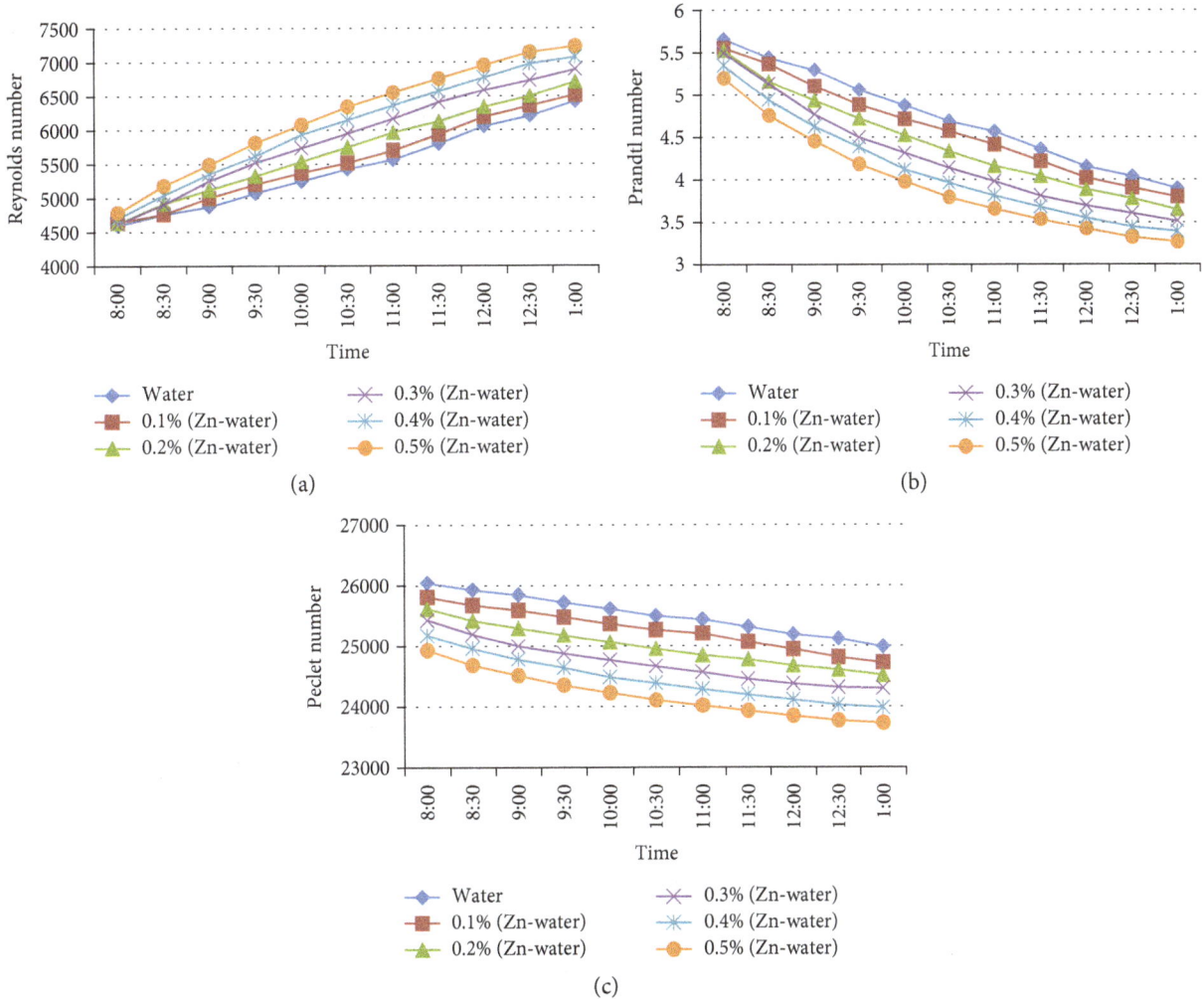

FIGURE 12: Effect of Zn-H_2O nanofluid at 2 L/min on (a) Reynolds number, (b) Prandtl number, and (c) Peclet number.

4.4. Physical Properties of Zn-H_2O Nanofluid.

All the physical properties of working fluid will change depending on the concentration ratio of nanoparticles such as density, specific heat, viscosity, and thermal conductivity. It was observed from the sketch that the variation of density of nanofluid is a function of volume concentration ratios and the density of water when increasing the temperature. Figure 11(a) represents the changes in specific heat of nanofluid with increasing the volume concentration ratios. This behavior is due to the change in density of nanofluid as a function of temperature. Figure 11(b) represents the viscosity of nanofluid as a function of volume concentration ratio. The results showed the decrease in nanofluid viscosity at all volume concentration ratios with rising temperature; this is explained by changing the physical properties of the water at rising temperature. Figure 11 shows the changes in thermal conductivity and thermal diffusivity of nanofluid, respectively, with increase in volume concentration ratios where the thermal conductivity is the most important in the physical properties of nanofluid, and it primarily depends on the temperature of fluid. It was observed that at higher temperature, there is greatest impact on these values and we noticed that with increasing the concentration ratios, the more heat are absorbed at the same time as compared with all values of volume concentration ratios, and this rise in temperatures leads to an increase in the thermal conductivity and thermal diffusivity of nanofluid, respectively. Figure 12 shows the influence of volume concentration ratios on Reynolds number; increasing concentration ratios led to increasing absorbing temperature, leading to increase in Reynolds number. This increase in Reynolds number is due to the reduction in viscosity of nanofluid and increase in density of nanofluid. Figure 12(c) represents the decreasing in Prandtl number with increasing in temperature for all volume concentration ratios; this is due to the increase in density and thermal diffusivity and the decline in viscosity with higher temperature. The influence of volume concentration ratios on Peclet number is shown in Figure 13. It is observed from the graph that the Peclet number decreased because of the increase in thermal diffusivity. The influence of volume concentration ratios as a function of temperature on the Nusselt number is shown in Figure 13(b). It was observed that the Nusselt number increased with volume concentration ratios (0.1,

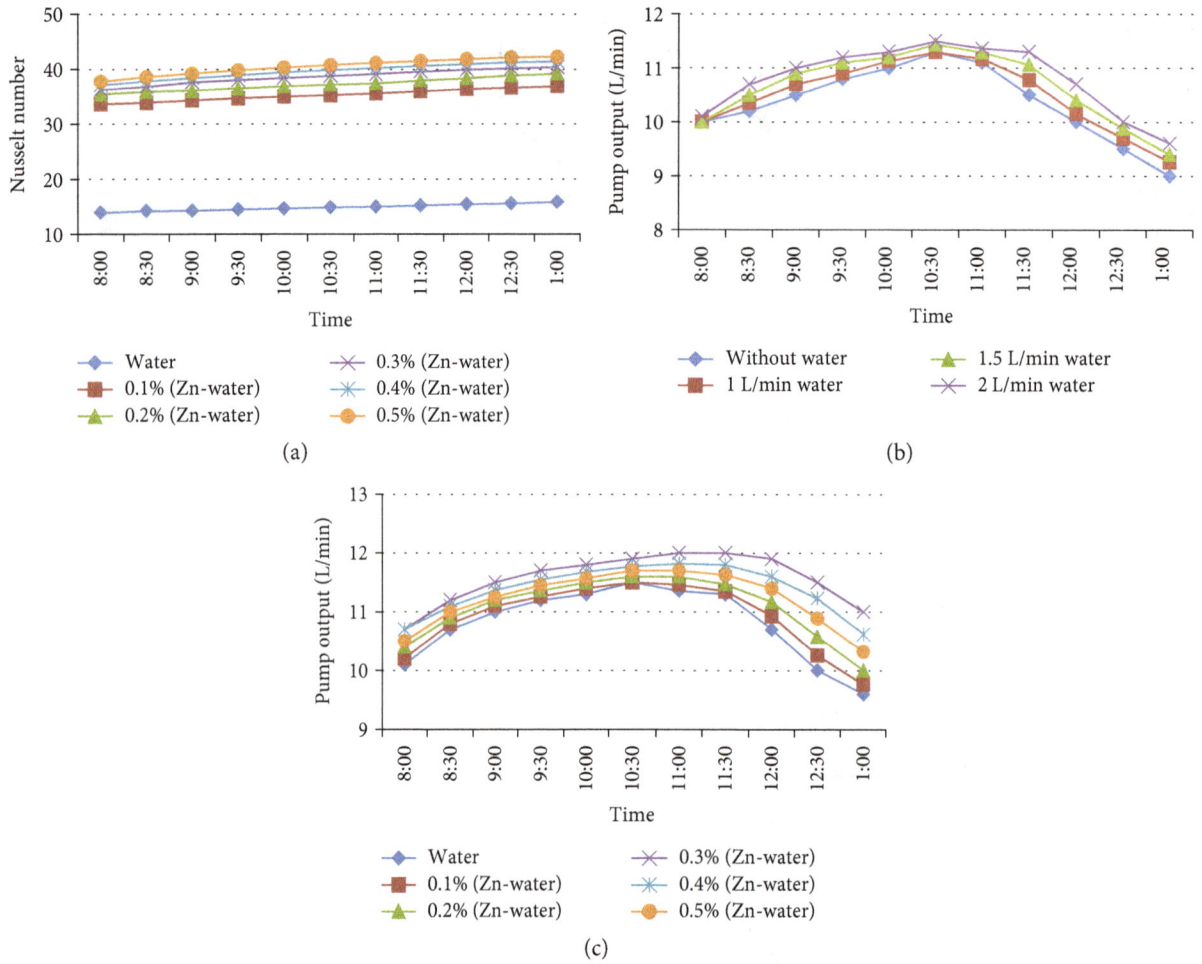

FIGURE 13: Effect of Zn-H_2O nanofluid at 2 L/min on (a) Nusselt number, (b) pump output at different mass flow rates, and (c) pump output at constant mass flow rate (2 L/min) with different concentration ratios.

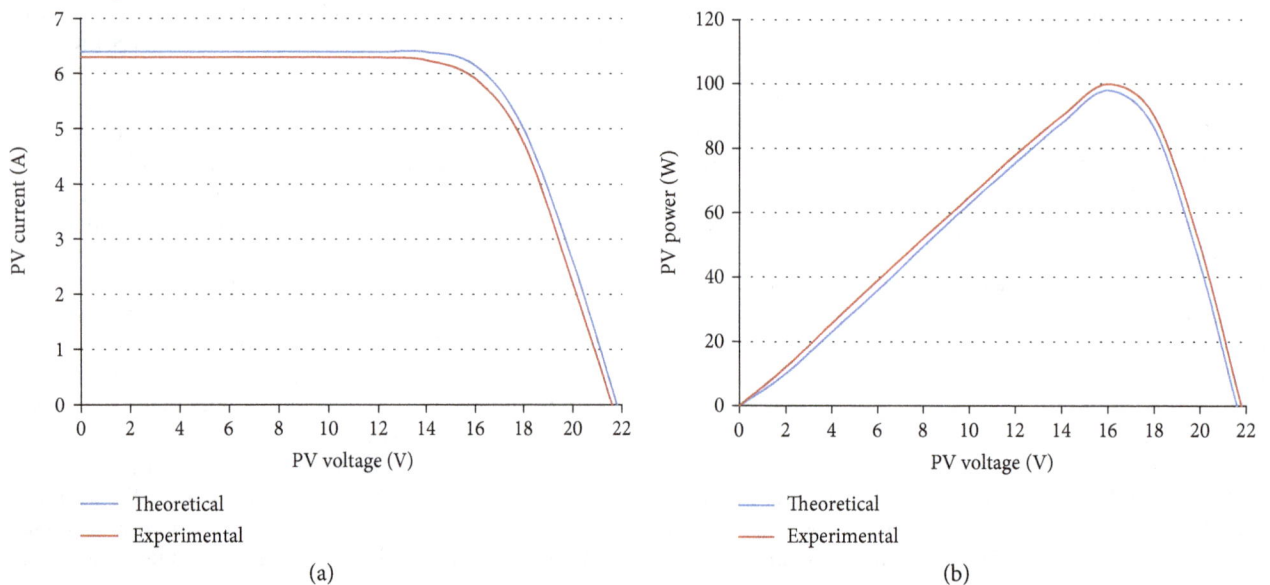

FIGURE 14: (a) Theoretical and experimental results comparison of I-V at 25°C, 1000 W/m². (b) Theoretical and experimental results comparison of P-V at 25°C, 1000 W/m².

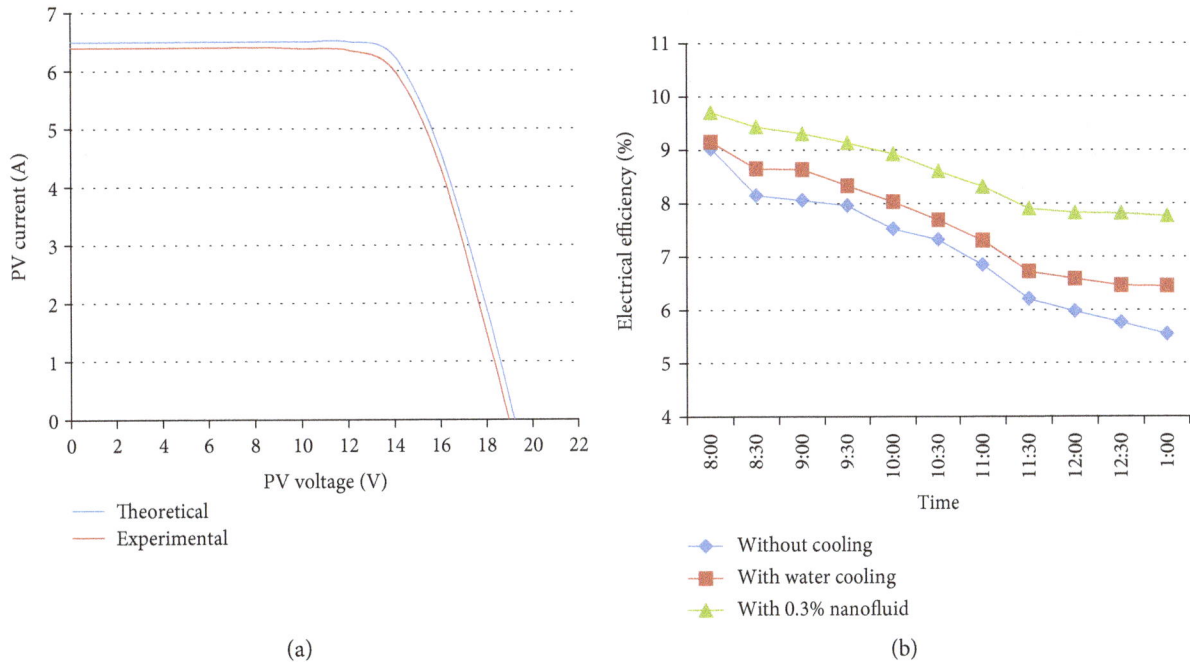

(a)

(b)

FIGURE 15: (a) Theoretical and experimental results comparison of I-V at 1000 W/m^2 and 70°C. (b) Comparison of electrical efficiency of PV/T without water, with water (2 L/min), and with Zn-H$_2$O nanofluid.

0.2, 0.3, 0.4, and 0.5%). This increase is due to the increase of the Reynolds and Prandtl numbers with the rising temperature, and with increasing concentration ratios, this leads to increasing the value of Nusselt number.

4.5. PV Performance of Water Pumping System. In this work, we have tested the operation of pumping systems designed to supply water for drinking or irrigation. The results show that the operating of pumping system depends deeply on the performance of the photovoltaic system and the peak power of the photovoltaic system.

The goal achieved via this study is the investigation of solar radiation changing effects on the pumping system performances. The obtained results show that due to increasing solar radiation, the pump flow increased. The DC voltage influences the speed of running motor. It is observed that low voltage generated from PV module due to high operating temperature lead to a decrease in the output of DC pump, while high voltage lead to an increase in the output of DC pump. It is observed that circulating the fluid through pipes at the photovoltaic cells' rear surface strongly enhances the performance of systems and subsystems, since motor pumps can receive most of the power of the cells by improving performance of the PV module as shown in Figures 13(b) and 13(c).

4.6. Results Comparison. When comparing between the experimental results which have been measured manually as shown in Figures 15(a) and 15(b) with theoretical results obtained from simulation using Matlab/Simulink for PV characteristics under the conditions (1) effect of solar radiation at constant temperature (25°C) and (2) effect of temperature at constant irradiation (1000 W/m^2). It can

be noticed from these figures that the difference between experimental and theoretical results is about less than 2% which is quite acceptable.

5. Conclusions

The variations in solar radiation mainly influence the output current, while the changes in temperature mainly affect the output voltage. Hybrid PV/T systems are one of the methods used to enhance the electrical efficiency of panel then improve the photovoltaic water pumping system performance. The electrical and thermal efficiencies of the hybrid system will increase with increasing mass flow rate of water. At optimum flow rate of 2 L/min, electrical efficiency was 6.5% and thermal efficiency was 60%. The results indicated that when nanofluid (Zn) is used at various concentration ratios (0.1, 0.2, 0.3, 0.4, and 0.5%) at 2 L/min flow rate, the cell temperature dropped more significantly from 76°C to 58°C at an optimum concentration ratio of 0.3% nanofluid; this led to an increase in the electrical efficiency of PV panel to 7.8%.

Nomenclature

A: Area of the PV module (m^2)
A_c: Area of collector (m^2)
C_{pf}:   Heat capacity of the base fluid (J/kg.c)
C_{pnf}:   Heat capacity of the nanofluid (J/kg.c)
G: Solar radiation (W/m^2)
I_m: Maximum current of PV (A)
I_{sc}:   Short-circuit current of solar cell (A)
K_f: Thermal conductivity of base fluid (W/m.c)

K_I: Cell's short-circuit current temperature coefficient (A/k)
K_{nf}: Thermal conductivity of the nanofluid (W/m.c)
K_ρ: Thermal conductivity of the nanoparticle (W/m.c)
m: Mass flow rate (kg/s)
ϕ: Volume concentration of the nanoparticles
T_{in}: Inlet temperature of the working fluid (°C)
T_0: Temperature of standard condition (25°C)
T_{out}: Outlet temperature of working fluid (°C)
V_m: Maximum voltage of PV (V)
V_{PV}: Output voltage (V)
β: Coefficient of silicon cell ($\beta = 0.0045$°C^{-1})
η_0: Nominal electrical efficiency at standard conditions
μ_{nf}: Nanofluid viscosity (kg/m.s)
μ_w: Water viscosity(kg/m.s)
ρ_{nf}: Density of the nanofluid (kg/m^3)
ρ_p: Density of the nanoparticles (kg/m^3).

Conflicts of Interest

The authors declare that they have no conflicts of interest.

References

[1] S. K. Firth, *Raising efficiency in photovoltaic systems: high resolution monitoring and performance analysis, [Ph.D. Thesis]*, De Montfort University, UK, 2006.

[2] A. Oi, *Design and simulation of photovoltaic water pumping system, [MSc. Thesis]*, California Polytechnic State University, San Luis Obispo, California, 2005.

[3] H. Teo, *Photovoltaic thermal (PV/T) system: effect of active cooling, [MSc. Thesis]*, National University of Singapore, Singapore, 2010.

[4] R. A. Taylor, P. E. Phelan, T. P. Otanicar et al., "Applicability of nanofluids in high flux solar collectors," *Renewable and Sustainable Energy*, vol. 3, no. 2, article 023104, 2011.

[5] M. Abdolzadeh and M. Ameri, "Improving the effectiveness of a photovoltaic water pumping system by spraying water over the front of photovoltaic cells," *Renewable Energy*, vol. 34, no. 1, pp. 91–96, 2009.

[6] W. A. EL-Basit, A. M. A. B. D. El-Maksood, and F. A. E.-M. S. Soliman, "Mathematical model for photovoltaic cells," *Leonardo Journal of Sciences*, vol. 23, pp. 13–28, 2013.

[7] S. Odeh and M. Behnia, "Improving photovoltaic module efficiency using water cooling," *Heat Transfer Engineering*, vol. 30, no. 6, pp. 499–505, 2009.

[8] H. Chaji, Y. Ajabshirchi, E. Esmaeilzadeh, S. Z. Heris, M. Hedayatizadeh, and M. Kahani, "Experimental study on thermal efficiency of flat plate solar collector using TiO2/water nanofluid," *Modern Applied Science*, vol. 7, no. 10, p. 60, 2013.

[9] H. Ben, "Study of electrical and thermal performance of a hybrid PVT collector," *International Journal of Electrical and Electronics*, vol. 3, no. 4, pp. 95–106, 2013.

[10] I. E. El-Seesy, T. Khalil, and M. T. Ahmed, "Experimental investigations and developing of photovoltaic/thermal system," *World Applied Sciences Journal*, vol. 19, no. 9, pp. 1342–1347, 2012.

[11] R. Darby, *Chemical Engineering Fluid Mechanics*, Marcel Dekker, Inc, New York, 2nd edition, 2001.

[12] J. Albadr, "Heat transfer through heat exchanger using Al2O3 nanofluid at different concentrations," *Case Studies in Thermal Engineering*, vol. 1, no. 1, pp. 38–44, 2013.

[13] A. M. Hussein, "Experimental measurements of nanofluids thermal properties," *International Journal of Automotive and Mechanical Engineering (IJAME)*, vol. 7, pp. 850–863, 2013.

8

Synergetic Control of Grid-Connected Photovoltaic Systems

Junjie Qian, Kaiting Li, Huaren Wu, Jianfei Yang, and Xiaohui Li

School of Electrical and Automation Engineering, Nanjing Normal University, Nanjing 210042, China

Correspondence should be addressed to Xiaohui Li; 61011@njnu.edu.cn

Academic Editor: Md. Rabiul Islam

It is important to improve the dynamic performance and the low-voltage ride-through (LVRT) capability of a grid-connected photovoltaic (PV) system. This paper presents synergetic control for the control of a grid-connected PV system. Modeling of a grid-connected PV system is described, and differential-algebra equations are obtained. Two control strategies are used in normal operation and during LVRT of a PV system. Practical synergetic controllers with two control strategies are synthesized. The mathematical expressions are derived for computing control variables. The design of the synergetic controllers does not require the linearization of the grid-connected PV system. A grid-connected PV system with synergetic controllers is simulated in Simulink surroundings. The control performance is studied in normal operation and during LVRT. Simulation results show that the synergetic controllers are robust and have good dynamic characteristics under different operation states.

1. Introduction

The world is faced with serious problems of energy depletion and environmental pollution. The research and development of photovoltaic (PV) technologies have become a hot topic in the world [1]. Solar PV is now used around the world as an important technology for the conversion of solar energy because of its cleanliness and security. The solar PV capacity increased 25% over 2014 to a record 50 GW, lifting the global total to 227 GW. The solar PV industry is one of the fastest growing high-tech industries [2].

The control of a PV system is an important and difficult task. A grid-connected PV system mainly includes maximum power point tracking (MPPT) and the control of the DC-AC converter. Much research on MPPT has been conducted, and various MPPT algorithms have been proposed. The maximum power point (MPP) may be located by the perturbation and observation (P&O) algorithms [3], the incremental conductance (InC) algorithm [4], and the artificial neural network algorithm [5].

Reference [6] proposes a modified InC algorithm. The algorithm eliminates the division calculations involved in its structure and improves the variable step size, which only depends on the PV power change. Reference [7] improves P&O algorithm. This approach combines ant colony optimization with the traditional P&O method to yield faster and efficient convergence. This improved P&O algorithm can recognize global MPP under partially shaded conditions. A new MPPT algorithm is proposed in [8]. This scheme uses a gray wolf optimization technique to track the global peak of a PV array under partial shading conditions. It can solve the problems such as lower tracking efficiency, steady-state oscillations, and transients as encountered in P&O.

A two-stage three-phase grid-connected PV system in [9] contains a DC-DC boost converter and a DC-AC VSC converter. Pulse width modulator (PWM) signals fire the converters. The DC-AC VSC control system has an external voltage control loop and an internal current control loop. The external control loop regulates the DC link voltage, and the internal control loop regulates grid currents. The grid currents are transformed into d-axis and q-axis currents to accomplish vector control. Two control loops adopt the proportional-integral (PI) algorithm. Reference [10] presents a novel sliding-mode (SM) control for grid-connected PV systems. A systematic adaptive procedure to calculate the band of the hysteresis comparators is developed to improve the performance of the SM controller. A vector controller can keep the maximum power delivery of the PV system. Reference [11] uses a probabilistic wavelet fuzzy neural network (PWFNN) to structure the reactive power controller for a grid-connected PV system. The balance of the active power between the PV array and the DC-AC converter during grid faults is controlled by the DC link voltage. The controller can improve the operation of the grid-connected PV system during LVRT. Reference [12] suggests injecting the maximum rated current to maximize the inverter power capability during LVRT. The strategy combines a proper balance between positive- and negative-current sequences. High- and low-power production scenarios limit the inverter output current to the maximum rated value and avoid active power oscillations. Reference [13] presents a new adaptive PI controller using the continuous mixed p-norm (CMPN) algorithm for enhancing the LVRT capability of grid-connected PV systems. The adaptive PI controller is used to control the DC-AC converter. The gains of the PI controller are changed by the CMPN algorithm online without the need to fine-tune or optimize.

Reference [14] reviews the general synergetic control design procedure. Synergetic control is applied to a DC-DC boost converter, deriving a basic control law. An adaptive control strategy gives better trade-off between large-signal stability and load step response time. Reference [15] introduces a practical synergetic controller to regulate the buck converters that coordinate pulse current charging of batteries. Simulation and experiment results show that the synergetic controller is robust for such nonlinear dynamic systems and achieves better performance than the standard PI controller.

This paper presents a design for the synergetic controllers for an internal current control loop of a three-phase grid-connected PV system. The theory of synergetic control is first described. Modeling of a grid-connected PV system is introduced, and differential-algebra equations are obtained. Synergetic controllers are derived in detail for normal operation and during LVRT of a PV system. Two control strategies are used during normal and LVRT operations of PV systems. The design of a synergetic controller does not require linearization of the PV system. The parameters T_1 and T_2 of the two synergetic controllers are the same, and T_1 equals T_2. The parameters of the synergetic controllers are easy to determine, and the proposed control schemes are easy to achieve. An example given in Matlab is adapted according to the synergetic controllers for assessing the performance of synergetic control of the grid-connected PV system. The synergetic control is chatter-free, and the simulation results demonstrate the effectiveness of the proposed control schemes.

2. Synergetic Control Theory

Synergetic control is a state space control method based on modern mathematics. Synergetic control is applicable to the control of nonlinear, dynamic, and high-dimensional systems. It can be perfectly analyzed by mathematical expressions.

The nonlinear state equation of a controlled system is

$$\dot{x} = f(x,u,t), \tag{1}$$

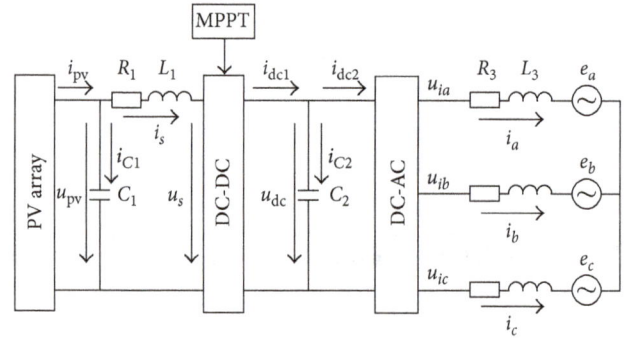

FIGURE 1: Main circuit of a grid-connected PV system.

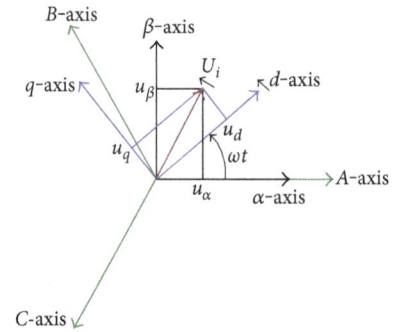

FIGURE 2: Reference frames.

where x is the state vector of the controlled system $\mathfrak{R}^n$, $f(\cdot)$ is a continuous nonlinear function, and u is the control vector of $\mathfrak{R}^m (m \leq n)$.

The macrovariables are defined for each input channel as a function of the state variables. The synergetic controller directs the system to move into the manifold from any initial motion point

$$\psi(x,t) = 0, \tag{2}$$

where ψ is the macrovector of $\mathfrak{R}^k$ $(0 < k \leq m)$.

The dynamic evolution of the macrovariable towards the manifolds is defined as follows [14]:

$$T\dot{\psi} + \psi = 0, \tag{3}$$

where T defines the rate of convergence of the system.

Substituting from (2) into (3) yields

$$T\frac{\partial \psi}{\partial x}\dot{x} + \psi = 0. \tag{4}$$

The control vector u can be acquired by substituting (1) into (4). The system can be controlled to stay in the desired manifold.

3. Modeling of the Grid-Connected PV System

The main circuit of the two-stage grid-connected PV system is described in Figure 1. A PV array is connected to a power

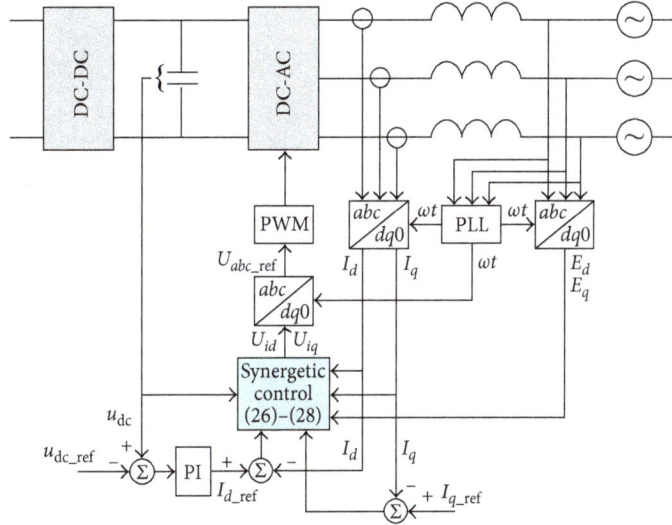

FIGURE 3: Synergetic control scheme of the PV system in normal operation.

grid via a DC boost converter and a three-phase voltage source converter (VSC) [9].

The differential equations (5), (6), (7), and (8) can be written according to Figure 1.

$$C_1 \frac{du_{pv}}{dt} = i_{pv} - i_s, \tag{5}$$

$$u_{pv} = R_1 i_s + L_1 \frac{di_s}{dt} + u_s, \tag{6}$$

$$C_2 \frac{du_{dc}}{dt} = i_{dc1} - i_{dc2}, \tag{7}$$

$$U_{iabc} - E_{abc} = R_3 I_{abc} + L_3 \frac{dI_{abc}}{dt}, \tag{8}$$

where

$$U_{iabc} = \begin{bmatrix} u_{ia} \\ u_{ib} \\ u_{ic} \end{bmatrix},$$

$$E_{abc} = \begin{bmatrix} e_a \\ e_b \\ e_c \end{bmatrix}, \tag{9}$$

$$I_{abc} = \begin{bmatrix} i_a \\ i_b \\ i_c \end{bmatrix},$$

E_{abc} is the grid voltage, U_{iabc} is the output voltage of the DC-AC VSC converter, and I_{abc} is the alternating current.

Equation (8) is based on a three-phase (abc) reference frame. abc, $\alpha\beta0$, and $dq0$ reference frames are shown in Figure 2 [16].

Equation (8) is transformed into (10) from the abc reference frame to the $dq0$rotating reference frame using the sinus-based Park transformation.

$$U_{idq0} - E_{dq0} = R_3 I_{dq0} + L_3 \frac{dI_{dq0}}{dt} + L_3 \begin{bmatrix} -\omega I_q \\ \omega I_d \\ 0 \end{bmatrix}, \tag{10}$$

where $U_{idq0} = PU_{iabc}, E_{dq0} = PE_{abc}, I_{dq0} = PI_{abc}$, and P is the Park transformation matrix given in (11). Consider

$$P = \frac{2}{3} \begin{bmatrix} \sin(\omega t) & \sin\left(\omega t - \frac{2\pi}{3}\right) & \sin\left(\omega t + \frac{2\pi}{3}\right) \\ \cos(\omega t) & \cos\left(\omega t - \frac{2\pi}{3}\right) & \cos\left(\omega t + \frac{2\pi}{3}\right) \\ \frac{1}{2} & \frac{1}{2} & \frac{1}{2} \end{bmatrix}. \tag{11}$$

Equation (10) may be written in (12) and (13). Consider

$$L_3 \frac{dI_d}{dt} = -R_3 I_d + \omega L_3 I_q - E_d + U_{id} = U_{3d} + U_{id}, \tag{12}$$

$$L_3 \frac{dI_q}{dt} = -R_3 I_q - \omega L_3 I_d - E_q + U_{iq} = U_{3q} + U_{iq}, \tag{13}$$

where

$$U_{3d} = -R_3 I_d + \omega L_3 I_q - E_d, \tag{14}$$

$$U_{3q} = -R_3 I_q - \omega L_3 I_d - E_q. \tag{15}$$

U_{id} and U_{iq} in (12) and (13) are control variables that control the DC-AC VSC converter.

The MPPT algorithm computes the duty cycle to control the DC boost converter. The relationship of inputs and outputs of the DC boost converter is depicted by (16) and (17). Consider

$$i_{dc1} = (1 - D)i_s, \tag{16}$$

$$u_s = (1 - D)u_{dc}, \tag{17}$$

where D is the duty cycle of the DC boost converter.

The instantaneous active and reactive powers are defined by (18) and (19). Consider

$$p = E_d I_d + E_q I_q, \tag{18}$$

$$q = -E_d I_q + E_q I_d. \tag{19}$$

Selecting $E_q = 0$, (18) and (19) change into the two following equations:

$$p = E_d I_d, \tag{20}$$

$$q = -E_d I_q. \tag{21}$$

The model above can be used to design the synergetic control of the grid-connected PV system.

4. Synergetic Control of a Grid-Connected PV System in Normal Operation

Control strategies of the DC boost converter and the DC-AC VSC converter in the grid-connected PV system must be made. MPPT is implemented in the DC boost converter in normal operation. There are a number of MPPT algorithms; however, this paper does not analyze them.

The DC-AC VSC converter is controlled using the synergetic control presented in this paper. The control variables U_{id} and U_{iq} in (12) and (13) are derived by synergetic control theory.

The reference value of I_d may be obtained from the external voltage control loop.

$$I_{d\_ref} = \left(K_P + \frac{K_i}{s}\right)(u_{dc} - u_{dc\_ref}) = K_P(u_{dc} - u_{dc\_ref}) + I_u, \tag{22}$$

where $I_u = K_i/s(u_{dc} - u_{dc\_ref})$; that is,

$$\frac{dI_u}{dt} = K_i(u_{dc} - u_{dc\_ref}), \tag{23}$$

where K_P is the gain of the proportional term, K_i is the gain of the integral term, u_{dc_ref} is the reference value of the DC voltage u_{dc}, and subscript ref denotes a reference value.

There are 2 control variables, and therefore, 2 macrovariables must be selected. The first macrovariable is

$$\psi_1 = I_{d\_ref} - I_d = K_P(u_{dc} - u_{dc\_ref}) + I_u - I_d. \tag{24}$$

Substituting (24) into (4), (25) is obtained:

$$T_1\left(K_P \frac{du_{dc}}{dt} + \frac{dI_u}{dt} - \frac{dI_d}{dt}\right) + \psi_1 = 0. \tag{25}$$

Substituting (12), (23), and (24) into (25), the first control variable is computed by (26) as follows:

$$U_{id} = K_P L_3 \frac{du_{dc}}{dt} + K_i L_3(u_{dc} - u_{dc\_ref}) + \frac{L_3}{T_1}(I_{d\_ref} - I_d) - U_{3d}. \tag{26}$$

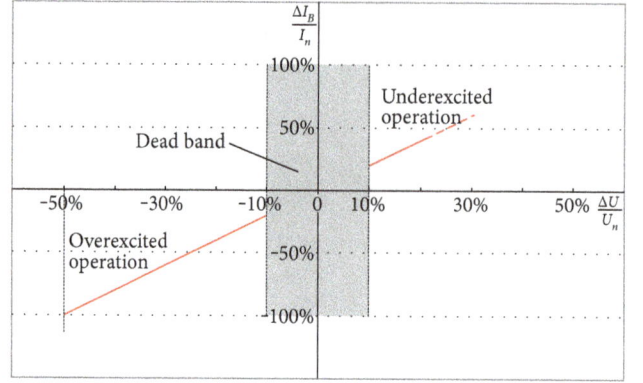

FIGURE 4: Principle of voltage support in the event of grid faults.

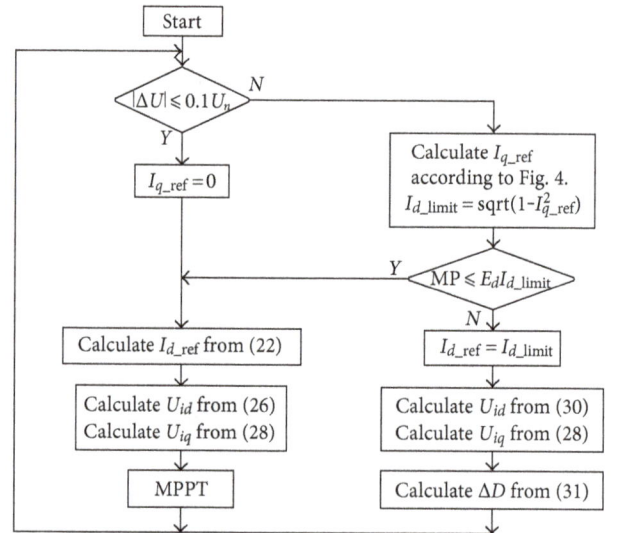

FIGURE 5: Synergetic control strategies of the PV system.

The second macrovariable is selected

$$\psi_2 = I_{q\_ref} - I_q, \tag{27}$$

where I_{q_ref} is a constant.

Substituting (27) into (4) and considering (13), the second control variable is as follows:

$$U_{iq} = \frac{L_3}{T_2}(I_{q\_ref} - I_q) - U_{3q}. \tag{28}$$

The control variables U_{id} and U_{iq} are computed by (26) and (28), respectively, to guarantee system stability in normal operation. The synergetic control scheme of the grid-connected PV system is shown in Figure 3.

5. Synergetic Control during LVRT

The PV system should stay connected and support the grid with reactive power during the voltage dip. Therefore, the reference value of I_q is $I_{q_ref} = $ const., depending on the voltage magnitude. Figure 4 depicts the principle of voltage support in the event of grid faults [17].

The abscissa in Figure 4 stands for $\Delta U/U_n$, and the ordinate is $\Delta I_B/I_n$. $\Delta U = U - U_0$ and $\Delta I_B = I_B - I_{B0}$, where U

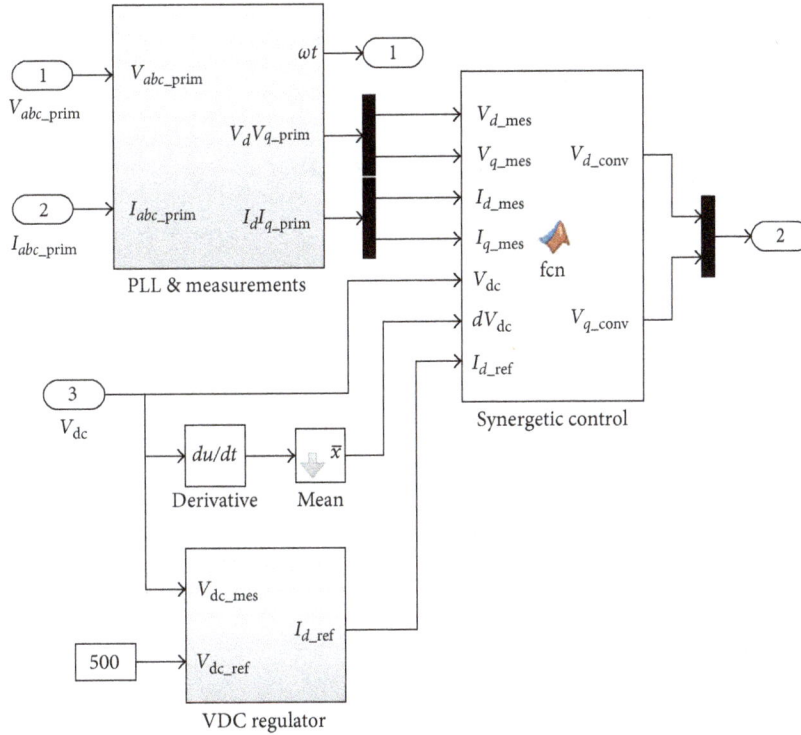

FIGURE 6: VSC main controller.

is the present voltage during the fault, U_n is the rated voltage, U_0 is voltage before the fault, I_B is the reactive current, I_n is the rated current, and I_{B0} is the reactive current before the fault. If a voltage dip is more than 10% of the rated voltage, the generator should provide a reactive current amounting to at least 2% of the rated current for each percent of the voltage dip within 20 ms after fault recognition [17]. If a voltage dip is more than 50% of the rated voltage, the generator must inject the grid with a reactive power of 100% of the rated current.

The limitation of I_d is $I_{d\_limit} = \sqrt{1 - I_{q\_ref}^2}$ pu, so the current will not be greater than the rated current.

If the maximum power of the PV array at MPP is less than the power $E_d I_{d\_limit}$, the maximum power can be injected into the grid with $I_d < I_{d\_limit}$. A MPPT is used during LVRT to obtain the maximum power and economic benefits, and the control strategies are the same as those in normal operation.

If the maximum power of the PV array at MPP is more than the power $E_d I_{d\_limit}$, $I_{d\_ref} = I_{d\_limit}$ is set, and the power output of the PV array equals $E_d I_{d\_limit}$ for the power balance. The following control strategies are used:

The first macrovariable is selected as

$$\psi_1 = I_{d\_ref} - I_d. \tag{29}$$

Substituting (29) into (4) and considering (12), the first control variable is computed as follows:

$$U_{id} = \frac{L_3}{T_1}(I_{d\_ref} - I_d) - U_{3d}. \tag{30}$$

Equation (28) is also used for the second control variable during LVRT.

A MPPT is not used, and the duty cycle D is determined by a PI controller to regulate the DC link voltage.

$$\Delta D = \left(K_{PD} + \frac{K_{iD}}{s} \right)(u_{dc\_ref} - u_{dc}), \tag{31}$$

$$D = D_0 + \Delta D, \tag{32}$$

where D_0 is the initial value of D.

The duty cycle of the DC boost converter is determined by (31) and (32) during LVRT. The DC-AC VSC converter is controlled on the basis of (28) and (30).

The control strategies described above are shown in Figure 5.

Equations (26) and (30) are derived according to the synergetic control algorithm. I_{d_ref} is variable and is computed by (22) in normal operation. Consequently, (26) is obtained for the control of the PV system in normal operation. I_{d_ref} is a constant, and (30) is derived for the second control strategy during LVRT. Equations (26) and (30) are used under different operating conditions of the grid-connected PV system.

The DC link voltage u_{dc} should remain stable to maintain good operation of the PV system. This requires a power balance in the PV system. If the power injected into the grid by the DC-AC converter is less than the output power of the PV array, u_{dc} will increase. If the output power of the converter is the same as the output power of the PV array, u_{dc} will not change. The two control strategies can satisfy the power balance in the PV system in normal operation and during a LVRT.

The PV array works at the MPP because of the MPPT in normal operation. If the output power of the converter is less than the output power of the PV array and u_{dc} increases, the DC voltage regulator will generate greater I_{d_ref} on the basis of (22). The synergetic control will result in more output power of the converter and achieve a power balance in the PV system after the regulation process.

The synergetic control maintains the output power $E_d I_{d\_limit}$ of the converter for the second control strategy during LVRT. If the output power of the PV array is greater than $E_d I_{d\_limit}$ and u_{dc} increases, the DC voltage regulator will give a smaller D according to (31) and (32). The voltage of the PV array will increase according to (17), and its output power will decrease due to the power-voltage characteristics of the PV array. The power balance in the PV system will be achieved once more, and u_{dc} will return to its reference value.

The DC-AC converter may be damaged due to the large current that passes through it. The two control strategies can prevent the converter from overcurrent in normal operation and during LVRT.

6. Case Studies

Matlab software provides an example titled *Detailed Model of a 100-kW Grid-Connected PV Array* [9]. The controllers of the example are adapted to assess the performance of synergetic control of the grid-connected PV system.

The example includes a PV array with an open-circuit voltage of 321 V. u_{dc_ref} is 500 V, and the rated AC voltage is 260 V. A distribution transformer has a voltage ratio of 25 kV/260 V.

6.1. Normal Operation Simulation. MPPT used in the case study is based on incremental conductance with an integral controller that can ensure that the system operates in MPP when the radiation intensity and temperature change rapidly.

The DC-AC converter is controlled using a synergetic control scheme. Equations (26) and (28) are rewritten according to the symbols in the Matlab example [9]:

$$V_{d\_conv} = K_P L_3 \frac{dV_{dc}}{dt} + K_i L_3 (V_{dc} - V_{dc\_ref})$$
$$+ \frac{L_3}{T_1}(I_{d\_ref} - I_d) + R_3 I_{d\_ref} - \omega L_3 I_{q\_ref} + V_{d\_mes},$$
(33)

$$V_{q\_conv} = \frac{L_3}{T_2}(I_{q\_ref} - I_q) + R_3 I_{q\_ref} + \omega L_3 I_{d\_ref} + V_{q\_mes}.$$
(34)

Figure 6 shows the VSC main controller containing synergetic control.

The VDC regulator in Figure 6 is constructed on the basis of (22). The inputs of the PLL and measurements block are primary voltages V_{abc_prim} and currents I_{abc_prim} of the distribution transformer. This block tracks the frequency

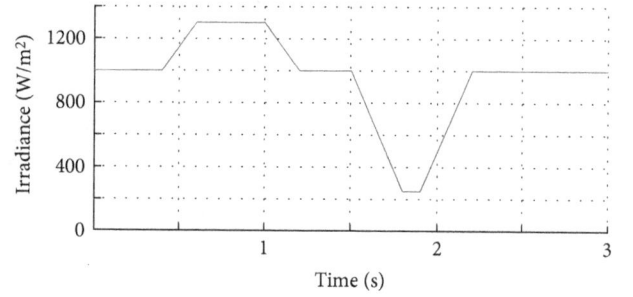

FIGURE 7: Sun irradiance.

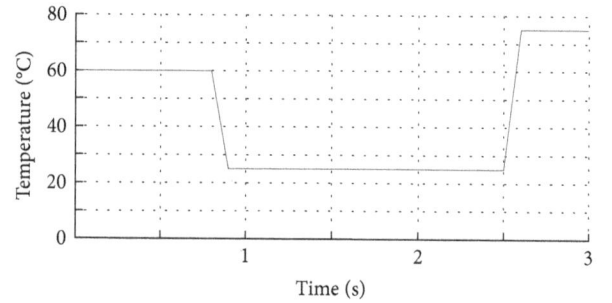

FIGURE 8: Temperature.

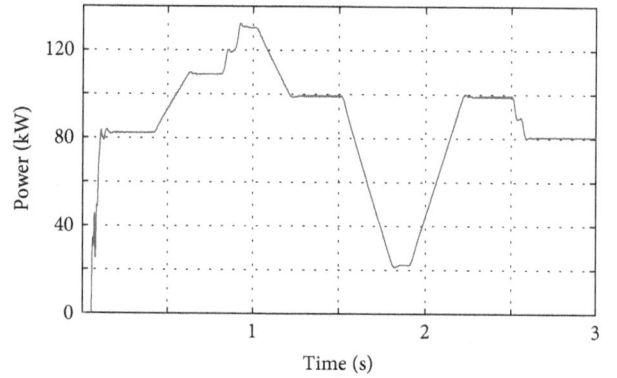

FIGURE 9: PV array output power.

and phase of a sinusoidal three-phase signal and performs Park transformation from a three-phase (*abc*) reference frame to a *dq*0 reference frame. The synergetic control block in Figure 6 is a Matlab function block. It includes Matlab code. The code computes 2 control variables, V_{d_conv} and V_{q_conv}, with (33) and (34). The control variables are transformed into the reference values of three-phase voltages to generate PWM and control the VSC converters.

The parameters $K_P = 0.2$, $K_i = 150$, $T_1 = 0.01$, and $T_2 = 0.01$ are selected and applied in (22), (33), and (34). The settings are $V_{d\_ref} = 500$ V and $I_{q\_ref} = 0$ for the simulation in normal operation. The standard test conditions are 1000 W/m² irradiance and 25°C temperature. The change of sun irradiance is shown in Figure 7. Figure 8 describes the change of temperature for the simulation.

The MPPT regulator changes the duty cycle to regulate the PV voltage for tracking maximum power. At $t = 0.1$ sec,

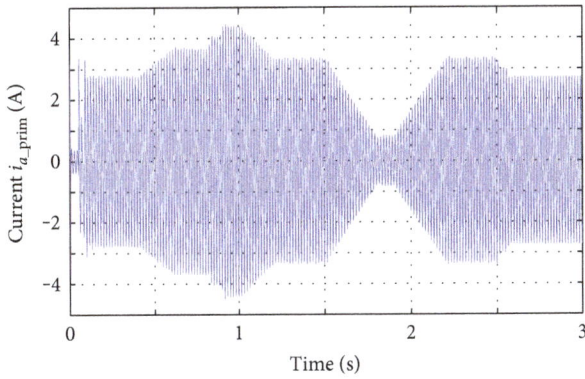

FIGURE 10: Primary current i_{a_prim}.

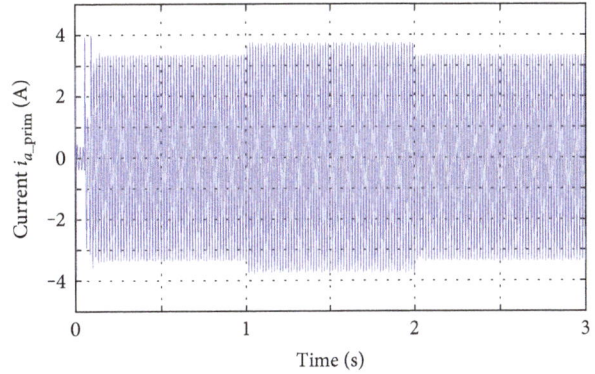

FIGURE 11: Bus voltage decrease.

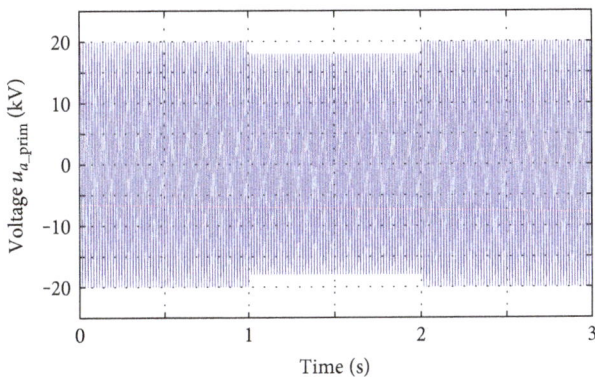

FIGURE 12: Current increase.

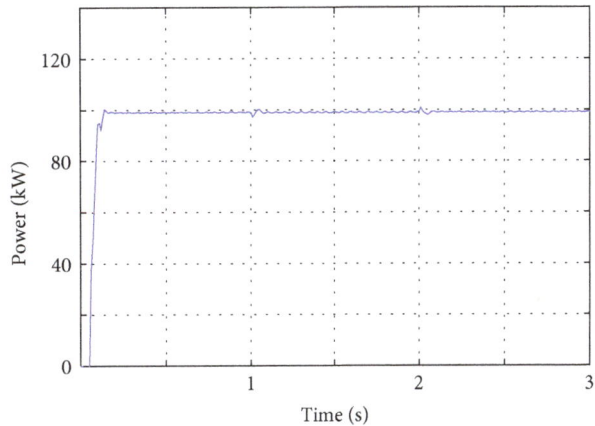

FIGURE 13: Power invariance.

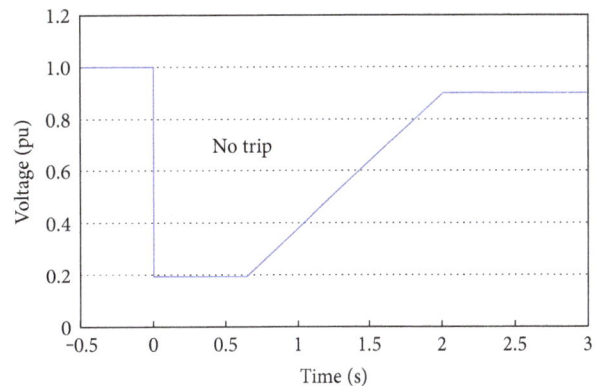

FIGURE 14: LVRT requirement.

the MPPT is enabled, and Figure 9 shows the PV array output power. Maximum power is 100.4 kW at the standard test conditions. VSC converters deliver the power to the grid by synergetic control.

The root mean square (RMS) of the primary voltage v_{a_prim} of the distribution transformer is constant. The RMS of the current i_{a_prim} is directly proportional to the power. Figure 10 shows the primary current i_{a_prim} of the distribution transformer.

Figures 9 and 10 indicate that the grid-connected PV system can track maximum power and deliver the power to the grid when the radiation and temperature change rapidly. The system operates stably.

The bus voltage may change in normal operation. Simulation of voltage fluctuation is performed at the standard test conditions. Figure 11 depicts a 25 kV bus voltage v_{a_prim}. The RMS of v_{a_prim} decreases by 10% at $t = 1$ sec.

The current i_{a_prim} is shown in Figure 12. i_{a_prim} increases when the bus voltage decreases.

The maximum power does not change because sun irradiance and temperature are constant at the standard test conditions. Figure 13 demonstrates that the power injected into the grid is almost constant during the grid voltage fluctuation. This means that the synergetic control performs well in normal operation.

6.2. LVRT Simulation. LVRT is the capability of electric generators to stay connected to the grid during short periods of voltage dip. LVRT is an important feature of the generator control system. There are several standards for LVRT requirements [17, 18]. Figure 14 shows the LVRT requirement in [19]. These requirements may be used for large solar power installations.

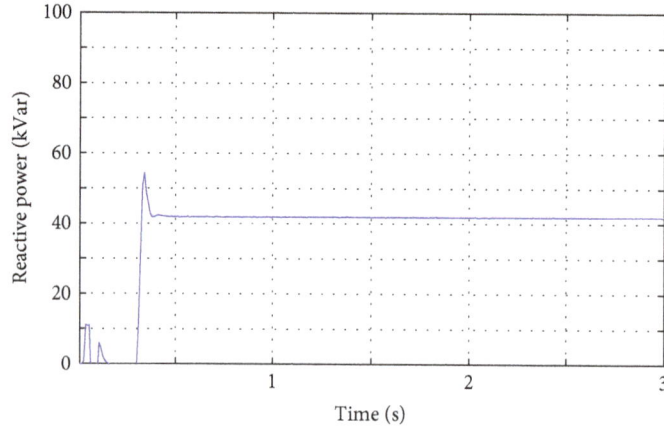

FIGURE 15: Reactive power injected into the grid at a 30% decrease of the bus voltage.

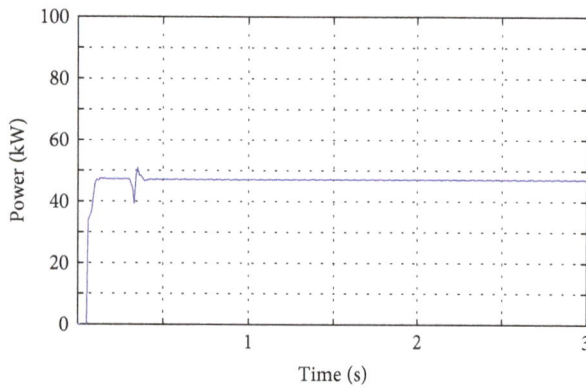

FIGURE 16: Active power injected into the grid at a 30% decrease of grid voltage.

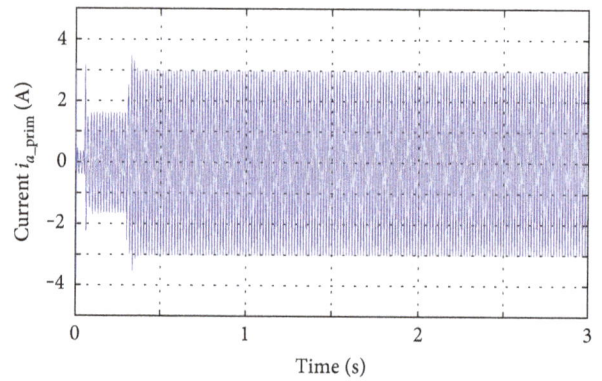

FIGURE 17: Primary current at a 30% decrease of grid voltage.

Three cases are used to test the performances of the synergetic control during a LVRT. The three-phase fault block in Simulink is connected in utility grid to simulate a three-phase short circuit with arc resistances.

The feature of the PV system is tested first when the bus voltage decreases by 30% due to a grid fault. The reference value of the reactive current I_q is $I_{q\_ref} = -0.6$ pu on the basis of Figure 4. The limitation of I_d is $I_{d\_limit} = 0.8$, and the power is $E_d I_{d\_limit} = 0.56$ pu. The maximum power of the PV array at MPP is 0.48 pu when it is simulated at an irradiance of $500\,\text{W/m}^2$ and a temperature of 25°C. The first control strategy is used because the maximum power of the PV array at MPP is less than $E_d I_{d\_limit}$. $I_{q\_ref} = -0.6$ is set; the other settings are equivalent to those used in the normal operation simulation. Equations (22), (33), and (34) and MPPT are used. The RMS of the voltage v_{a_prim} decreases by 30% at $t = 0.3$ s, and the simulation results are shown in Figures 15–17.

Figure 15 depicts the reactive power injected into the grid. The PV system provides the grid with a reactive power of 42 kVar during the LVRT.

The active power injected into the grid is shown in Figure 16. The injected active power is 47.2 kW when the reactive power injected into the grid is 42 kVar at a 30% decrease of the grid voltage. The maximum power of the

PV array at MPP is delivered to the grid, and the solar energy is fully utilized.

The rated primary current is 3.3 A. Figure 17 shows that the primary current is less than its rated value. Therefore, the PV system may stay connected to the grid and provide the grid with 47.2 kW and 42 kVar at a 30% decrease of the grid voltage due to grid fault.

Then, simulations will test the performances of the synergetic control when the grid voltage decreases to 0.2 pu. The second control strategy is used because $I_{q\_ref} = -1$ according to Figure 4 and $I_{d\_limit} = 0$. The duty cycle D is computed according to (31) and (32). Equation (30) is rewritten using the symbols in [9]:

$$V_{d\_conv} = \frac{L_3}{T_1}(I_{d\_ref} - I_d) + R_3 I_{d\_ref} - \omega L_3 I_{q\_ref} + V_{d\_mes}.$$

(35)

Equations (31), (32), (34), and (35) are used to compute the control variables for controlling the DC boost converter and the VSC converters during LVRT. The parameters are $K_{PD} = 0.01, K_{iD} = 0.1, T_1 = 0.01$, and $T_2 = 0.01$. The settings are $V_{d\_ref} = 500\,\text{V}, I_{q\_ref} = -1$, and $I_{d\_ref} = 0$ pu.

A three-phase short circuit occurs in the power distribution system at $t = 0.3$ sec for the LVRT simulation. The bus voltage decreases to 0.2 pu during short circuit. v_{a_prim} and

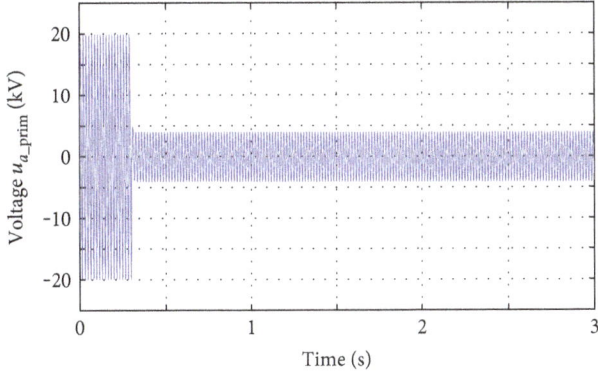

FIGURE 18: 25 kV bus voltage during LVRT.

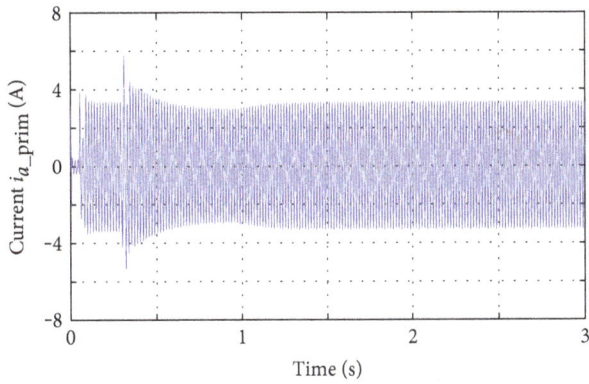

FIGURE 19: Primary current during LVRT.

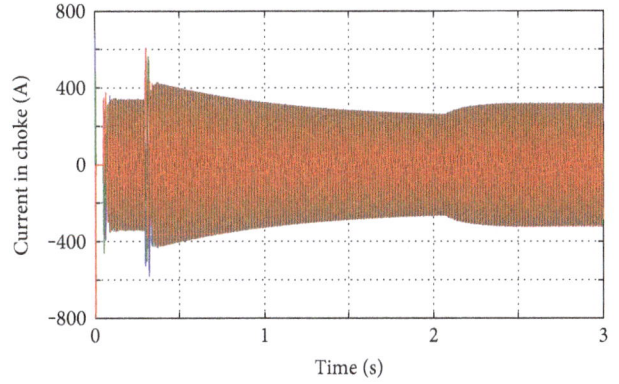

FIGURE 20: Currents I_d and I_q during LVRT.

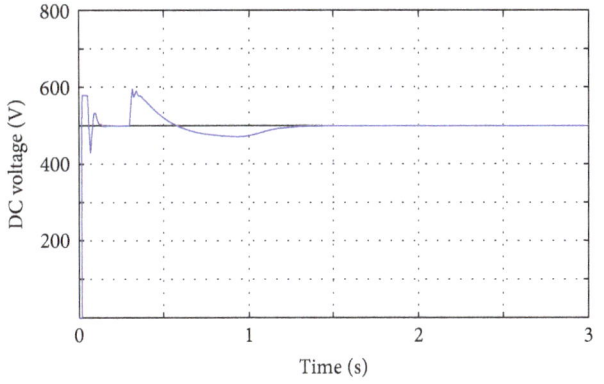

FIGURE 21: DC link voltage V_{dc}.

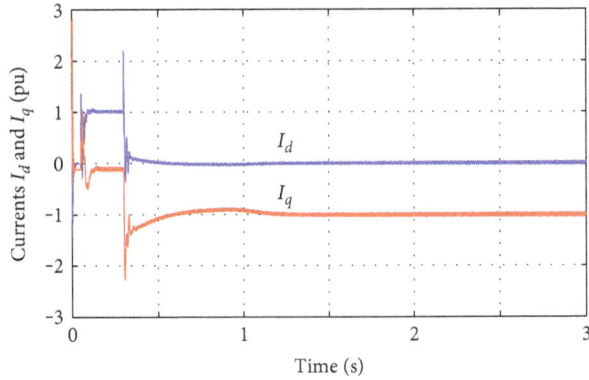

FIGURE 22: Three-phase currents flowing in the choke.

i_{a_prim} are shown in Figures 18 and 19, respectively. The 25 kV bus voltage in Figure 18 decreases to 0.2 pu after $t = 0.3$ sec.

Small fluctuations of primary current arise after the voltage dip. The peak value of i_{a_prim} during LVRT is less than double of that in normal operation.

Currents I_d and I_q are depicted in Figure 20. I_d and I_q are regulated to 0 and −1 pu, respectively, after the dip. The PV system stays connected and supports the grid with reactive power.

The DC link voltage V_{dc} is shown in Figure 21. It fluctuates slightly and then remains at 500 V after the dip.

Another simulation is made for testing the performance of the synergetic control at a bus voltage of 0.05 pu. The second control strategy is used. A three-phase short circuit is applied in the bus at $t = 0.3$ sec, and the bus voltage decreases to 0.05 pu to simulate the voltage of the fault arc. The current settings are $I_{d\_ref} = 0$ and $I_{q\_ref} = -0.9$ pu, to avoid a large current. The reference value of current I_q is $I_{q\_ref} = -1$ pu, to provide the grid with reactive power after $t = 0.35$ sec. Figure 22 shows three-phase currents flowing in the choke. The maximum value of the current is less than 2.0 pu. Synergetic control can limit the current peak during LVRT.

The example in [9] has the current regulator with PI control and simulates only the normal operation of the PV system. The regulator from (31) and (32) is added to the example to simulate the LVRT of the PV system. $I_{d\_ref} = 0$ and $I_{q\_ref} = -1$ pu are fixed. Two parameters of the current regulator with PI control are changed from 0.3 and 20 (for normal operation) to 0.03 and 0.002, respectively, for LVRT. Therefore, parameter tuning of the current regulator with PI control is difficult. Chatter has been the main obstacle for sliding-mode control systems [10]. Probabilistic wavelet fuzzy neural networks (PWFNNs) include a membership layer, probabilistic layer, wavelet layer, and rule layer [11]. PWFNNs are very complicated, such that it is difficult to determine the PWFNN parameters. The gains of the PI controller in [13] must be changed online using the CMPN algorithm for improving the LVRT capability of grid-connected PV systems.

The parameters of the synergetic controller described in this paper are invariant during normal operations and during LVRT, and parameter T_1 equals T_2. There is no need to change the parameters of the synergetic controller with a complicated algorithm. Therefore, parameter tuning of the synergetic controller is easy, and the controller is sufficiently robust for use in the PV system. The design of the synergetic controller is simple and does not require linearization of the PV system. In addition, the proposed control schemes are easy to realize. The synergetic control is chatter-free and displays good static and dynamic performance.

Asymmetric short circuit faults may occur in power systems. Unbalanced grid voltages are comprised of positive, negative, and zero sequence voltage components. The zero sequence is not considered here because of three-wire systems. Unbalanced grid voltage sags will cause performance deterioration of the converter. Negative sequences result in DC-link voltage ripples and harmonic power. The positive and negative sequence currents should be controlled simultaneously to improve control performance. Reference [20] used separate current controllers for positive and negative sequences. Synergetic control may be used in two current controllers to control positive and negative sequence currents separately. It is possible to use synergetic control for improving the performance of PV systems during unbalanced voltage sags.

7. Conclusions

Solar PV is an important renewable energy technology and does not generate pollution. PV systems are developing rapidly, and most PV systems are grid-connected. Research on the control of grid-connected PV systems contributes to the improvement of the operation of the distribution network and the PV system.

Synergetic control can be used for the control of a grid-connected PV system. The design of a synergetic controller does not require the linearization of the PV system. The mathematical expressions for computing control variables can be derived according to the synergetic control algorithm and the mathematical model of a grid-connected PV system. Two control strategies are used in normal operation and during LVRT. The parameters T_1 and T_2 of the two synergetic controllers are the same. The parameters of the synergetic controllers are easy to determine, and results indicate that the synergetic controllers are robust. The grid-connected PV system can obtain a maximum power point and inject the power into the grid when the radiation and temperature change rapidly. The DC link voltage and AC currents are limited, and the DC-AC VSC converters are not damaged during LVRT. The PV system can run with $I_q = -1$ pu continuously and supply reactive power to the grid when the grid voltage decreases to 0.05 pu.

Synergetic control has good dynamic characteristics in normal operation and during LVRT and is the alternative solution for grid-connected PV systems.

Conflicts of Interest

The authors declare that there is no conflict of interests regarding the publication of this paper.

Acknowledgments

This work was financially supported by the National Natural Science Foundation of China (51177074, 51407095) and the Jiangsu Province Natural Science Foundation (BK20151548).

References

[1] A. Rosa, *Fundamentals of Renewable Energy Processes*, Academic Press, Boston, MA, USA, 2012.

[2] REN21, *Renewables 2016 Global Status Report*, REN21 Secretariat, Paris, France, 2016, http://www.ren21.net/wp-content/uploads/2016/06/GSR_2016_Full_Report_REN21.pdf.

[3] N. Femia, G. Petrone, G. Spagnuolo, and M. Vitelli, "Optimization of perturb and observe maximum power point tracking method," *IEEE Transactions on Power Electronics*, vol. 20, no. 4, pp. 963–973, 2005.

[4] K. H. Hussein, I. Muta, T. Hoshino, and M. Osakada, "Maximum photovoltaic power tracking: an algorithm for rapidly changing atmospheric conditions," *IEE Proceedings-Generation, Transmission and Distribution*, vol. 142, no. 1, pp. 59–64, 1995.

[5] P. Q. Dzung, L. D. Khoa, H. H. Lee, L. M. Phuong, and N. T. D. Vu, "The new MPPT algorithm using ANN based PV," in *Proceedings of the International Forum on Strategic Technology*, pp. 402–407, Ulsan, South Korea, 13–15 October 2010.

[6] N. Zakzouk, M. Elsaharty, A. Abdelsalam, A. Helal, and B. Williams, "Improved performance low-cost incremental conductance PV MPPT technique," *IET Renewable Power Generation*, vol. 10, no. 4, pp. 561–574, 2016.

[7] K. Sundareswaran, V. Vigneshkumar, P. Sankar, S. Simon, P. Nayak, and S. Palani, "Development of an improved P&O algorithm assisted through a colony of foraging ants for MPPT in PV system," *IEEE Transactions on Industrial Informatics*, vol. 12, no. 1, pp. 187–200, 2016.

[8] S. Mohanty, B. Subudhi, and P. Ray, "A new MPPT design using grey wolf optimization technique for photovoltaic system under partial shading conditions," *IEEE Transactions on Sustainable Energy*, vol. 7, no. 1, pp. 181–188, 2016.

[9] MathWorks, *Detailed Model of a 100-kW Grid-Connected PV Array*, MathWorks, Natick, MA 01760-2098, USA, 2015, http://www.mathworks.com/examples/simpower/mw/sps_product-power_PVarray_grid_det-detailed-model-of-a-100-kw-grid-connected-pv-array.

[10] N. Kumar, T. Saha, and J. Dey, "Sliding-mode control of PWM dual inverter-based grid-connected PV system: modeling and performance analysis," *IEEE Journal of Emerging Selected Topics in Power Electronics*, vol. 4, no. 2, pp. 435–444, 2016.

[11] F. Lin, K. Lu, and T. Ke, "Probabilistic wavelet fuzzy neural network based reactive power control for grid-connected three-phase PV system during grid faults," *Renewable Energy*, vol. 92, pp. 437–449, 2016.

[12] J. Sosa, M. Castilla, J. Miret, J. Matas, and Y. Al-Turki, "Control strategy to maximize the power capability of PV three-phase inverters during voltage sags," *IEEE Transactions on Power Electronics*, vol. 31, no. 4, pp. 3314–3323, 2016.

[13] H. M. Hasanien, "An adaptive control strategy for low voltage ride through capability enhancement of grid-connected photovoltaic power plants," *IEEE Transactions on Power Apparatus and Systems*, vol. 31, no. 4, pp. 3230–3237, 2016.

[14] E. Santi, A. Monti, D. Li, K. Proddutur, and R. Dougal, "Synergetic control for dc-dc boost converter: implementation options," *IEEE Transactions on Industry Applications*, vol. 39, no. 6, pp. 1803–1813, 2003.

[15] Z. Jiang and R. Dougal, "Synergetic control of power converters for pulse current charging of advanced batteries from a fuel cell power source," *IEEE Transactions on Power Electronics*, vol. 19, no. 4, pp. 1140–1150, 2004.

[16] MathWorks, *abc to dq0, dq0 to abc*, MathWorks, Natick, MA 01760-2098, USA, 2015, http://www.mathworks.com/help/physmod/sps/powersys/ref/abctodq0dq0toabc.html.

[17] *Grid Code High and Extra High Voltage*, E.ON Netz GmbH, Bayreuth, Germany, 2006.

[18] BDEW, *Technical Guideline: Generating Plants Connected to the Medium-Voltage Network*, BDEW, Berlin, Germany, 2008, http://www.bdew.de.

[19] DEIF, *LVRT Capability-Test Results*, http://www.deifwindpower.com/wind-turbine-solutions/control-systems/lvrt-test-results.

[20] H. Chong, R. Li, and B. Jim, "Unbalanced-grid-fault ride-through control for a wind turbine inverter," *IEEE Transactions on Industry Applications*, vol. 44, no. 3, pp. 845–856, 2008.

9

Influence of Front and Back Contacts on Photovoltaic Performances of p-n Homojunction Si Solar Cell: Considering an Electron-Blocking Layer

Md. Feroz Ali[1] and Md. Faruk Hossain[2]

[1]*Department of Electrical and Electronic Engineering (EEE), Pabna University of Science & Technology (PUST), Pabna 6600, Bangladesh*
[2]*Electrical & Electronic Engineering, Rajshahi University of Engineering & Technology (RUET), Rajshahi 6204, Bangladesh*

Correspondence should be addressed to Md. Feroz Ali; feroz071021@gmail.com

Academic Editor: Yatendra S. Chaudhary

In this simultion work, the effect of front and back contacts of p-n homojunction Si solar cell with an electron-blocking layer (EBL) has been studied with the help of a strong solar cell simulator named AMPS-1D (analysis of microelectronic and photonic structures one dimensional). Without the effect of these contact parameters, low solar cell efficiency has been observed. Fluorine-doped tin oxide (FTO) with high work function (5.45 eV) has been used as the front contact to the proposed solar cell. Zinc (Zn) metal which has a work function of 4.3 eV has been used as the back contact of the proposed model. With FTO as the front contact and Zn as the back contact, the optimum efficiency of 29.275% (Voc = 1.363 V, Jsc = 23.747 mA/cm^2, FF = 0.905) has been observed. This type of simple Si-based p-n homojunction solar cell with EBL of high efficiency has been proposed in this paper.

1. Introduction

Because of the increasing trend of price of fossil fuels and some of their drastic and dangerous effects on greenhouse, the world is now looking for green energy like solar cells [1]. For its green power, low cost, and availability, renewable energy plays an important role in the world energy, especially solar photovoltaic cell which has a great contribution to the world's electrical energy. Solar energy [2] is another increasingly hot topic in recent years due to the inevitable exhaustion of fossil and mineral energy sources in the next fifty years [3]. Although solar cell has a great disadvantage of higher initial cost, after payback, it is still the best option for clean energy. For example, the calculated energy payback for the current PV systems is 3-4 years, which depends on the type of PV panel (thin film technology or multicrystalline silicon), but this time may be expected to be reduced to 1-2 years as manufacturing techniques improve [4]. People all over the world have investigated different types of silicon solar cells for many years [5]. One of the main reasons that silicon is the choice for semiconductor material in the field of microelectronics is that it forms a unique oxide on the surface when heated to high temperatures which facilitates device fabrication for two reasons: (i) it neutralizes defects on the silicon surface and (ii) it allows for straightforward planar processing [6]. The performance of solar photovoltaic cells depends on their design, material properties, and fabrication technology; that is why researchers present improved cells over periods of time, although the overall process is not only quite complex but also expensive and time-consuming [7]. One of the best methods for simulating solar cell is the numerical approach which helps the researchers to find out a design optimization. There are many major objectives of numerical modeling and simulation in solar cell research such as testing the validity of the proposed physical structures and geometry on solar cell performance and fitting the modeling output to experimental results, which have become indispensable tools for designing a high-efficiency solar cell [7]. The front and back contacts have great influence on efficiency as well as performance of silicon solar cell. Ni back contact gives high performance compared with other metals [8]. The efficiency of solar cell also depends on the work function [9] of the front and back contacts. The work function, Φ, is the energy required to remove an electron from the highest filled level in the stationary Fermi distribution of a solid, a point in a field-free zone just outside the solid, at absolute zero. The relatively high evaporation rates can be achieved with low work function materials and must be operated at lower temperatures [10]. High-performance electrodes must exhibit a low work function, φ, uniformly over the electrode surface. The electron source community has identified several promising materials that can uniformly reduce φ through the use of surface adsorbents and/or bandgap modification [10]. The theoretical efficiency limits of solar energy conversion are strongly dependent not only on the range but also on a number of different bandgaps or effective bandgaps that can be incorporated into a solar cell [11].

In this work, p-n homojunction Si solar cell with an electron-blocking layer (EBL) [1] has been studied with the variation of front and back contact parameters such as the work function (eV) and reflection coefficient. For the simulation of this device, a strong solar cell simulator named AMPS-1D (analysis of microelectronic and photonic structures one dimensional) has been used. Other parameters/data of the solar cell device have been adopted from various practical references [12]. Although some have carried out this topic before [1], this study is the first and unique one to consider the work function and reflection coefficient of the front and back contacts to improve the efficiency of Si homojunction solar cell. Authors have been tried to simulate this solar cell, and the efficiency of 29.275% has been observed at the thickness of 6000 nm for each p-layer Si and n-layer Si and 50 nm of EBL with 2.10 eV bandgap along with the work functions of the front and back contacts which are 1.5 eV and 0.5 eV, respectively.

2. Simulation Model

In AMPS software, the physics of device transport can be captured in three governing equations: Poisson's equation, the continuity equation for free holes, and the continuity equation for free electrons [13]. Determining transport characteristics then becomes a task of solving these three coupled nonlinear differential equations subject to appropriate boundary conditions. These three equations and the corresponding boundary conditions, along with the numerical solution technique used to solve them. AMPS software assumes that the material system under examination is in steady state. That is, it is assumed that there is no time dependence. It follows that the terminal characteristics generated by AMPS are the quasi-static characteristics. Table 1 shows the boundary conditions of the front and back contacts of AMPS. PHIBO is the difference

TABLE 1: Boundary conditions of the front and back contacts of AMPS.

Contact parameters	Description
PHIBO = Φ_{bo} (front contact)	$E_c - E_f$ in at $x = 0$ (eV)
PHIBL = Φ_{bL} (back contact)	$E_c - E_f$ in at $x = L$ (eV)

between the work function of the front contact and electron affinity of the semiconductor in contact. Similarly, PHIBL is the difference between the work function of the back contact and electron affinity of the semiconductor in contact.

PHIBO is the difference between the work function of the front contact and the electron affinity of the associated semiconductor. Similarly, the PHIBL is the difference between the work function of the back contact and the electron affinity of the associated semiconductor. Table 2 shows the surface recombination speed of AMPS. Here, for the simulation, the SNO, SPO, SNL, and SPL were all selected as 1e7 cm/sec.

Table 3 shows the reflection coefficient for light impinging on the front and back surfaces. In this case, RF 0.1 and RB 0.9 have been selected for the simulation. It means that the front contact of the device can reflect only 10% of the incident light and the back contact of the device can reflect 90% of the incident light.

3. The Proposed Model with Front and Back Contact Parameters

Figure 1 shows the p-n homojunction solar cell with EBL and the front and back contact parameters. The spectral response (SR) has been observed with this model.

In the proposed device, p-n solar cell with p-type Si of 6000 nm in thickness and also with n-type Si solar cell of 6000 nm in thickness has been incorporated along with the proper front and back contacts in order to enhance efficiency. As some researches have been carried out on this homojunction Si solar cell, this is the first research with AMPS to consider the proper front and back contacts. A 50 nm thick electron-blocking layer (EBL) has been embodied on the top of the p-n diode where light is immersed. It has been observed that the higher the thickness of p-n Si layer, the higher the efficiency. But with p-n Si layer of 6000 nm thickness, the optimum efficiency is 29.275%, comparatively higher than conventional Si solar cell. For this solar cell simulation, a PHIBO (work function of the front contact) of 1.6 eV and reflection coefficient of 0.1 and a PHIBL (work function of the back contact) of 0.5 eV and reflection coefficient of 0.9 have been selected. The simulation result shows that the efficiency and performance of the solar cell device are optimum; hence, these are the best solar cell contacts with the corresponding parameters for this device.

4. Simulation Parameters

The parameters included in Table 4 and Table 5 are used to simulate the solar cell in AMPS-1D. The temperature of 300 K is used as the default, and to get the all results,

TABLE 2: Surface recombination speed of AMPS.

Contact parameters	Description
SNO (recombination speed of the electron of the front contact)	Electrons at $x = 0$ interface (cm/sec)
SPO (recombination speed of the hole of the front contact)	Hole at $x = 0$ interface (cm/sec)
SNL (recombination speed of the electron of the back contact)	Electrons at $x = L$ interface (cm/sec)
SPL (recombination speed of the hole of the front contact)	Hole at $x = L$ interface (cm/sec)

TABLE 3: Reflection coefficient for light impinging on the front and back surfaces.

Contact parameters	Description
RF	Reflection coefficient at $x = 0$ (front surface)
RB	Reflection coefficient at $x = L$ (back surface)

AM 1.5 illuminations are used. AM means air mass coefficient. The air mass coefficient is defined as the direct optical path length through the Earth's atmosphere, expressed as the ratio relative to the path length vertically upwards, that is, at the zenith [14].

5. Simulation Result and Discussion

Figure 2 shows the PHIBO versus efficiency plot of the proposed solar cell's front contact. We have tried to simulate the front contact PHIBO from 1.2 eV to 2.2 eV. It has been observed that, at 1.6 eV PHIBO, the efficiency is 29.275%. Before the value of 1.6 eV, which has a drastic effect on efficiency, and beyond the value of 1.6 eV, the efficiency has a negligible increase. That is why the PHIBO is 1.6 eV which is the optimum value of the proposed solar cell. Figure 3 shows the PHIBO versus Voc plot of the proposed solar cell's front contact. It has been observed that, at PHIBO 1.6 eV, the open circuit voltage is 1.363 V and it is the optimum value as Figure 3 describes. Figure 4 shows the PHIBO versus Jsc plot of the proposed solar cell's front contact. The plot describes that the short circuit current of the solar cell increases with the increase in PHIBO but the changes are quite small. Figure 5 shows the PHIBO versus fill factor plot of the proposed solar cell's front contact. It has been observed that the fill factor is optimum (0.905) at PHIBO 1.6 eV.

Figure 6 shows the PHIBL versus efficiency plot of the proposed solar cell's back contact. We have tried to vary the PHIBL from 0.1 eV to 0.8 eV. It has been observed that the optimum efficiency (29.275%) has been achieved at PHIBL 0.5 eV. After 0.5 eV, the efficiency has been drastically affected as shown in the plot. Figure 7 shows the PHIBL versus Jsc plot of the proposed solar cell's back contact. Here, the short circuit current has no effect (almost the same, i.e., 23.747 mA/cm^2) on PHIBL as shown in the plot. Figure 8 shows the PHIBL versus fill factor (FF) plot of the proposed solar cell's back contact. The optimum FF (0.905) has been observed at PHIBL 0.5 eV. FF has been decreasing sharply after PHIBL 0.5 eV which is depicted in Figure 8. Figure 9

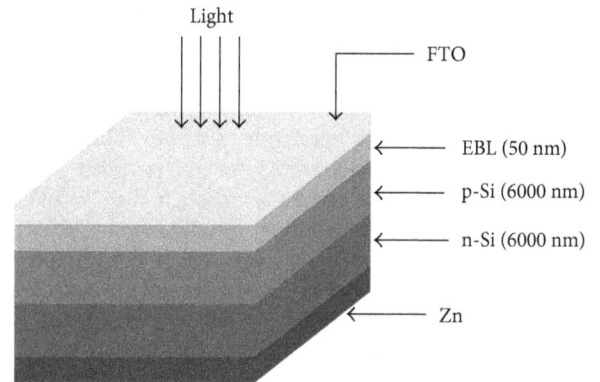

FIGURE 1: The proposed schematic diagram of p-n homojunction solar cell with EBL.

shows the PHIBL versus Voc plot of the proposed solar cell's back contact. The optimum open circuit voltage (1.363 V) has been observed at PHIBL 0.5 eV as shown in the graph.

Figure 10 shows the J-V and P-V plots of the proposed solar cell. Here, the maximum power of 29.28434 mW/cm^2 has been observed.

Figure 11 shows the band diagram of the proposed solar cell. This figure shows the difference between the conduction and valence band as well as the difference between the Fermi labels of the proposed solar cell's front and back contacts.

6. Conclusion

To conclude, for the front contact of the solar cell, a PHIBO of 1.6 eV, SNO and SPO of 1.0×10^7 cm/sec, and reflection coefficient (RF) of 0.1 (10%) have been observed as the optimum values which correspond to FTO with high work function (5.45 eV). Also, for the back contact of the solar cell, a PHIBL of 0.6 eV, SNL and SPL of 1.0×10^7 cm/sec, and reflection coefficient (RB) of 0.9 (90%) have been observed as the optimum values which correspond to metal Zn (4.3 eV). With these front and back contact parameters, the solar cell has the following performance parameters: Voc 1.363 V, Jsc 23.747 mA/cm^2, FF 0.905, efficiency 29.275%, and maximum power 29.28434 mW/cm^2. By controlling the bandgap and band state parameters, the efficiency can also be improved. So, this kind of p-n homojunction Si solar cell with the following proposed front contact (FTO) and back contact (Zn) can be fabricated in the laboratory and can be compared with the simulation result in the future.

TABLE 4: Different layers of electronic properties used in the APMS simulation.

Electronic properties	EBL	p-Si	n-Si
Relative permittivity, ε_r	11.9	11.9	11.9
Electron mobility, μ_n (cm^2/v-s)	40.0	20.0	20.0
Hole mobility, μ_p (cm^2/v-s)	4.0	2.0	2.0
Acceptor & donor concentration (cm^{-3})	$N_A = 1.0 \times 10^{18}$	$N_A = 1.0 \times 10^{18}$	$N_D = 1.0 \times 10^{18}$
Bandgap (eV)	2.10	1.82	1.82
Effective density of states in the conduction band (cm^{-3})	2.5×10^{20}	2.5×10^{20}	2.5×10^{20}
Effective density of states in the valence band (cm^{-3})	2.5×10^{20}	2.5×10^{20}	2.5×10^{20}
Electron affinity (eV)	3.85	3.80	3.80

TABLE 5: Optimum values of the front and back contact parameters.

Front contact	Back contact
PHIBO = 1.60 eV	PHIBL = 0.50 eV
SNO = 1.0×10^7 cm/sec	SNL = 1.0×10^7 cm/sec
SPO = 1.0×10^7 cm/sec	SPL = 1.0×10^7 cm/sec
RF = 0.10	RB = 0.90

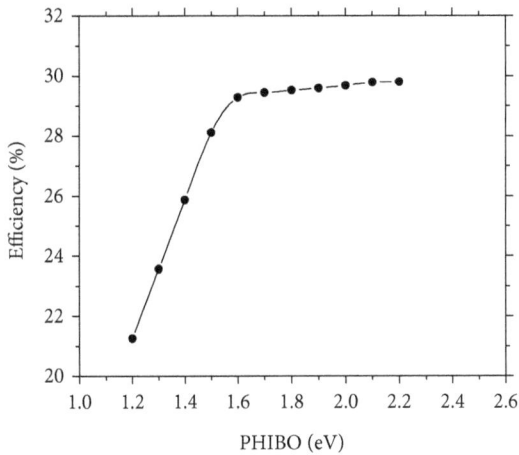

FIGURE 2: PHIBO versus efficiency plot of the proposed solar cell's front contact.

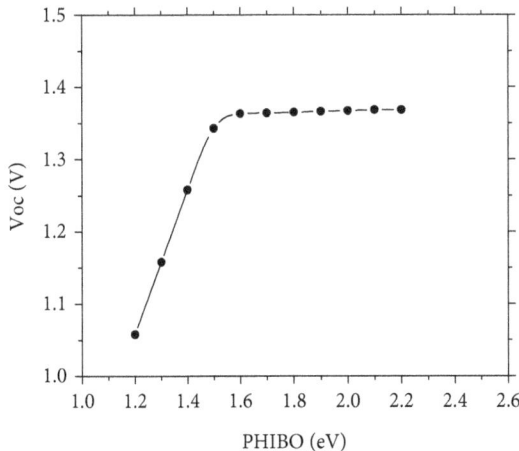

FIGURE 4: PHIBO versus Jsc plot of the proposed solar cell's front contact.

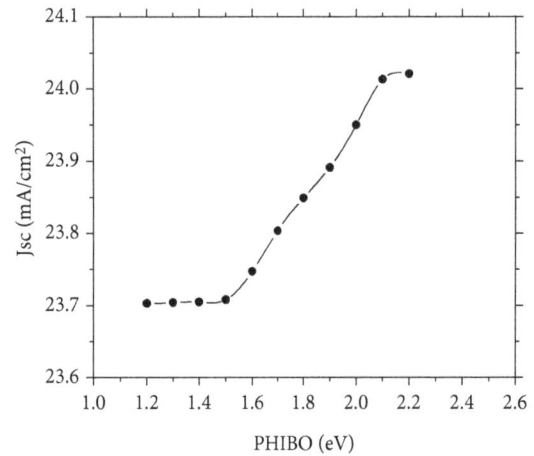

FIGURE 3: PHIBO versus Voc plot of the proposed solar cell's front contact.

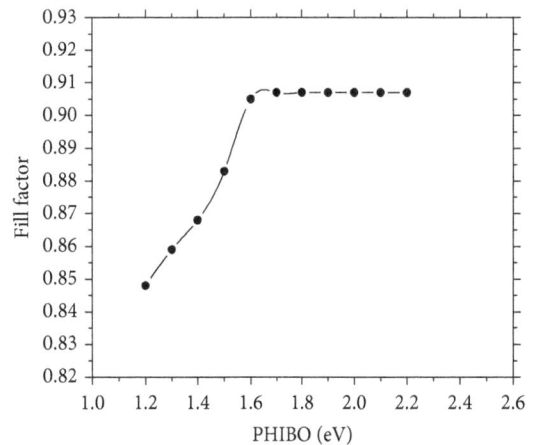

FIGURE 5: PHIBO versus fill factor plot of the proposed solar cell's front contact.

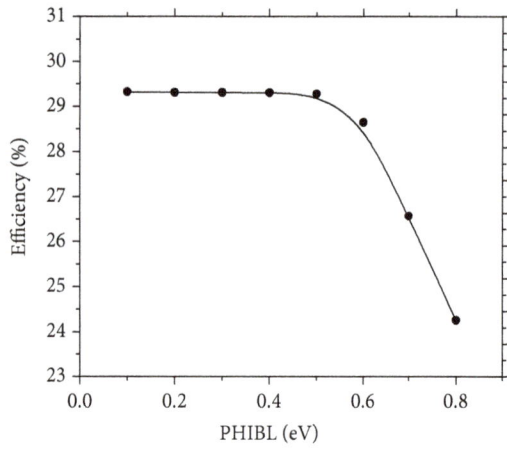

FIGURE 6: PHIBL versus efficiency plot of the proposed solar cell's back contact.

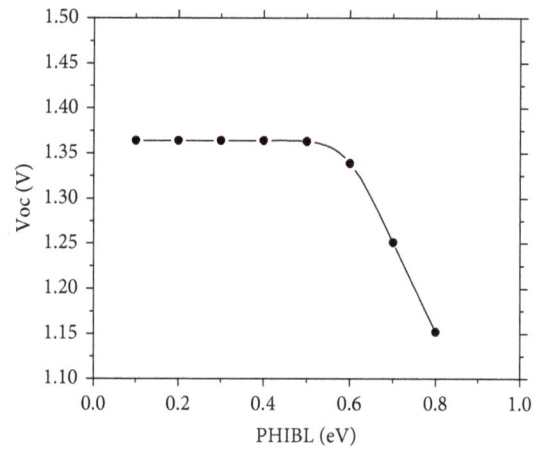

FIGURE 7: PHIBL versus Jsc plot of the proposed solar cell's back contact.

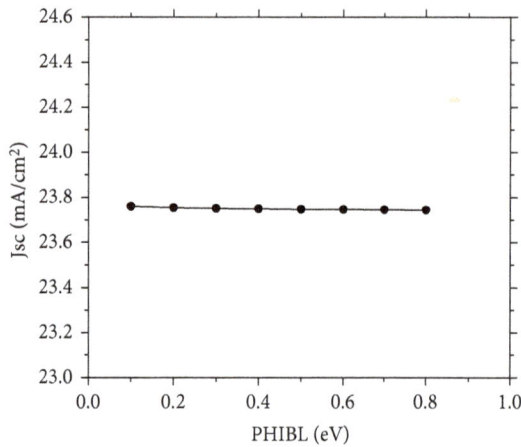

FIGURE 8: PHIBL versus fill factor plot of the proposed solar cell's back contact.

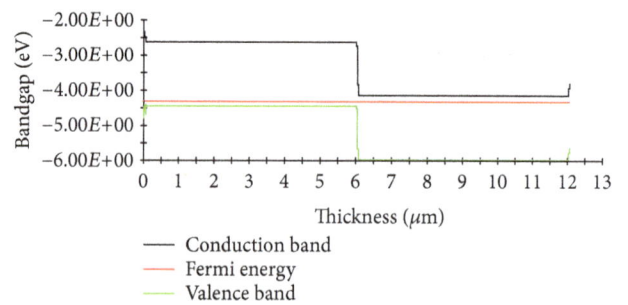

FIGURE 9: PHIBL versus Voc plot of the proposed solar cell's back contact.

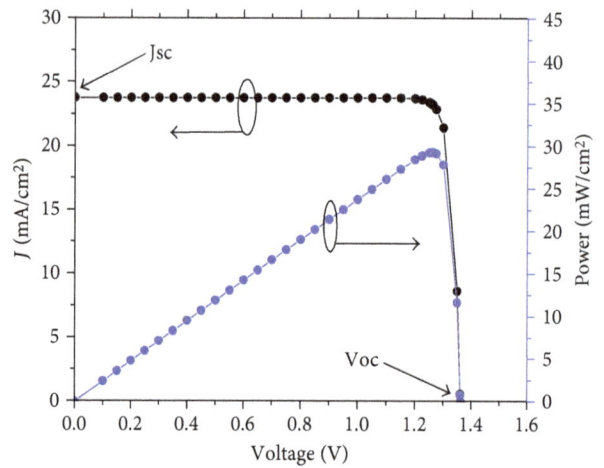

FIGURE 10: J-V and P-V plots of the proposed solar cell.

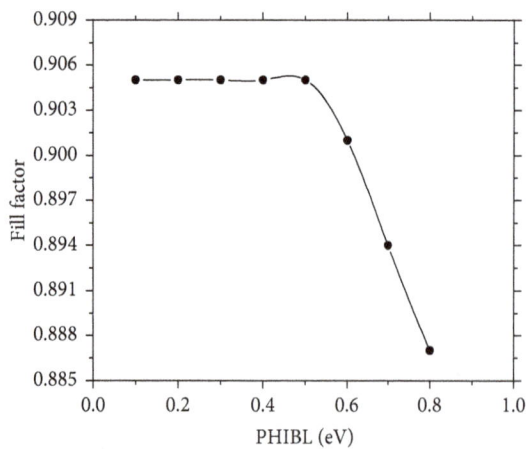

FIGURE 11: Band diagram of the proposed solar cell.

Conflicts of Interest

The authors declare that they have no conflicts of interest.

Acknowledgments

The authors would like to thank the Pennsylvania State University, USA, for providing them the analysis of microelectronic and photonic structures one-dimensional (AMPS-1D) device simulation package.

References

[1] M. F. Ali and M. F. Hossain, "Effect of bandgap of EBL on efficiency of the p-n homojunction Si solar cell from numerical analysis," *International Conference on Electrical & Electronic Engineering (ICEEE), IEEE Explored*, pp. 245–248, 2015, http://ieeexplore.ieee.org/document/7428268/.

[2] M. F. Ali and M. F. Hossain, "Simulation and observation of efficiency of p-n homojunction Si solar cell with defects and EBL by using AMPS-1D," *International Journal of Engineering and Applied Sciences (IJEAS)*, vol. 2, no. 12, pp. 137–140, 2015.

[3] M. F. Ali, R. Islam, N. Afrin, M. Firoj Ali, S. C. Motonta, and M. F. Hossain, "A new technique to produce electricity using solar cell in aspect of Bangladesh: dye-sensitized solar cell (DSSC) and it's prospect," *American Journal of Engineering Research*, vol. 3, pp. 35–40, 2014.

[4] J. A. Turner, "A realizable renewable energy future," *Science*, vol. 285, no. 5428, pp. 687–689, 1999.

[5] M. F. Ali and M. F. Hossain, "Improving efficiency of an amorphous silicon (pa-SiC: H/ia-Si: H/na-Si: H) solar cell by affecting bandgap and thickness from numerical analysis," *International Journal of Engineering and Applied Sciences (IJEAS)*, vol. 2, no. 12, 2015.

[6] May 2016, http://berc.berkeley.edu/why-are-solar-cells-made-of-silicon_1/.

[7] M. I. Kabir, S. A. Shahahmadi, V. Lim, S. Zaidi, K. Sopian, and N. Amin, "Amorphous silicon single-junction thin-film solar cell exceeding 10% efficiency by design optimization," *International Journal of Photoenergy*, vol. 2012, Article ID 460919, p. 7, 2012.

[8] F. T. Zohora, M. A. M. Bhuiyan, and S. Saimoom, "Simulation and optimization of high performance CIGS solar cells," *International Conference on Mechanical Engineering and Renewable Energy*, 2015.

[9] D. Rached and R. Mostefaoui, "Influence of the front contact barrier height on the indium tin oxide/hydrogenated p-doped amorphous silicon heterojunction solar cells," *Thin Solid Films*, vol. 516, no. 15, pp. 5087–5092, 2008.

[10] K. Robert, C. C. Battaile, A. C. Marshall, D. B. King, and D. R. Jennison, *Low Work Function Material Development for the Microminiature Thermionic Converter*, Sandia National Laboratories, 2004, http://prod.sandia.gov/techlib/access-control.cgi/2004/040555.pdf.

[11] M. A. Hossain, J. Mondal, M. Feroz Ali, and M. A. A. Humayun, "Design of high efficient InN quantum dot based solar cell," *International Journal of Scientific Engineering and Technology*, vol. 3, no. 4, pp. 346–349, 2014.

[12] J. Arch, S. V. FonAsh, J. Cuiffi et al., *A Manual for AMPS-1D for Windows 95/NT: A One-Dimensional Device Simulation, Program for the Analysis of Microelectronic and Photonic Structures, and Pennsylvania State University 1997, USA.*

[13] S. Banik and M. S. K. Shekh, *Design and Simulation of Ultra-Thin CdS-CdTe Thin-Film Solar Cell*, East West University, 2014, http://dspace.ewubd.edu/bitstream/handle/123456789/1308/Sowrabh_Banik.pdf?sequence=1.

[14] May 2016, https://en.wikipedia.org/wiki/Air_mass_%28solar_energy%29.

10

Non-Toxic Buffer Layers in Flexible Cu(In,Ga)Se₂ Photovoltaic Cell Applications with Optimized Absorber Thickness

Md. Asaduzzaman, Md. Billal Hosen, Md. Karamot Ali, and Ali Newaz Bahar

Department of Information and Communication Technology (ICT), Mawlana Bhashani Science and Technology University (MBSTU), Santosh, Tangail 1902, Bangladesh

Correspondence should be addressed to Md. Asaduzzaman; asaduzzaman.mbstu@gmail.com

Academic Editor: Meenakshisundaram Swaminathan

Absorber layer thickness gradient in $Cu(In_{1-x}Ga_x)Se_2$ (CIGS) based solar cells and several substitutes for typical cadmium sulfide (CdS) buffer layers, such as ZnS, ZnO, ZnS(O,OH), $Zn_{1-x}Sn_xO_y$ (ZTO), ZnSe, and In_2S_3, have been analyzed by a device emulation program and tool (ADEPT 2.1) to determine optimum efficiency. As a reference type, the CIGS cell with CdS buffer provides a theoretical efficiency of 23.23% when the optimum absorber layer thickness was determined as 1.6 μm. It is also observed that this highly efficient CIGS cell would have an absorber layer thickness between 1 μm and 2 μm whereas the optimum buffer layer thickness would be within the range of 0.04–0.06 μm. Among all the cells with various buffer layers, the best energy conversion efficiency of 24.62% has been achieved for the ZnO buffer layer based cell. The simulation results with ZnS and ZnO based buffer layer materials instead of using CdS indicate that the cell performance would be better than that of the CdS buffer layer based cell. Although the cells with ZnS(O,OH), ZTO, ZnSe, and In_2S_3 buffer layers provide slightly lower efficiencies than that of the CdS buffer based cell, the use of these materials would not be deleterious for the environment because of their non-carcinogenic and non-toxic nature.

1. Introduction

Solar cells based on thin-film CuInSe₂ (CIS) and $Cu(In_{1-x}Ga_x)Se_2$ (CIGS) absorbers have major contributions in photovoltaic technology due to their lower cost, flexible modules, and high energy conversion efficiency of more than 21% [1–4]. The absorber works as a p-type doped area into a CIS and CIGS model with a typical thickness of 1-2 μm [5]. For the CIS absorber layer, the band gap is about 1 eV (0.98–1.04 eV) [6]. By introducing the gallium content, [Ga]/([Ga + In]), into the CIS absorber, the band gap of CIGS can be varied from 1.04 to 1.7 eV [4] although the optimal band gap of CIGS ranges from 1.16 eV to 1.38 eV due to the Ga grading in the CIS absorber [7–10]. To show the effects of absorber band gap on the properties of the CIGS solar cell, in this study, the CIGS band gap is optimized as 1.25 eV for [Ga]/([Ga + In]) = 0.45 [11]. While growing CdS buffer on the CIGS absorber, the critical thickness of the CdS buffer layer was found to be around 50 nm.

Besides this, the chemical bath-deposited (CBD) ZnS buffer layer having a wider band gap ($E_g = 3.68$ eV) has shown the second highest efficiency in the thin-film technology for the replacement of CdS [12]. Atomic layer-deposited (ALD) $Zn_{1-x}Sn_xO_y$ (ZTO) has also been used for the replacement of CdS, and the best cell efficiency of the CIGS solar cell with a ZTO buffer layer has been recorded as 18.2% [11]. The wider band gap of ZTO is used as it permits transmitting the photons having lower wavelength into the absorber. Hence, the current generation is also increased owing to using ZTO as a buffer. Zinc oxide, ZnO, is also used as an alternative to the CdS buffer layer as it has a band gap of 3.30 eV, which is 0.88 eV wider than that of CdS ($E_g = 2.42$ eV). But with the ZnO buffer layer, the cell does not show the light-soaking effect [13]. During the heteroepitaxial growth of the buffer layer using the atomic layer deposition (ALD) technique, the existence of the effect is realized. This effect is primarily associated with the surface properties of the CIGS absorber. To subside the light-soaking effect, the CIGS

TABLE 1: Material parameters for CIGS solar cell simulation.

Parameters	ZnO : Al	i-ZnO	CdS	CIGS
Thickness, τ (μm)	0.2	0.02	0.05	1
Dielectric constant, E_{ps}	7.8	7.8	12	13.6
Refractive index, N_{dx}	2	2	3.15	3.67
Band gap, E_g (eV)	3.30	3.30	2.42	1.25
Electron affinity, χ_e (eV)	4.6	4.6	3.74	4.19
Donor concentration, N_d (cm^{-3})	5×10^{16}	1×10^{17}	3×10^{16}	0
Acceptor concentration, N_a (cm^{-3})	0	0	0	2×10^{16}
Electron mobility, μ_n (cm^2 V^{-1} s^{-1})	100	100	120	110
Hole mobility, μ_p (cm^2 V^{-1} s^{-1})	25	25	40	35
Conduction band effective density of states, N_c (cm^{-3})	2.2×10^{18}	2.2×10^{18}	2.2×10^{18}	2.2×10^{18}
Valence band effective density of states, N_v (cm^{-3})	1.8×10^{19}	1.8×10^{19}	1.8×10^{19}	1.8×10^{19}

TABLE 2: Base parameters of the substitute buffer layers.

Parameters	ZnO	ZnS	ZnS(O,OH)	$Zn_{1-x}Sn_xO_y$	ZnSe	In_2S_3
τ (μm)	0.03	0.04	0.05	0.05	0.04	0.05
E_g (eV)	3.30	3.68	3.80	3.74	2.71	2.90
χ_e (eV)	4.60	4.13	4.24	4.06	4.09	3.85
N_d (cm^{-3})	1×10^{17}	5×10^{16}	1×10^{17}	1×10^{17}	2×10^{17}	1×10^{17}
μ_n (cm^2 V^{-1} s^{-1})	100	250	200	160	60	400
μ_p (cm^2 V^{-1} s^{-1})	30	70	60	40	20	120

surface is doped or etched with Zn, and hence, the solar cell with the ALD technique yielded an efficiency of 13.9% [14]. Besides the CdS, ZnS, and ZnO based buffer layers, we also investigate the properties of ZnSe, In_2S_3, and ZnS(O,OH) buffer layers and their effects on the CIGS solar cell performance.

2. Experimental

ADEPT 2.1 (a device emulation program and tool: version 2.1) is a one-dimensional online simulator to analyze the electrical and optical characteristics of silicon based solar cells, CIGS and CdTe based thin films, GaAs solar cells, and so on [15]. In this paper, ADEPT 2.1 was used to simulate and investigate CIGS solar cell properties with different buffer layers. Current-voltage characteristics in light and dark conditions can be achieved from the simulation carried out by ADEPT 2.1 simulation. For all the solar cell structures, quantum efficiency, J-V characteristics, electric field, current generation, saturation current, energy band gap profile, and so on are measured as a function of light bias, voltage, or temperature. Moreover, from the ADEPT 2.1 simulation, recombination profiles, carrier concentration, electric field distributions, and carrier current densities can be obtained as a function of thickness. However, the properties of the conventional CdS and its alternative materials such as ZnO, ZnS, ZnS(O,OH), ZTO, ZnSe, and In_2S_3 and their effects on the performance of the n-ZnO/i-ZnO/buffer/CIGS solar cell are emphasized in this study. The

analytical aspects are observed due to the changes in efficiency (η), open-circuit voltage (V_{oc}), short-circuit current (J_{sc}), and fill factor (FF). The effects of CdS buffer layer thickness, the various buffer layer material parameters, and the CIGS absorber thickness are also taken into account during the simulation. The base parameters for the CIGS cell structure with CdS buffer used for the simulation are shown in Table 1 [1, 3, 4, 11, 16, 17]. The most important parameters of different buffer layer materials needed for the simulations are depicted in Table 2 [4, 17–31].

3. Results and Discussions

3.1. Effects of Absorber Thickness on CIGS Cell Performance. To make the simulation reasonable, the conventional CIGS thin-film structure has been ascertained in terms of Cu(In,Ga)Se$_2$ absorber. Firstly, to determine the optimum thickness of the CIGS cell structure with the most commonly used CdS buffer, the thickness of the CIGS absorber ($\tau_{absorber}$) was varied. In Figure 1(a), it is shown that the open-circuit voltage (V_{oc}) is increasing with varying $\tau_{absorber}$. Figure 1(b) shows that the short-circuit current density (J_{sc}) of the solar cell is also increasing with the increase of $\tau_{absorber}$. Because of being in the p-type area in the cell, V_{oc} and J_{sc} are increased with enhancement of the CIGS absorber layer thickness, $\tau_{absorber}$. This phenomenon permits the photons with longer wavelengths to be accumulated, which in turn leads to generation of electron-hole pair (EHP). It is also easily realized that if $\tau_{absorber}$ is decreased, the value of

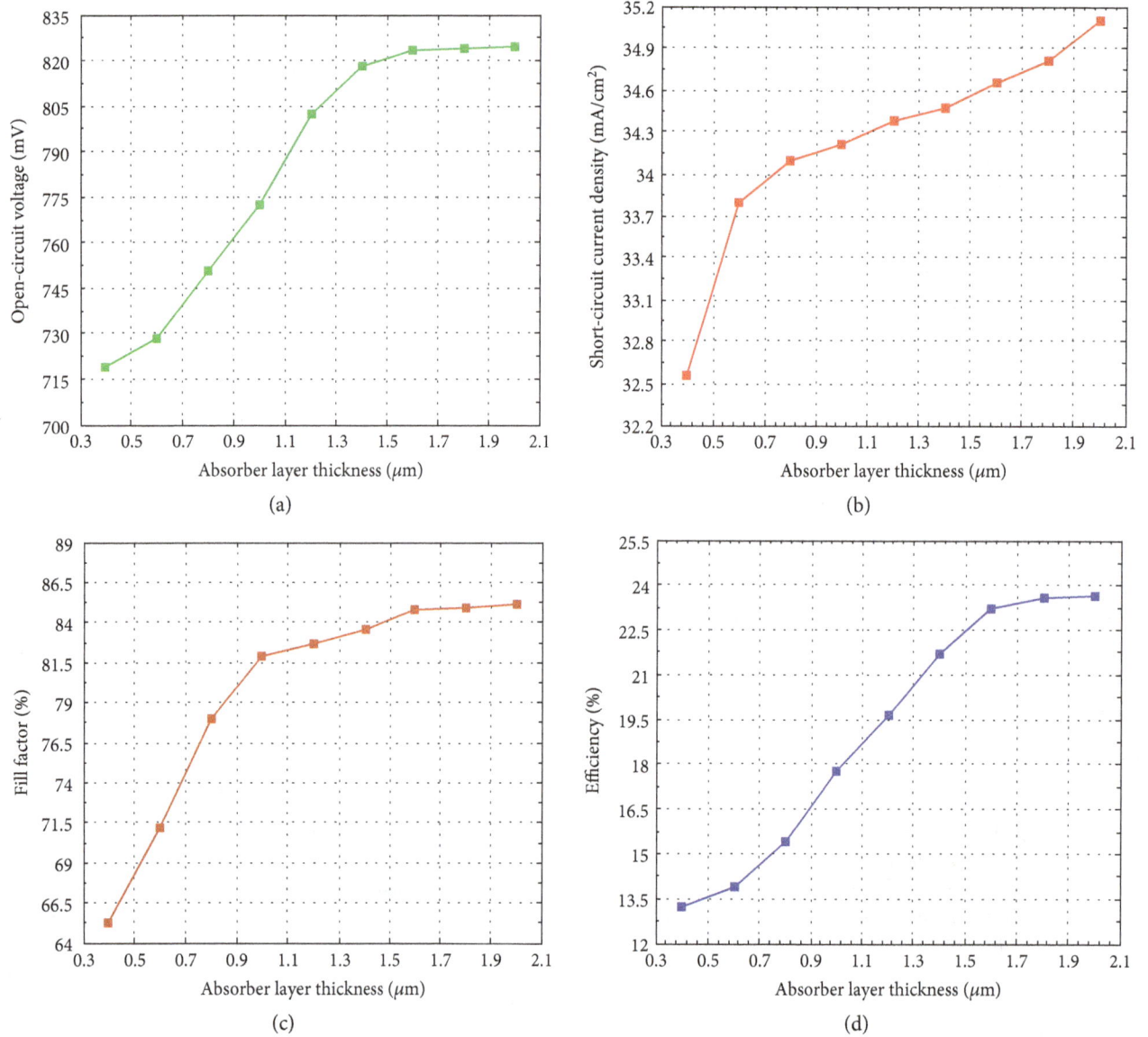

FIGURE 1: Performance variation due to variable thickness of the CIGS absorber layer: (a) open-circuit voltage (V_{oc}); (b) short-circuit current density (J_{sc}); (c) fill factor; (d) efficiency.

both V_{oc} and J_{sc} will be decreased. It is seen from Figure 1(c) that, after $1\,\mu m$, the fill factor (FF) of this simulated CIGS solar cell is almost constant with the $\tau_{absorber}$ variation. From Figure 1(d), it is clearly observed that the solar cell efficiency (η) increases with the increase of the CIGS absorber thickness, but over $1.6\,\mu m$, the efficiency variation seems to be very slow. $\tau_{absorber}$ would be optimum around 1.6–$2\,\mu m$. The efficiencies were recorded as 17.76% and 23.67% for the thicknesses of $1\,\mu m$ and $2\,\mu m$, respectively. By comparing these results, it is observed that a 5.47% increase in efficiency was found due to the increase of $0.6\,\mu m$ (from $1\,\mu m$) in $\tau_{absorber}$. And an enhancement of $0.4\,\mu m$ from $1.6\,\mu m$ in $\tau_{absorber}$ results in only 0.44% increase in efficiency.

The recombination process at the back contact of the cell may cause the enhancement in the values of J_{sc} and V_{oc}. The

back contact will go very close to the depletion area while reducing the absorber layer thickness. Therefore, the back contact captures the electrons for the recombination in an easy way. Hence, a smaller number of electrons contribute to the collection efficiency, and thus, the values of J_{sc} and V_{oc} decrease. Figure 2 shows the J-V characteristic curve with the performance parameters such as V_{oc}, J_{sc}, FF, and efficiency computed after conducting the simulation.

Figure 3 shows the variation in quantum efficiency with respect to photons' wavelengths according to the variation in absorber layer thickness, $\tau_{absorber}$. It is observed that if the thickness of the absorber layer increases, then the quantum efficiency of the solar cell is increased. While increasing the thickness of the absorber layer, a large number of photons are absorbed, including photons

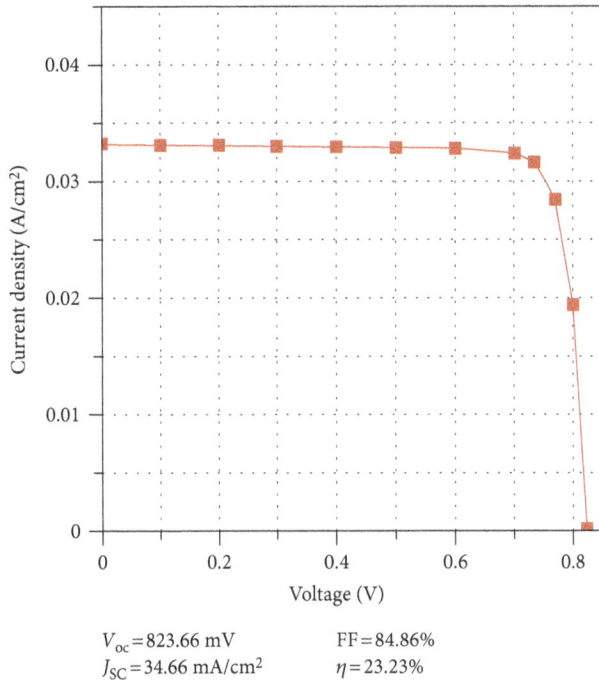

$V_{oc} = 823.66$ mV   FF $= 84.86\%$
$J_{SC} = 34.66$ mA/cm^2 $\eta = 23.23\%$

FIGURE 2: J-V characteristic curve with photovoltaic parameters.

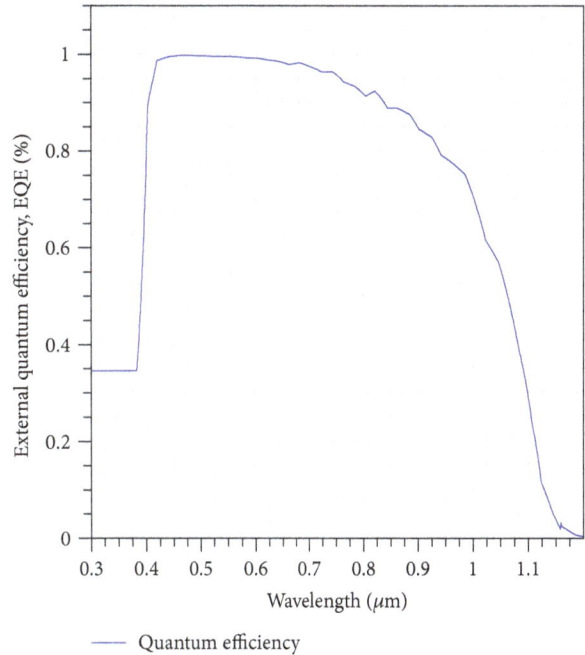

FIGURE 3: Spectral response of CIGS photovoltaic cell.

having longer wavelengths. The absorbed photons produce a larger number of electron-hole pairs (EHP) in this way. Thus, the increased $\tau_{absorber}$ substantiates the increase in external quantum efficiency (EQE).

3.2. Effects of Various Buffer Layers on Cell Efficiency. The six potential buffer layers which have been investigated besides the conventional CdS are ZnS, ZnS(O,OH), ZnSe, ZnO, $Zn_{1-x}Sn_xO_y$, and In_2S_3. For all cases, the thickness has been varied from $0.01\,\mu$m to $0.1\,\mu$m. More photons are absorbed by the CdS buffer layer while increasing the thickness of the layer. The optimum thickness of CdS buffer is determined as $0.05\,\mu$m since J_{sc} decreased dramatically after $0.05\,\mu$m [32]. However, the ZnS(O,OH) based CIGS solar cell is typically not affected by the thickness of the buffer layer because of the wider band gap of ZnS. It is seen that the efficiencies of the ZnO, ZnSe, and In_2S_3 based solar cells have been decreased when the thicknesses of these layers cross beyond $0.05\,\mu$m. Hence, the optimum thickness of the mentioned buffer layers would be suggested as $0.05\,\mu$m. On the other hand, after the buffer layer thickness of $0.06\,\mu$m, the efficiency of the $Zn_{1-x}Sn_xO_y$ (ZTO) based CIGS cell increases. Figure 4 gives a summary of the simulated J-V characteristic curves with different performance parameters for the CIGS cell based on various buffer layers.

From the simulated results, it is suggested that the ZnO buffer would be a promising alternative to the CdS buffer layer in the CIGS cell as the ZnO and ZnS based cells have reached efficiency levels of 23.67% and 24.62%, respectively. Consequently, ZnO and ZnS are proposed to be very potential materials for conventional CdS replacement because of having wider band gap than that of CdS and higher efficiency than that of the CdS based cell. Although ZTO, ZnS(O,OH), ZnSe, and In_2S_3 based cells provide slightly lower efficiencies than CdS based CIGS solar cells, these materials are nontoxic in nature while the CdS is categorized as carcinogenic and toxic [33]. As a result, these materials can be used as substitute buffer layers for CIGS absorber-based chalcopyrite thin-film photovoltaic devices and will be environmentally friendly enough compared to that of the CdS based cell. The comparisons among different performance parameters of CIGS cells with different substitute buffer layers are shown in Figure 5.

Finally, the comparisons of the improved cell efficiencies for different buffer layers with the relevant experimental results in the literature [1, 11, 12, 17, 34–36] are shown in Table 3.

4. Conclusions

At first, the performance measurements due to the variation in absorber layer thickness in the CIGS solar cell with a CdS buffer layer have been investigated, and thus, the optimum thickness has been figured out from the analysis. Afterwards, various buffer layer materials such as ZnO, ZnS, ZnS(O,OH), ZnSe, ZTO, and In_2S_3 have been used into the CIGS cells. The optimum thicknesses of all the buffer layer materials studied are in the range of 0.04–$0.06\,\mu$m whereas the absorber layer thickness was optimized as $1.6\,\mu$m. While comparing the performances of the cells with different buffer layers, it has been observed that the cell with a ZnS buffer layer reveals the highest efficiency of 24.62% among all the cells studied. The second highest efficiency of 23.67% was found from the simulation result of the cell having a ZnO buffer layer. Besides, the comparative analysis of the cells with ZnS(O,OH), ZnSe, ZTO, and In_2S_3 buffer layers also shows a higher efficiency of more than 18%, which asserts the possible replacement of the conventional CdS buffer layer material in CIGS thin-film device structures.

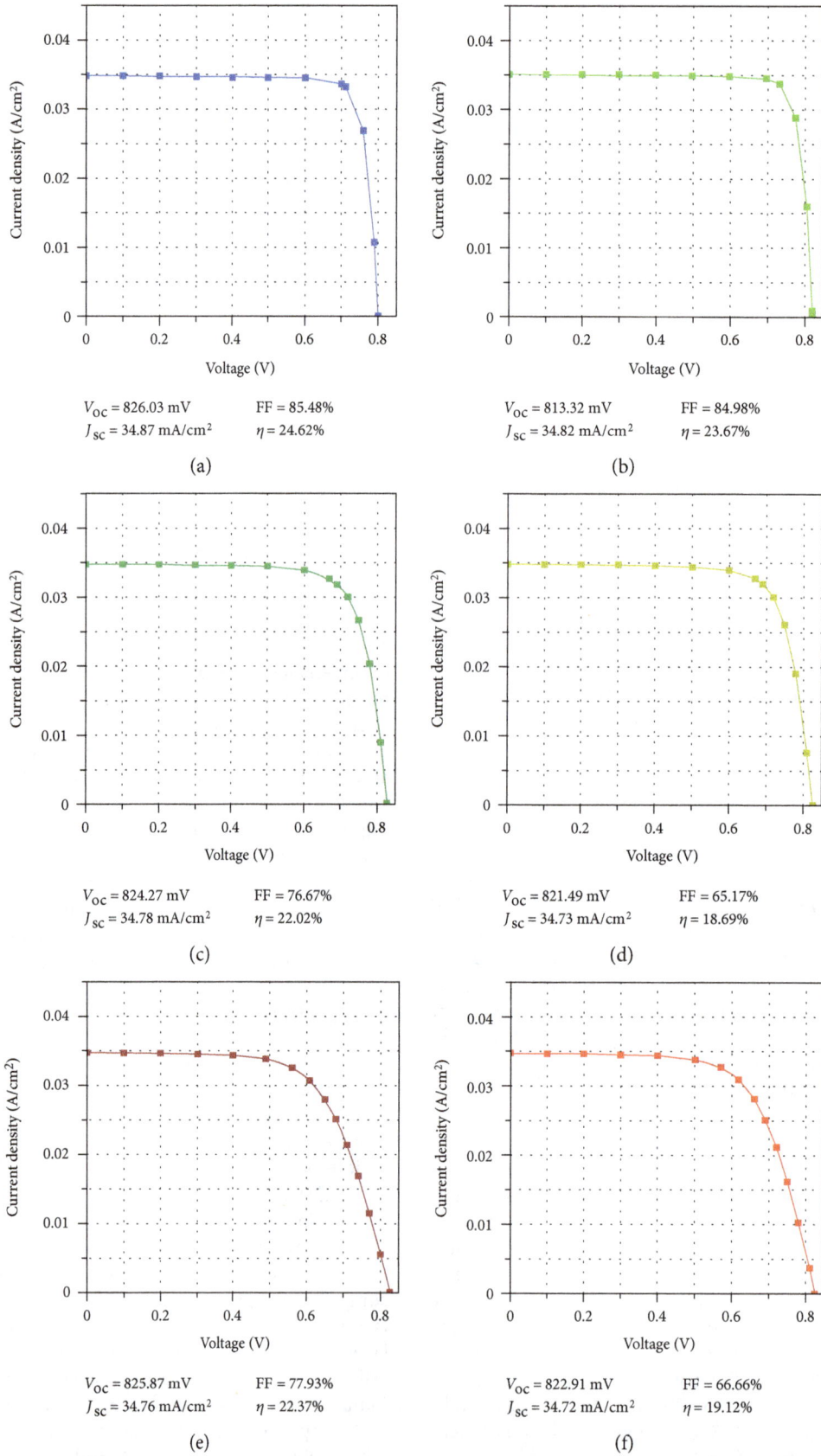

FIGURE 4: *J-V* characteristic curves for CIGS cell with different buffer layers: (a) ZnS; (b) ZnO; (c) ZnS(O,OH); (d) ZnSe; (e) ZTO; (f) In$_2$S$_3$.

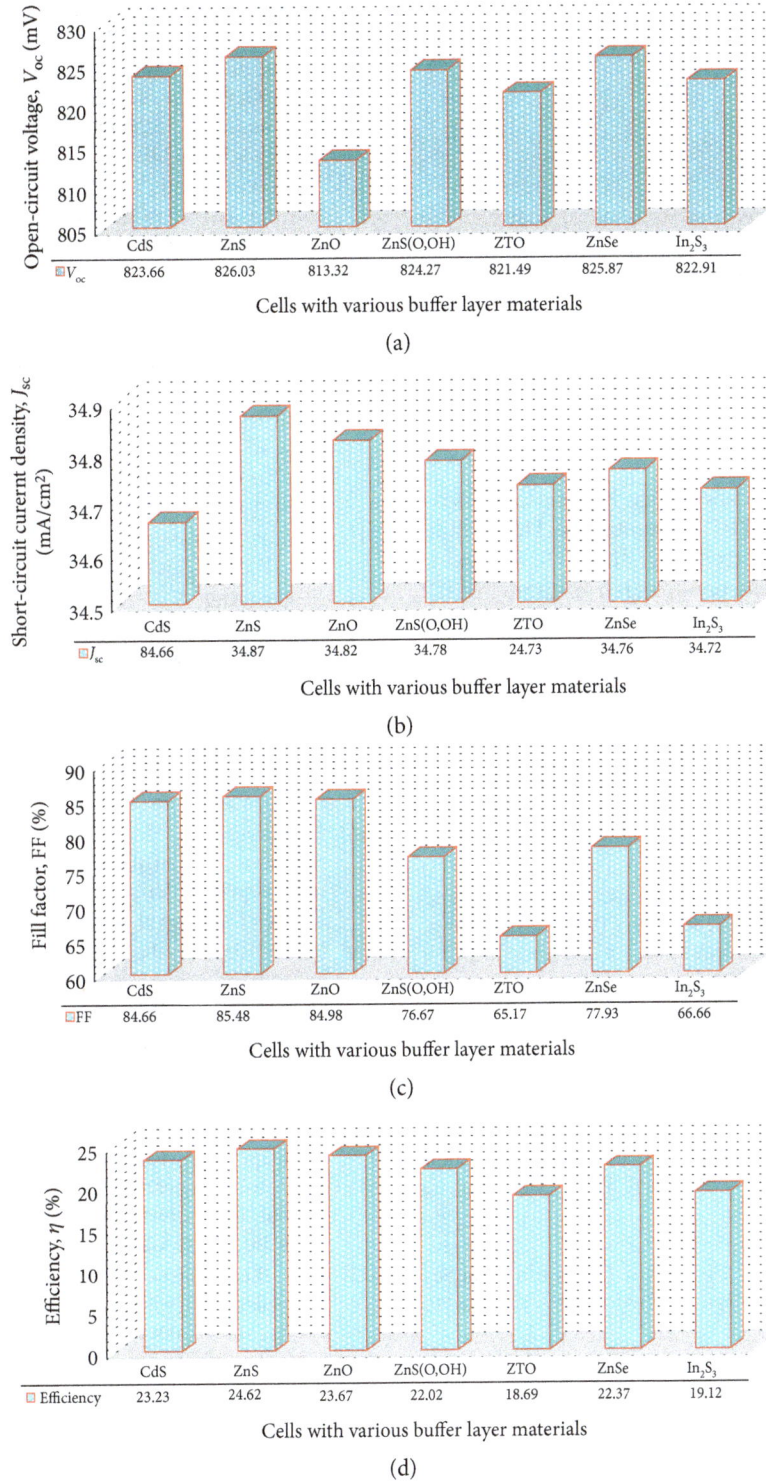

| ☐V_{oc} | 823.66 | 826.03 | 813.32 | 824.27 | 821.49 | 825.87 | 822.91 |
| | CdS | ZnS | ZnO | ZnS(O,OH) | ZTO | ZnSe | In$_2$S$_3$ |

Cells with various buffer layer materials

(a)

| ☐J_{sc} | 84.66 | 34.87 | 34.82 | 34.78 | 24.73 | 34.76 | 34.72 |
| | CdS | ZnS | ZnO | ZnS(O,OH) | ZTO | ZnSe | In$_2$S$_3$ |

Cells with various buffer layer materials

(b)

| ☐FF | 84.66 | 85.48 | 84.98 | 76.67 | 65.17 | 77.93 | 66.66 |
| | CdS | ZnS | ZnO | ZnS(O,OH) | ZTO | ZnSe | In$_2$S$_3$ |

Cells with various buffer layer materials

(c)

| ☐ Efficiency | 23.23 | 24.62 | 23.67 | 22.02 | 18.69 | 22.37 | 19.12 |
| | CdS | ZnS | ZnO | ZnS(O,OH) | ZTO | ZnSe | In$_2$S$_3$ |

Cells with various buffer layer materials

(d)

FIGURE 5: Performance comparison among CIGS cells with different buffer layers: (a) open-circuit voltage, V_{oc}; (b) short-circuit current density, J_{sc}; (c) fill factor, FF; (d) efficiency, η.

TABLE 3: Comparisons of the improved efficiencies with the relevant experimental results.

Cells	CdS	ZnS	ZnO	ZnS(O,OH)	ZTO	ZnSe	In$_2$S$_3$
Proposed cells	23.23	24.62	23.67	22.02	18.69	22.37	19.12
Reference cells	22.60 [1]	18.10 [12]	20.80 [34]	18.40 [35]	18.20 [11]	19.15 [17]	16.40 [36]

Conflicts of Interest

The authors declare that there is no conflict of interest regarding the publication of this paper.

Authors' Contributions

Md. Asaduzzaman conducted the simulation, analyzed the simulation results, and prepared and submitted the manuscript. Md. Billal Hosen and Md. Karamot Ali assisted in conducting the simulation, analyzing the results, and preparing the manuscript. Ali Newaz Bahar supervised the study. All authors finally approved the manuscript for submission.

Acknowledgments

The authors would like to acknowledge Purdue University, USA, for the use of the ADEPT 2.1 simulator. They would also like to express their extreme gratitude to Professor Dr. Nowshad Amin, Solar Energy Research Institute (SERI), Universiti Kebangsaan Malaysia (UKM), Malaysia, for providing incredible support for this study.

References

[1] P. Jackson, R. Wuerz, D. Hariskos, E. Lotter, W. Witte, and M. Powalla, "Effects of heavy alkali elements in $Cu(In,Ga)Se_2$ solar cells with efficiencies up to 22.6%," *Physica Status Solidi (RRL)*, vol. 10, no. 8, pp. 583–586, 2016.

[2] M. A. Green, K. Emery, Y. Hishikawa, W. Warta, and E. D. Dunlop, "Solar cell efficiency tables (version 48)," *Progress in Photovoltaics: Research and Applications*, vol. 24, no. 7, pp. 905–913, 2016.

[3] T. M. Friedlmeier, P. Jackson, A. Bauer et al., "Improved photocurrent in $Cu(In,Ga)Se_2$ solar cells: from 20.8% to 21.7% efficiency with CdS buffer and 21.0% Cd-free," *IEEE Journal of Photovoltaics*, vol. 5, no. 5, pp. 1487–1491, 2015.

[4] M. Asaduzzaman, M. Hasan, and A. N. Bahar, "An investigation into the effects of band gap and doping concentration on $Cu(In,Ga)Se_2$ solar cell efficiency," *SpringerPlus*, vol. 5, no. 1, 578 pages, 2016.

[5] A. O. Pudov, A. Kanevce, H. A. Al-Thani, J. R. Sites, and F. S. Hasoon, "Secondary barriers in $CdS–CuIn_{1-x}Ga_xSe_2$ solar cells," *Journal of Applied Physics*, vol. 97, no. 6, 064901 pages, 2005.

[6] K. Ramanathan, M. A. Contreras, C. L. Perkins et al., "Properties of 19.2% efficiency $ZnO/CdS/CuInGaSe_2$ thin-film solar cells," *Progress in Photovoltaics: Research and Applications*, vol. 11, no. 4, pp. 225–230, 2003.

[7] A. Han, Y. Sun, Y. Zhang, X. Liu, F. Meng, and Z. Liu, "Comparative study of the role of Ga in CIGS solar cells with different thickness," *Thin Solid Films*, vol. 598, pp. 189–194, 2016.

[8] O. Lundberg, M. Edoff, and L. Stolt, "The effect of Ga-grading in CIGS thin film solar cells," *Thin Solid Films*, vol. 480–481, pp. 520–525, 2005.

[9] S. H. Song and S. A. Campbell, "Heteroepitaxy and the performance of CIGS solar cells," *IEEE 39th Photovoltaic Specialists Conference (PVSC)*, pp. 2534–2539, 2013.

[10] O. Lundberg, M. Bodegård, J. Malmström, and L. Stolt, "Influence of the $Cu(In,Ga)Se_2$ thickness and Ga grading on solar cell performance," *Progress in Photovoltaics: Research and Applications*, vol. 11, no. 2, pp. 77–88, 2003.

[11] J. Lindahl, U. Zimmermann, P. Szaniawski et al., "Inline $Cu(In,Ga)Se$ Co-evaporation for high-efficiency solar cells and modules," *IEEE Journal of Photovoltaics*, vol. 3, no. 3, pp. 1100–1105, 2013.

[12] T. Nakada and M. Mizutani, "18% efficiency Cd-free $Cu(In, Ga)Se_2$ thin-film solar cells fabricated using chemical bath deposition (CBD)-ZnS buffer layers," *Japanese Journal of Applied Physics*, vol. 41, no. 2B, pp. L165–L167, 2002, no. Part 2.

[13] R. Mikami, H. Miyazaki, T. Abe, A. Yamada, and M. Konagai, "Chemical bath deposited (CBD)-ZnO buffer layer for CIGS solar cells," *Proceedings of 3rd World Conference on Photovoltaic Energy Conversion, 2003*, vol 1, pp. 519–522, 2003.

[14] S. Chaisitsak, A. Yamada, and M. Konagai, "Comprehensive study of light-soaking effect in $ZnO/Cu(InGa)Se_2$ solar cells with Zn-based buffer layers," *MRS Online Proceedings Library Archive*, vol. 668 H9–H10, 2001.

[15] J. Gray, X. Wang, R. V. K. Chavali, X. Sun, A. Kanti, and J. R. Wilcox, "Adept 2.1," 2015.

[16] M. Gloeckler, A. L. Fahrenbruch, and J. R. Sites, "Numerical modeling of CIGS and CdTe solar cells: setting the baseline," *Proceedings of 3rd World Conference on Photovoltaic Energy Conversion, 2003*, vol 1, pp. 491–494, 2003.

[17] P. Chelvanathan, M. I. Hossain, and N. Amin, "Performance analysis of copper–indium–gallium–diselenide (CIGS) solar cells with various buffer layers by SCAPS," *Current Applied Physics*, vol. 10, no. 3, Supplement, pp. S387–S391, 2010.

[18] T. Nakada, K. Furumi, and A. Kunioka, "High-efficiency cadmium-free $Cu(In,Ga)Se_2$ thin-film solar cells with chemically deposited ZnS buffer layers," *IEEE Transactions on Electron Devices*, vol. 46, no. 10, pp. 2093–2097, 1999.

[19] M. M. Islam et al., "CIGS solar cell with MBE-grown ZnS buffer layer," *Solar Energy Materials and Solar Cells*, vol. 93, no. 6–7, pp. 970–972, 2009.

[20] M. Nguyen, K. Ernits, K. F. Tai et al., "ZnS buffer layer for $Cu_2ZnSn(SSe)_4$ monograin layer solar cell," *Solar Energy*, vol. 111, pp. 344–349, 2015.

[21] C. Zhang, S. Pan, T. Heng et al., "Stable inverted low-bandgap polymer solar cells with aqueous solution processed low-temperature ZnO buffer layers," *International Journal of Photoenergy*, vol. 2016, no. 2016, Article ID 3675036, 7 pages, 2016.

[22] R. N. Bhattacharya and K. Ramanathan, "$Cu(In,Ga)Se_2$ thin film solar cells with buffer layer alternative to CdS," *Solar Energy*, vol. 77, no. 6, pp. 679–683, 2004.

[23] J. Lindahl, J. Keller, O. Donzel-Gargand, P. Szaniawski, M. Edoff, and T. Törndahl, "Deposition temperature induced conduction band changes in zinc tin oxide buffer layers for $Cu(In,Ga)Se_2$ solar cells," *Solar Energy Materials and Solar Cells*, vol. 144, pp. 684–690, 2016.

[24] M. Asaduzzaman, A. N. Bahar, M. M. Masum, and M. M. Hasan, "Cadmium free high efficiency $Cu_2ZnSn(S,Se)_4$ solar cell with $Zn_{1-x}Sn_xO_y$ buffer layer," *Alexandria Engineering Journal*, 2017, article in press.

[25] D. Hariskos, S. Spiering, and M. Powalla, "Buffer layers in $Cu(In,Ga)Se_2$ solar cells and modules," *Thin Solid Films*, vol. 480–481, pp. 99–109, 2005.

[26] N. Khoshsirat, N. A. M. Yunus, M. N. Hamidon, S. Shafie, and N. Amin, "Analysis of absorber and buffer layer band gap grading on CIGS thin film solar cell performance using

SCAPS," *Pertanika Journal of Science and Technology*, vol. 23, no. 2, pp. 241–250, 2015.

[27] L. Wang, X. Lin, A. Ennaoui, C. Wolf, M. C. Lux-Steiner, and R. Klenk, "Solution-processed In_2S_3 buffer layer for chalcopyrite thin film solar cells," *EPJ Photovoltaics*, vol. 7, 70303 pages, 2016.

[28] F. Engelhardt, L. Bornemann, M. Köntges et al., "$Cu(In,Ga)Se_2$ solar cells with a ZnSe buffer layer: interface characterization by quantum efficiency measurements," *Progress in Photovoltaics: Research and Applications*, vol. 7, no. 6, pp. 423–436, 1999.

[29] N. A. Okereke and A. J. Ekpunobi, "ZnSe buffer layer deposition for solar cell application," *Journal of Non-Oxide Glasses*, vol. 3, no. 1, pp. 31–36, 2011.

[30] J. Wang and M. Isshiki, "Wide-bandgap II–VI semiconductors: growth and properties," in *Springer Handbook of Electronic and Photonic Materials*, S. Kasap and P. Capper, Eds., pp. 325–342, Springer US, 2006.

[31] T.-H. Yeh, C.-H. Hsu, W.-H. Ho, S.-Y. Wei, C.-H. Cai, and C.-H. Lai, "An ammonia-free chemical-bath-deposited ZnS(O,OH) buffer layer for flexible $Cu(In,Ga)Se_2$ solar cell application: an eco-friendly approach to achieving improved stability," *Green Chemistry*, vol. 18, no. 19, pp. 5212–5218, 2016.

[32] Y. Yamamoto, K. Saito, K. Takahashi, and M. Konagai, "Preparation of boron-doped ZnO thin films by photo-atomic layer deposition," *Solar Energy Materials and Solar Cells*, vol. 65, no. 1–4, pp. 125–132, 2001.

[33] M. P. Waalkes, "Cadmium carcinogenesis in review," *Journal of Inorganic Biochemistry*, vol. 79, no. 1–4, pp. 241–244, 2000.

[34] P. Jackson, D. Hariskos, R. Wuerz, W. Wischmann, and M. Powalla, "Compositional investigation of potassium doped $Cu(In,Ga)Se_2$ solar cells with efficiencies up to 20.8%," *Physica Status Solidi (RRL)*, vol. 8, no. 3, pp. 219–222, 2014.

[35] T. Kobayashi, K. Yamauchi, and T. Nakada, "Comparison of cell performance of ZnS(O,OH)/CIGS solar cells with UV-assisted MOCVD-ZnO:B and sputter-deposited ZnO:Al window layers," *IEEE Journal of Photovoltaics*, vol. 3, no. 3, pp. 1079–1083, 2013.

[36] M. A. Mughal, R. Engelken, and R. Sharma, "Progress in indium (III) sulfide (In_2S_3) buffer layer deposition techniques for CIS, CIGS, and CdTe-based thin film solar cells," *Solar Energy*, vol. 120, pp. 131–146, 2015.

11

Design and Optimization of Elliptical Cavity Tube Receivers in the Parabolic Trough Solar Collector

Fei Cao,[1,2] **Lei Wang,**[1] **and Tianyu Zhu**[1]

[1]*College of Mechanical and Electrical Engineering, Hohai University, Changzhou 213022, China*
[2]*Sunshore Solar Energy Company Limited, Nantong 226300, China*

Correspondence should be addressed to Fei Cao; yq.cao@hotmail.com

Academic Editor: Stoian Petrescu

The nonfragile cavity receiver is of high significance to the solar parabolic trough collector (PTC). In the present study, light distributions in the cavity under different tracking error angles and PTC configurations are analyzed. A new elliptical cavity geometry is proposed and analyzed. It is obtained from this study that light distribution on the tube receiver is asymmetrical when tracking error occurs. On increasing the tracking error angle, more lights are sheltered by the cavity outer surface. The PTC focal distance has negative correlation with the cavity open length, whereas the PTC concentration ratio has positive correlation with the cavity open length. Increasing the tracking error angle and increasing the PTC focal distance would both decrease the cavity blackness. Introducing a flat plate reflector at the elliptical cavity open inlet can largely increase the cavity darkness.

1. Introduction

Solar energy has been recognized as one of the most important energy sources at present and in the further energy structure. Due to the discontinuous, low energy flux, periodicity, and unsteady characteristics of solar energy, solar concentration is commonly utilized in solar engineering fields, among which the solar parabolic trough collectors (PTC) are the most widely accepted solar concentration style [1]. The light ray tracing [2–4] and structural analysis [5–7] for the PTC have been widely discussed in the literature.

Serrano-Aguilera et al. studied the continuous linear reflectors for flat plate receivers with Inverse Monte Carlo ray-tracing method, where a quasi-planar-concentrated flux distribution is required [2]. Cheng et al. made a 3D numerical study of heat transfer characteristics in the receiver tube of PTC [3], and they then carried out a comparative analysis for PTC with a detailed Monte Carlo ray-tracing optical model [4]. They concluded that the ideal characteristics and optical performance of the PTCs were very different from some critical points determined by the divergence

phenomenon of the nonparallel solar beam [4]. Giannuzzi et al. defined a guideline for steel structures' design and assessment of the components of PTC. Their codes were developed for practical usage and were evaluated under some specific conditions [5]. Liang et al. compared three optical models for the PTC and optimized the geometric parameters according to their models [6]. Cheng et al. optimized the geometrical structure of the PTC based on the particle swarm optimization algorithm and the Monte Carlo ray-tracing method, which found a balance between the calculating speed and result accuracy [7].

With respect to the heat receiver of the PTC, some structures are designed for the PV or PV/T in PTCs [3]. However, tube receivers are more commonly used in PTC combined with thermodynamic cycles [8–10]. In general, two kinds of tubes are commonly utilized in the PTC as the receiver, namely, the vacuum tube [2, 4–10] and the cavity tube [11–13]. The main advantages of the vacuum tube are the high thermal maintenances and low module cost, whereas its structure is frangible. Though a kind of "metal inner and vacuum glass outer" tube is proposed, the connection area of metal and glass is also frangible due to

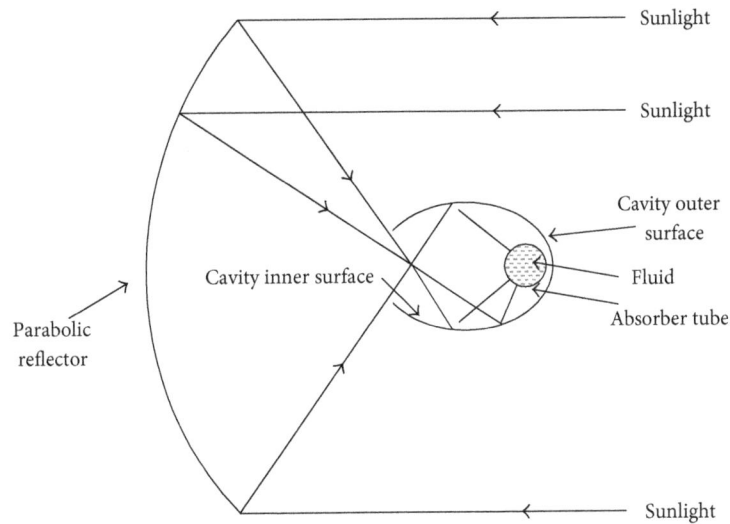

FIGURE 1: Schematic of a cavity tube receiver in the PTC.

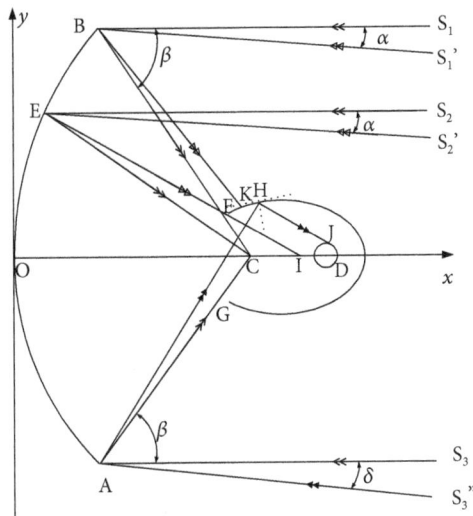

FIGURE 2: Sunlight trace in a PTC with an elliptical cavity tube receiver by considering the tracking error angle α under the rectangular coordinate system.

TABLE 1: Parameters of the PTC, cavity, and tube absorber.

Parameter/unit	Value
Major axis/mm	100
Minor axis/mm	60
PTC focal distance/mm	800
Elliptic focal distance/mm	80
PTC open width/mm	850
Absorber tube radius/mm	18
Elliptic cavity open length/mm	12

parameters of the PTC with the cavity tube receiver are then discussed. A new geometry is finally proposed according to the parameters' analysis results.

2. Mathematical Models

To track the solar ray in the PTC, the cross section of a PTC with an elliptical cavity tube receiver is established under the rectangular coordinate system as shown in Figure 2. The major axis, minor axis, and the focal distance of the elliptical cavity are A, B, and C in Figure 2, respectively. The open length of the elliptical cavity is L_{el}. The focal distance and open length of the parabolic reflector are f and L_{pr}, respectively. The radius of the tube receiver is r_{tube}. The left focal point of the elliptical cavity is located at the focal point of the PTC. In ideal condition, the sunlight is reflected by the parabolic reflector and concentrated at point C. When the sun tracking error occurs, there is a tracking error angle α in the PTC, which causes part of the sunlight not to be reflected into the cavity, that is, Light S_1'BK. There are three representative groups of light paths which can be reflected into the elliptical cavity in Figure 2, namely, the marginal light of the parabolic reflector (i.e., Light S_1BC, S_2EC, and S_3AC), the marginal light of the tube receiver (i.e., Light

uninterrupted and periodical thermal stress from the inner and outer sides. The other kind of solar receiver is the cavity tube, whose schematic is shown in Figure 1. The outer cover of the receiver is the cavity, with an open inlet towards the parabolic trough reflector. As there is no connection area between the cavity outer cover and the inner absorber tube, this structure is nonfrangible. The incoming lights are reflected by the parabolic reflector and entered the cavity through its open inlet. Sunlight is then reflected for several times and finally reached the tube receiver at the ellipse focus. There are some studies on optical performance of the solar cavity receivers [10–13]. Very limited studies have been presented to the parameter analysis and optimization of the elliptical cavity tube receiver. Considering this, the PTC with the cavity tube receiver is simulated through using the Monte Carlo method. The

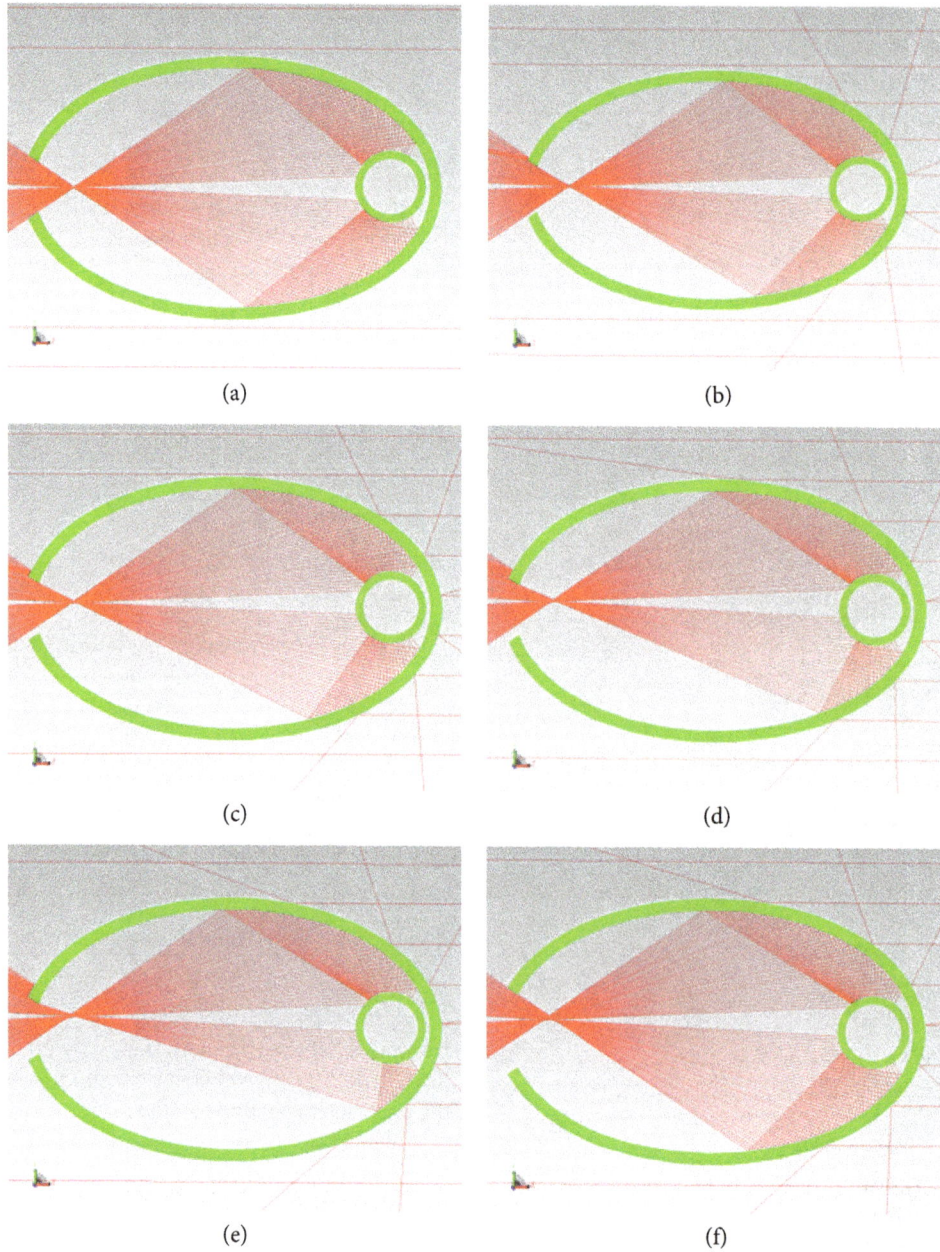

FIGURE 3: Light distribution in the elliptical cavity when (a) under ideal condition; (b) $L_{el} = 12$ mm and $\alpha = 0.05°$; (c) $L_{el} = 12$ mm and $\alpha = 0.1°$; (d) $L_{el} = 12$ mm and $\alpha = 0.15°$; (e) $L_{el} = 12$ mm and $\alpha = 0.2°$; and (f) $L_{el} = 19$ mm and $\alpha = 0.2°$.

S_3"AHJ), and the light with a tracking error angle of α (i.e., Light S_2'EFI).

2.1. Tube Receiver. According to Figure 2, the elliptic cavity and the parabolic reflector can be expressed as

$$\frac{(x-f-c)^2}{a^2} + \frac{y^2}{b^2} = 1, \tag{1}$$

$$y^2 = 4fx. \tag{2}$$

The marginal light AC can be expressed as

$$y = k_1(x - f), \tag{3}$$

where

$$k_1 = \frac{-B/2}{(B^2/16) - f}. \tag{4}$$

The angle β between the Light S_3A and Light AC is

$$\beta = \arctan k_1. \tag{5}$$

The marginal light of the tube receiver AH can be expressed as

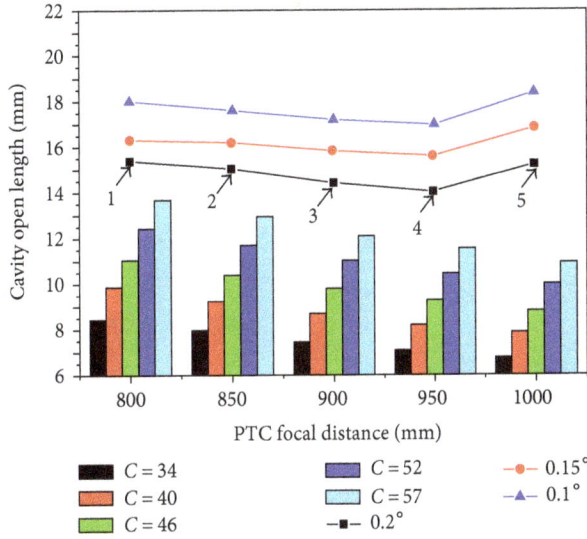

FIGURE 4: Relationship between the PTC focal distance and cavity open length under different concentration ratios and tracking error angles.

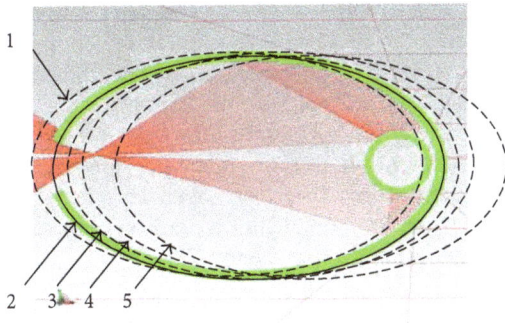

FIGURE 5: Elliptical cavity under different PTC focal lengths when $\alpha = 0.2°$.

$$y = k_2 \left(x - \frac{L_{pr}^2}{16f} \right) - \frac{L_{pr}}{2}, \qquad (6)$$

where

$$k_2 = \tan(\delta + \beta). \qquad (7)$$

Define point H as (m, n) and the values of m and n can be obtained through (1) and (6). The slope of the tangent light for the tube receiver HJ, which is reflected by the cavity inner surface, is

$$k_4 = \frac{k_2 - k_3 + k_3(1 + k_2 k_3)}{1 + k_2 k_3 - (k_2 - k_3)k_3}, \qquad (8)$$

where k_3 is the slope of the tangent light:

$$k_3 = -\frac{b^2(m - f - c)}{a^2 n}. \qquad (9)$$

Correspondingly, in order to reach the absorber tube inside the cavity, the relationship between the absorber tube radius and the light slopes is

$$\frac{|k_4(f + 2c) - k_4 m + n|}{\sqrt{k_4^2 + 1}} \leq r_{\text{tube}}. \qquad (10)$$

2.2. Tracking Error Angle. The Light EC can be expressed as

$$y = k_5(x - f). \qquad (11)$$

According to (1) and the open length of the elliptical cavity, the Light EFI with a tracking error angle of α is

$$y = k_6 \left(x - \sqrt{a^2 \left(1 - \frac{L_{el}^2}{4b^2} \right)} - f - c \right) + \frac{1}{2} L_{el}, \qquad (12)$$

where k_5 and k_6 are the slopes of Light EFI and Light EC, respectively.

Correspondingly, in order to enter the elliptical cavity, the relationship between the tracking error angle and the light slopes is

$$|\alpha| \leq \arctan \frac{k_5 - k_6}{1 + k_5 k_6}. \qquad (13)$$

The concentration ratio in the PTC with the elliptical cavity tube receiver is defined as

$$C = \frac{L_{pr} - 2b}{2\pi r}. \qquad (14)$$

The Monte Carlo method is utilized to simulate the light distribution in the elliptical cavity. Equations (1)–(13) are then converted into the Fortran codes to determine the configuration sizes of the elliptical cavity and the tube receiver.

3. Results and Discussion

3.1. Light Distribution in the Elliptical Cavity. Dimensions of the cavity tube receiver are summarized in Table 1. Light distribution in the elliptical cavity is shown in Figure 3. Figure 3(a) shows the light distribution in the cavity under ideal condition. It is found that the light distribution is longitudinally symmetrical. Taking the above half section as an example, some of the incident sunlight is sheltered by the cavity itself, which causes no light reaching the tube receiver at the left of the tube. The other light is reflected by the parabolic reflector, enters the cavity, and reaches the absorber tube, leading to the increase in the heat flux on the tube. After that, the direct light from the cavity open inlet and the reflected light from the elliptic inner surface are merged, generating a peak on the tube. The other reflected sunlight from the elliptic inner surface then reaches the rest tube surface. No light can reach the range at the right of the tube receiver due to the shelter of the tube itself.

Tracking error usually occurs in practical control of PTCs. Four tracking error angles are then considered in Figures 3(b), 3(c), 3(d), and 3(e), respectively. Due to the

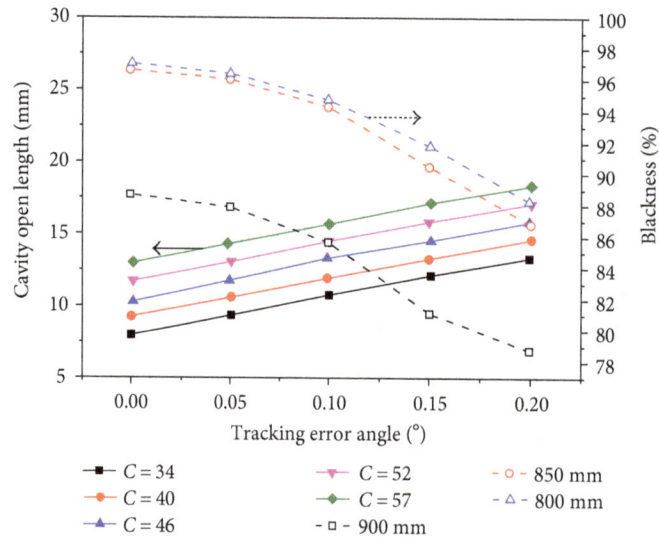

FIGURE 6: The cavity blackness under different tracking error angles, cavity open lengths, concentration ratios, and PTC focal distances.

tracking error, the light distribution on the tube is asymmetrical. Some lights are sheltered by the cavity outer surface. However, the light distribution tendency under ideal condition can also be found in Figures 3(b), 3(c), 3(d), and 3(e), namely, the merged lights directly from the cavity open inlet and reflected from the elliptic inner surface also generate peaks on the above and below half sections of the tubes. On increasing the tracking error angle, more lights are sheltered by the cavity outer surface and less lights reach the below half section of the tube receivers.

The elliptical cavity open length is increased in Figure 3(f). Comparing Figures 3(e) and 3(f), it is found that increasing the cavity open length can allow more lights to enter the cavity. As more sunlight enters, the light distribution on the tube receiver is more close to the ideal condition as shown in Figure 3(a).

3.2. Parameter Analysis.
PTC focal distance and cavity open length are two core parameters of the PTC and the cavity tube receiver. Their relationship is discussed in Figure 4. It is found that the PTC focal distance and the tracking error angle have negative correlations with the cavity open length. But the PTC concentration ratio has positive correlation with the cavity open length. When tracking error occurs, on increasing the PTC distance, the cavity open length first decreases and then increases to maintain the concentration ratio. This can be explained by Figure 5. Taking the condition of $\alpha = 0.2°$ as an example, five elliptical cavity locations are shown in Figure 5. It is found that on increasing the PTC focal distance, the elliptical cavity is moved toward right. As the reflected light is first concentrated toward the focal point and then diverges in the cavity, the cavity open length needs to first decrease then increase to allow the light to enter the cavity.

For the cavity tube receiver, the blackness, which is the percentage of the sunlight on the tube receiver surface to the total incident sunlight, is proposed to evaluate the cavity performance. The cavity open length and cavity blackness

under different tracking error angles are shown in Figure 6. It is found that, to a specific PTC and cavity tube receiver, increasing the tracking error angles would decrease the cavity blackness; and in order to maintain the concentration ratio, the cavity open length needs to be enlarged. Under a specific tracking error angle, increasing the PTC focal distance would decrease the cavity blackness. The reason is that more sunlight is sheltered by the cavity outer surface as indicated in Figures 3(b), 3(c), 3(d), and 3(e). Also, when the PTC concentration ratio increases, for example the PTC width increases, the cavity open length needs to be increased to allow more light to transfer into the cavity as indicated by angle β in Figure 2.

3.3. Elliptical Cavity Optimization.
According to the discussion above, a flat plate reflector is added at the cavity inlet to enhance the cavity tube receiver performance. The geometry of the proposed cavity is shown in Figure 7(a), and the top flat plate reflector can be described as

$$y = -\tan\beta(x + c). \tag{15}$$

The light distribution in the cavity is shown in Figure 7(b). It is found that the sheltered sunlight is reflected by the flat plate, which can finally reach the tube receiver. The cavity darkness of the new cavity receiver under different tracking error angles and PTC focal distances is shown in Figure 8. According to Figures 3, 6, 7(b), and 8, it is found that introducing the flat plate can largely increase the cavity darkness. The cavity darkness decreases when the tracking error angle increases. But introducing the flat plate reflector breaks the monotonic relationships of the cavity darkness under different PTC focal distances. Moving the elliptical cavity along the long axis direction leads to the incident sunlight reflected by the flat plate under different incident angles, which generates different multireflections inside the cavity, and finally leads to the curves in Figure 8.

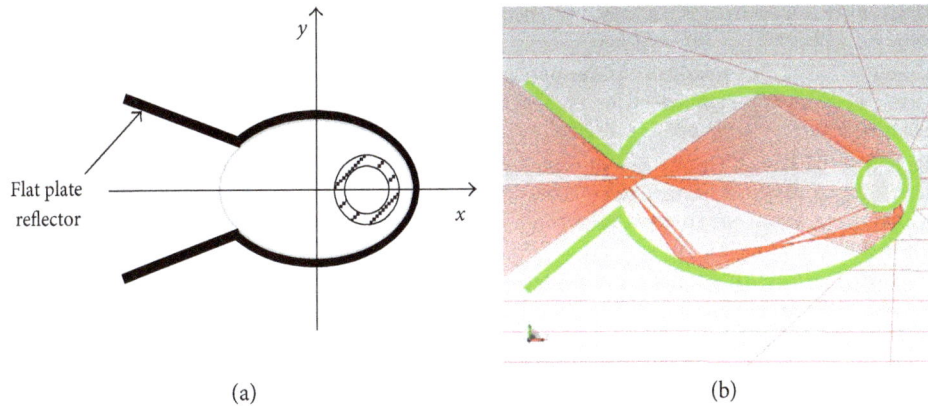

FIGURE 7: (a) Newly proposed elliptical cavity geometry and (b) light distribution in the cavity.

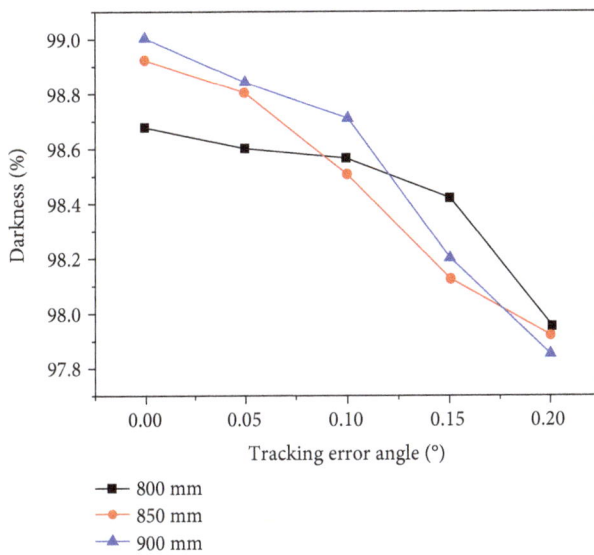

FIGURE 8: Darkness of the newly proposed cavity receiver under different tracking error angles and PTC focal distances.

4. Conclusions

Solar parabolic trough collectors (PTC) are the most widely accepted solar concentration style. The PTC with an elliptical cavity tube receiver has not been well discussed in the literature. In the present study, the light distribution in the cavity under different tracking error angles and PTC configurations are analyzed. A new elliptical cavity geometry is proposed and analyzed. The following conclusions are obtained through this study:

(1) The light distribution on the tube receiver is asymmetrical. On increasing the tracking error angle, more lights are sheltered by the cavity outer surface and less lights reach the below section of the tube receivers.

(2) The PTC focal distance and the tracking error angle have negative correlations with the cavity open length, whereas the PTC concentration ratio has positive correlation with the cavity open length. On increasing the PTC focal distance, the cavity open length needs to first decrease and then increase to maintain the concentration ratio.

(3) Increasing the tracking error angles would decrease the cavity blackness. Increasing the PTC focal distance would decrease the cavity blackness. Introducing a flat plate reflector at the elliptical cavity open inlet leads to multireflections inside the cavity, which can largely increase the cavity darkness.

Conflicts of Interest

The authors declare that they have no conflicts of interest.

Acknowledgments

This research was funded by the National Natural Science Foundation of China (no.: 51506043), the Fundamental Research Funds for the Central Universities (no.: 2014B19714), and the Applied Basic Research Plan of Changzhou City (no.: CJ20160043).

References

[1] V. K. Jebasingh and G. M. J. Herbert, "A review of solar parabolic trough collector," *Renewable and Sustainable Energy Reviews*, vol. 54, pp. 1085–1091, 2016.

[2] J. J. Serrano-Aguilera, L. Valenzuela, and J. Fernandez-Reche, "Inverse Monte Carlo Ray-Tracing method (IMCRT) applied to line-focus reflectors," *Solar Energy*, vol. 124, pp. 184–197, 2016.

[3] Z. D. Cheng, Y. L. He, J. Xiao, Y. B. Tao, and R. J. Xu, "Three-dimensional numerical study of heat transfer characteristics in the receiver tube of parabolic trough solar collector," *International Communications in Heat and Mass Transfer*, vol. 37, no. 7, pp. 782–787, 2010.

[4] Z. D. Cheng, Y. L. He, F. Q. Cui, B. C. Du, Z. J. Zheng, and Y. Xu, "Comparative and sensitive analysis for parabolic trough solar collectors with a detailed Monte Carlo ray-tracing optical model," *Applied Energy*, vol. 115, no. 4, pp. 559–572, 2014.

[5] G. M. Giannuzzi, C. E. Majorana, A. Miliozzi, V. A. Salomoni, and D. Nicolini, "Structural design criteria for steel components

of parabolic-trough solar concentrators," *Journal of Solar Energy Engineering*, vol. 129, no. 4, pp. 382–390, 2007.

[6] H. Liang, S. You, and H. Zhang, "Comparison of three optical models and analysis of geometric parameters for parabolic trough solar collectors," *Energy*, vol. 96, pp. 37–47, 2016.

[7] Z. Cheng, Y. He, and D. Baochun, "Geometric optimization on optical performance of parabolic trough solar collector systems using particle swarm optimization algorithm," *Applied Energy*, vol. 148, pp. 282–293, 2015.

[8] P. M. Zadeh, T. Sokhansefat, A. B. Kasaeian, F. Kowsary, and A. Akbarzadeh, "Hybrid optimization algorithm for thermal analysis in a solar parabolic trough collector based on nanofluid," *Energy*, vol. 82, pp. 857–864, 2015.

[9] G. Kumaresan, R. Sridhar, and R. Velraj, "Performance studies of a solar parabolic trough collector with a thermal energy storage system," *Energy*, vol. 47, no. 1, pp. 395–402, 2012.

[10] V. S. Reddy, S. C. Kaushik, and S. K. Tyagi, "Exergetic analysis and performance evaluation of parabolic trough concentrating solar thermal power plant (PTCSTPP)," *Energy*, vol. 39, no. 1, pp. 258–273, 2012.

[11] F. Chen, M. Liu, P. Zhang, and X. Luo, "Thermal performance of a novel linear cavity absorber for parabolic trough solar concentrator," *Energy Conversion and Management*, vol. 90, pp. 292–299, 2015.

[12] F. Chen, M. Liu, R. H. E. Hassanien et al., "Study on the optical properties of triangular cavity absorber for parabolic trough solar concentrator," *International Journal of Photoenergy*, vol. 2015, Article ID 895946, 9 pages, 2015.

[13] W. Gao, G. Q. Xu, T. T. Li, and H. W. Li, "Modeling and performance evaluation of parabolic trough solar cavity-type receivers," *International Journal of Green Energy*, vol. 12, no. 12, pp. 1263–1271, 2015.

12

Organic Dyes Containing Coplanar Dihexyl-Substituted Dithienosilole Groups for Efficient Dye-Sensitised Solar Cells

Ciaran Lyons,[1] **Neelima Rathi,**[2] **Pratibha Dev,**[1] **Owen Byrne,**[1] **Praveen K. Surolia,**[1] **Pathik Maji,**[1] **J. M. D. MacElroy,**[1] **Aswani Yella,**[3] **Michael Grätzel,**[3] **Edmond Magner,**[2] **Niall J. English,**[1] **and K. Ravindranathan Thampi**[1]

[1]*SFI Strategic Research Cluster in Solar Energy Conversion, UCD School of Chemical and Bioprocess Engineering, University College Dublin, Dublin 4, Ireland*
[2]*SFI Strategic Research Cluster in Solar Energy Conversion, Department of Chemical Sciences and Bernal Institute, University of Limerick, Limerick, Ireland*
[3]*Laboratoire de Photonique et Interfaces (LPI), Ecole Polytechnique Fédérale de Lausanne, 1015 Lausanne, Switzerland*

Correspondence should be addressed to Edmond Magner; edmond.magner@ul.ie, Niall J. English; niall.english@ucd.ie, and K. Ravindranathan Thampi; ravindranathan.thampi@ucd.ie

Academic Editor: Bill Pandit

A chromophore containing a coplanar dihexyl-substituted dithienosilole (CL1) synthesised for use in dye-sensitised solar cells displayed an energy conversion efficiency of 6.90% under AM 1.5 sunlight irradiation. The new sensitiser showed a similar fill factor and open-circuit voltage when compared with N719. Impedance measurements showed that, in the dark, the charge-transfer resistance of a cell using CL1 in the intermediate-frequency region was higher compared to N719 (69.8 versus 41.3 Ω). Under illumination at AM 1.5G-simulated conditions, the charge-transfer resistances were comparable, indicative of similar recombination rates by the oxidised form of the redox couple. The dye showed instability in ethanol solution, but excellent stability when attached to TiO$_2$. Classical molecular dynamics indicated that interactions between ethanol and the dye are likely to reduce the stability of CL1 in solution form. Time-dependent density functional theory studies were performed to ascertain the absorption spectrum of the dye and assess the contribution of various transitions to optical excitation, which showed good agreement with experimental results.

1. Introduction

Dye-sensitised solar cells (DSSCs) [1] have the distinct advantage of being responsive to low and diffuse light levels, as well as to the light incident under acute irradiation angles. This renders the technology particularly suitable for indoor situations and other similar applications where the incident light is confined to a select band in the visible spectrum, which may be the characteristic output of a certain light source. There, it is possible to tune the spectral sensitivity of DSSC as the light-harvesting function of the cell is separated from the semiconductor, unlike in inorganic solar cells, and is taken up by a sensitiser. The standard and most studied sensitisers for making DSSC are ruthenium dyes with

a reported maximum efficiency of 11.9% [2]. Although the DSSC is optimal for room interiors and vertical façade positions, its certified solar-to-electric power conversion efficiency (PCE) under standard air mass 1.5 (AM 1.5) reporting conditions (1000 W/m^2 solar light intensity and 298 K) is still a factor of 2 below that of Si solar cells. Much effort has been expended in developing more efficient dyes [3, 4]. However, Ru is expensive, has limited availability, and has an undesirable environmental impact when used in large amounts. A newer and increasingly emerging area of solar research lies in perovskite solar cells. One key challenge for perovskite commercialisation is stability. The light-sensitive material in these devices dissolves in the presence of water and decomposes at high temperature. Scientists must also address the possibility of lead contamination before these cells can be commercialised at large scale [3].

The development of new organic dyes has been the subject of much interest lately, especially metal-free dyes [5–7]. Organic dyes can be more versatile in their light absorption properties, owing to a larger number of molecular structure variations possible and in principle cheaper for mass scale production. They also possess higher molar extinction coefficients allowing efficient light harvesting using thinner layers when compared to Ru dyes. Porphyrins and phthalocyanines containing inexpensive metal atoms also have attracted considerable attention in the recent years [8, 9]. Recent reports have described a porphyrin-based dye in combination with a Co-complex containing electrolyte, a redox mediator exhibiting higher reduction potentials than that of I_3^-, which has displayed the highest efficiency to date of 13%, as well as over 14% from a DSSC cophotosensitised with an alkoxysilyl-anchor dye and a carboxy-anchor organic dye [5–7]. Another reason for developing organic dyes is the industrial and architectural preferences for specific dye colours such as green, bright red, golden, and blue, when conceiving new buildings with integrated photovoltaics and designing newer interior designs. Indeed, industry is even willing to compromise slightly on solar cell efficiency for desired colour characteristics. This confers to organic dye research a rather compelling new impetus and scope.

In general, organic dyes consist of three segments: a donor, a linker, and an acceptor group. Much research has been performed on altering the nature of each of these groups with the aim of tuning the dyes' absorption spectrum in the visible spectral region. Synthetic chemists increasingly turn to molecular structures based on the current understanding of dyes' structure-property relationships. The introduction of long-chain alkyloxy groups in the dye structure is suggested to generally retard the charge recombination process [10]. Donor-π-bridge-acceptor (D-π-A) sensitisers, endowed with such groups, recently reached open-circuit voltage (V_{oc}) values exceeding 0.8 V when used with Co(II/III)tris(bipyridyl) redox electrolytes. However, the energy conversion efficiencies of these dyes remained in the 6.7 to 9.6% range because of their insufficient solar light harvesting, resulting in low photocurrents [10, 11]. The quest for newer dyes is therefore very challenging. In fact, dyes with narrower spectral bands still may find use in DSSC devices destined for indoor applications, where the available spectrum is usually influenced by indoor lighting systems.

Five-membered heterocycles containing silicon have also attracted attention recently, in particular, silole derivatives with a 2,2'-bithiophene group connected through a silicon atom. Si, an abundant tetravalent element like C and widely used in organic synthesis, is thus a natural choice for organic sensitisers. Dithienosilole derivatives have been used in many optoelectronic devices, such as light emitting diodes [12], whilst dithienosilole-based polymers have been used in polymer solar cells [13]. There have been a few reports on the use of these materials in DSSCs, too. Ko et al. described the synthesis of silole-spaced triarylamine derivatives containing combinations of phenyl and methyl groups attached to silicon, with a PCE of 7.50% [14, 15]. Lin et al. reported

a coplanar diphenyl-substituted dithienosilole dye with an efficiency of 7.60%. The additional O-hexyl donor groups were added to the triphenylamine donor group to provide increased electron donor capabilities [14, 15]. However, these compounds possessed increased conjugation and lowered LUMO (lowest unoccupied molecular orbital) levels. Coplanarity of the π-spacer was obtained by bridging the two thiophene units with a silicon atom, leading to more effective rates of electron transfer. To our knowledge, there have been no reports to date on the effects of adding large alkyl chains to the silicon group.

Here, we report a dithienosilole dye (Scheme 1) containing alkyl chains to help prevent aggregation of the dye sensitiser. The brightly red-coloured dye showed a good power conversion efficiency of 6.90% under 1 sun, in the absence of any additional electron donor groups. Electrochemical impedance spectroscopy (EIS) was used to study the charge-transfer properties of the dye-sensitised solar cell [16–20]. Time-dependent density functional theory (TD-DFT) was used to calculate the absorption spectra and compare it with experimental spectra. On exposure to ethanol for a week in a dye bath, the dye showed a decrease in efficiency to 5.83% and the colour of the dye bath went from the bright fluorescent red to a dull brown. Cell performance and UV-Vis absorption tested over time are shown in Supporting Information available online at https://doi.org/10.1155/2017/7594869. However, DSSCs made using this dye and a standard iodide-tri-iodide electrolyte still showed a stable incident photon-to-current efficiency (IPCE) value even after a month. Computational studies using classical molecular dynamics (MD) were performed to provide an insight into solvent effects and investigate hydrogen bonding between the dye and the ethanol medium.

2. Experimental Section

2.1. Materials. The following materials were used as received: 1-butyl-3-methyl imidazolium iodide (BMII, Merck 4.90187.0100) and guanidinium thiocyanate (GuSCN, Merck 8.20613.0250). Ethanol was purchased from Lennox, Dublin. All other chemicals were purchased and used as received from Sigma-Aldrich. Toluene was distilled under sodium benzophenone and stored under nitrogen. Dimethylformamide was dried using barium oxide. Titanium dioxide paste (Ti-Nanooxide D20), electrolyte (Iodolyte AN-50), Surlyn film (Meltonix 1170-60), and fluorine-doped tin oxide glass slides (TCO30-8, ~8 ohm/square) were purchased from Solaronix SA (Aubonne, Switzerland). A second "light scattering" paste of particle size 150–250 nm was purchased from DyeSol Ltd. (WER 2-0).

Three types of TiO_2 pastes were used to make cells, including the commercially purchased samples. The first was a "transparent" type containing 20 nm particles of TiO_2 prepared from Evonik P25 powder using a standard fabrication procedure [21]. Ethyl cellulose (Fluka, #46080) and anhydrous terpineol (Fluka, #46070) were used as received to make a paste. A second paste of particle size 150–250 nm and purchased from DyeSol Ltd. (WER 2-0) was used to form a light scattering layer. Both pastes were screen printed with a

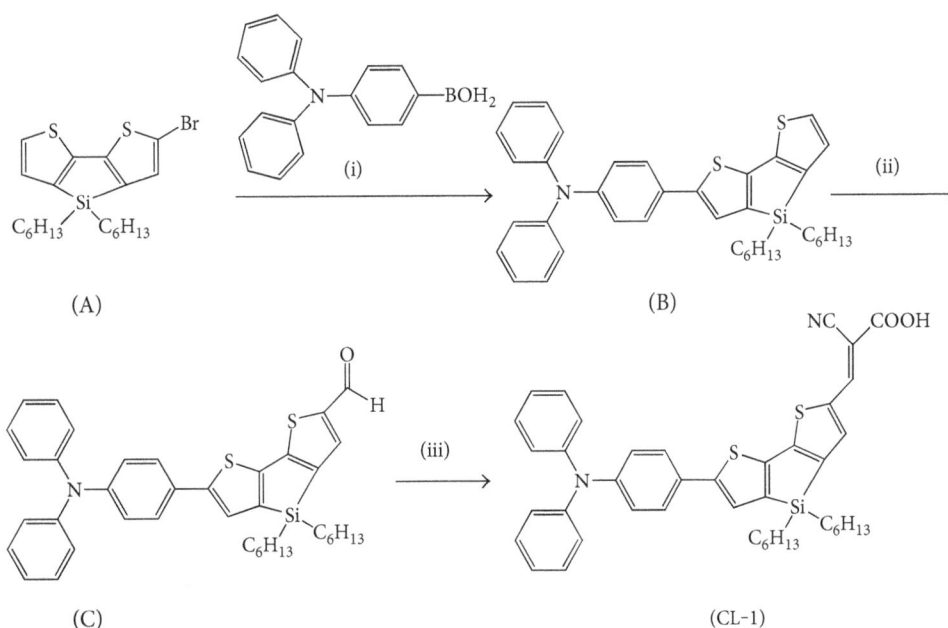

SCHEME 1: Reagents and conditions: (i) 4-(diphenylamino) phenylboronic acid, palladium(II) acetate, cesium carbonate, toluene, reflux, 24 h, (ii) POCl$_3$, DMF, 90°C, 24 h, and (iii) 2-cyanoacetic acid, piperidine, acetonitrile, reflux, 4 h.

90T mesh to yield electrodes of approximately 15 μm thickness (10 μm transparent layer and a 5 μm scattering layer). The standard liquid electrolyte used was labelled E1 [22] and was comprised of 0.03 M iodine, 0.1 M guanidinium thiocyanate, 0.5 M 4-tert-butylpyridine, and 0.6 M BMII in an acetonitrile : valeronitrile solvent mixture (85 : 15 by volume). The Solaronix paste was used without light scattering layers in certain experiments, where cell efficiencies were not considered the primary objective.

2.2. Dye Characterisation. Absorption and emission spectra of the dye were obtained, both in the dissolved form as well as in the chemisorbed state on TiO$_2$ surface. For experiments with TiO$_2$, a mixture was prepared using 100 μl Ti-Nanooxide D20, 35 μl polyethylene glycol (PEG), and 100 μl of 1% Tween-80 in water, which were mixed together to obtain a gel. The gel was spread onto an FTO glass slide and left to dry for 30 min after which the electrode was calcined at 450°C for 30 minutes. The adsorption of CL1 dye was performed by immersing the TiO$_2$ electrodes in CL1 (0.57 mM) solution in ethanol overnight (18 h). The electrodes were then rinsed with ethanol and dried in air. Solution-based absorption and emission spectra were recorded on a Varian Cary 300 UV-Visible and a Cary eclipse fluorescence spectrophotometer, respectively, using quartz cuvettes. The absorption spectrum of the anchored dye was obtained on a UV-1800 Shimadzu spectrophotometer.

A Perkin Elmer Spectrum 100 FT-IR spectrometer was used to record ATR-FTIR spectra at a resolution of 2 cm^{-1}. The spectra reported represent averages of 100 scans. A CHI800 potentiostat was used for cyclic-voltammetry measurements. Dye-immobilised TiO$_2$, Pt wire, and Ag/AgCl/ KCl were used as the working, counter, and reference electrodes, respectively. Solutions were bubbled with N$_2$ gas and kept under an N$_2$ atmosphere during experiments.

Electrochemical impedance spectroscopy was performed using an impedance analyser (Solartron analytical, 1260) connected to a potentiostat (Solartron Analytical, 1287). EIS spectra were measured under illumination and at applied bias voltage equivalent to the open-circuit voltage (V_{oc}) of the device in the dark over the frequency range of 0.1 to 10^5 Hz using a 10 mV amplitude AC signal. Impedance spectra were fitted to an equivalent circuit model using ZView software (Scribner Associates Inc.).

^{1}H and ^{13}C NMR spectra were recorded using Bruker 300 and 400 MHz instruments using the residual signals $\delta = 7.26$ ppm and 77.0 ppm for CDCl$_3$ and $\delta = 2.50$ ppm and 39.4 ppm for [D$_6$]-DMSO.

2.3. General Synthetic Procedure. The synthetic procedure for the preparation of compound A has been described previously [23]. The synthesis of B was achieved through Suzuki coupling between the bromo-silole derivative and triphenylamine-boronic acid. Aldehyde addition was performed using Vilsmeier formylation using phosphorus oxychloride to yield C. Claisen condensation between cyano-2-acetic and the aldehyde in the presence of piperidine yielded CL-1 as a purple solid with a yield of 41%.

2.4. 5-[N,N-Bis(phenylamino)phenyl]-3,3'-dihexyllsilylene-2,2'-bithiophene (B). Compound A (0.59 g, 1.34 mmol), 4-(diphenylamino) phenylboronic acid (0.469 g, 1.61 mmol), palladium(II) acetate (0.030 g, 0.134 mmol), and cesium carbonate (2.638 g, 8.09 mmol) were dissolved in toluene (75 ml). The reaction mixture was refluxed overnight. The organic layer was extracted using CH$_2$Cl$_2$ and washed with H$_2$O and then dried over Na$_2$SO$_4$. The solvent was removed by rotary evaporation. The crude product obtained was then purified using column chromatography (petroleum ether/CH$_2$Cl$_2$ = 6/1) to yield compound B as a yellow wax

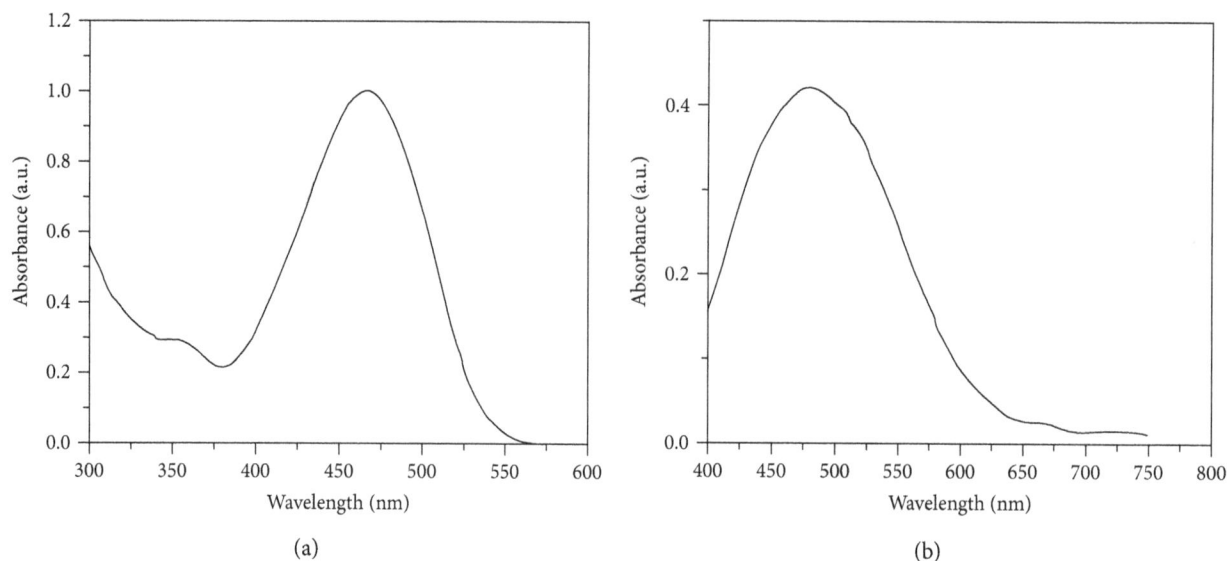

FIGURE 1: Absorption spectrum of CL1 dye showing an absorption band at 468 nm (a) in ethanol (0.057 mM) and (b) adsorbed on a TiO$_2$ film showing a red shift indicative of interactions between the dye molecule and the semiconductor.

with 61% yield. ^{1}H NMR (CDCl$_3$, 400 MHz): δ ppm 7.46 (2H, $J = 12$ Hz, d), 7.29–6.97 (15H, m), 1.45–1.16 (16H, m), 0.95–0.87 (m, 4H), and 0.87–0.79 (6H, m); ^{13}C NMR ([D$_6$]-DMSO, 100 MHz): δ ppm 149.42, 147.61, 147.51, 146.89, 144.78, 143.12, 141.05, 129.63, 129.27, 129.17, 128.90, 126.38, 124.95, 124.85, 124.39, 124.14, 123.90, 122.98, 122.64, 32.85, 31.42, 24.16, 22.55, 14.07, and 11.91.

2.5. 5-[N,N-Bis(phenylamino)phenyl]-5′-formyl-3,3′-hexylsilylene-2,2′-bithiophene (C).
Compound B (0.50 g, 0.825 mmol) was dissolved in DMF (50.00 ml). At 0°C, POCl$_3$ (0.184 ml, 1.956 mmol) was added to this solution and the mixture was stirred at 90°C overnight. The organic layer was extracted using CH$_2$Cl$_2$ and washed with H$_2$O and before drying over Na$_2$SO$_4$. The solvent was removed by rotary evaporation. The crude product was then purified using column chromatography (petroleum ether/CH$_2$Cl$_2$ = 4/1) to yield compound C as an orange wax in 74% yield. ^{1}H NMR (CDCl$_3$, 400 MHz): δ ppm 9.85 (1H, s), 7.69 (1H, s), 7.47 (2H, $J = 8$ Hz, d), 7.30–7.21 (5H, m), 7.14–7.03 (8H, m), 1.44–1.19 (16H, m), 0.99–0.91 (4H, m), and 0.88–0.79 (6H, m); ^{13}C NMR (CDCl$_3$, 100 MHz): δ ppm 181.41, 158.06, 148.15, 147.04, 146.77, 146.29, 145.03, 143.20, 140.68, 138.49, 132.94, 128.34, 126.75, 125.72, 124.00, 123.70, 122.34, 122.31, 119.94, 31.77, 30.34, 23.06, 21.49, 13.02, and 10.70.

2.6. 3-[5-[N,N-Bis(phenylamino)phenyl]-3,3′-dihexylsilylene-2,2′-bithiophene-5′-yl]-2-cyanoacrylic Acid (CL-1).
At first, compound C (0.05 g, 0.883 mmol), 2-cyanoacetic acid (0.02 g, 0.258 mmol), and a drop of piperidine were dissolved in acetonitrile (10 ml). This mixture was then refluxed for 2.5 h. After removal of the solvent, the crude product was purified using column chromatography (CH$_2$Cl$_2$/MeOH = 30/1) to yield Cl-1 as a purple solid with 41% yield. ^{1}H NMR (CDCl$_3$, 400 MHz): δ ppm 8.05 (1H, s), 7.67 (1H, s), 7.59 (2H, $J = 12$ Hz, d), 7.52 (1H, s), 7.38–7.28

(4H, m), 7.12–6.94 (8H, m). 1.39–1.10 (16H, m), 0.99–0.89 (4H, m), and 0.84–0.73 (6H, m); HRMS (TOF-MS-ESI) m/z: 699.2535 [M+]; calculated for C$_{42}$H$_{43}$N$_2$O$_2$S$_2$Si [M+]: 699.2535.

2.7. Cell Fabrication and Characterisation.
DSSCs were manufactured as described previously [24]. Screen printing was used to deposit layers of TiO$_2$ on a fluorine-doped tin oxide (FTO) conducting transparent glass substrate. In all cases, a nonporous, dense blocking underlayer of TiO$_2$ was deposited first on the FTO substrate via TiCl$_4$ treatment [21] in order to reduce charge recombination, prior to screen printing. The TiO$_2$ paste was then printed on the TiCl$_4$ treated glass using a Tiflex Ltd., France, screen printer and involved several cycles. After deposition of each layer, the films were kept in an ethanol saturated chamber for 6 min followed by drying at 125°C for 6 min, whilst the final sintering involved gradual heating in an oven at 325°C (5 min), 375°C (5 min), 450°C (15 min), and 500°C (30 min). After sintering, a layer of TiCl$_4$ was deposited followed by sintering at 500°C for 30 min. The TiO$_2$ active area was 0.283 cm^2 (6 mm diameter circular spot). The sintered electrodes were placed in a dye bath of N719 (benchmark dye as supplied by Dyesol Ltd., without further purification) dissolved in an acetonitrile : tert-butyl alcohol: THF mixture (vol 4.5 : 4.5 : 1) or Si dye (CL-1) in ethanol at a concentration of 200 μM for 16–20 hours. The counter electrode was prepared with a thin film of Pt catalyst deposited via a drop of H$_2$PtCl$_6$ solution (2 mg Pt content in 1 ml ethanol) and heat treated at 400°C for 15 minutes. The dye-coated TiO$_2$ electrode and Pt-coated counter electrode were sandwiched together and sealed using a Bynel polymer gasket (50 μm thick). Electrolyte was filled into the space between the two electrodes through a hole in the counter electrode via the vacuum back-filling method. The back hole was then heat sealed with a thin piece (0.1 mm thick) of glass, again with Bynel.

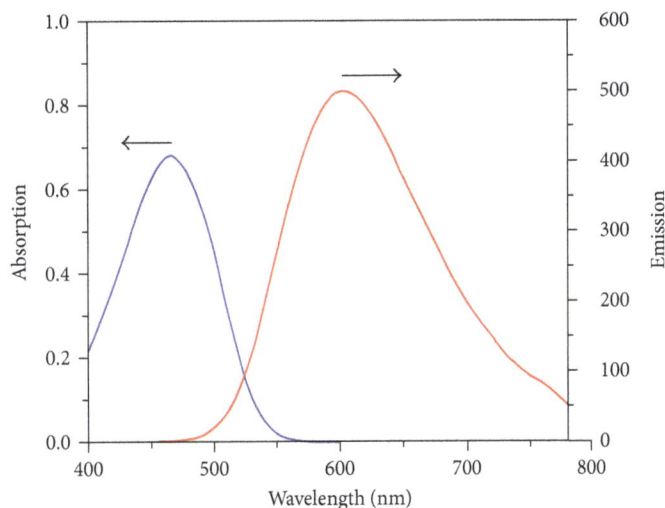

FIGURE 2: Absorption (blue line) and emission (red line) spectra of CL1 in ethanol in arbitrary units showing values of 468 nm and 618 nm, respectively.

2.8. Electrooptical Characterisation.

2.8. Electrooptical Characterisation. Current/voltage (I-V) curves, open-circuit voltage (V_{oc}), short-circuit current density (J_{sc}), and the fill factor (FF) were measured using a Newport 91195A-1000 solar simulator and Newport 69920 Arc Lamp Power Supply. A Newport 81088A air mass filter was placed before the output of the solar simulator to simulate the AM 1.5 spectrum. I-V measurements were recorded with a GAMRY Instruments potentiostat. Spectral response and incident photon-to-current efficiency (IPCE) measurements were made with using a solar cell spectral response/QE/IPCE measurement system (Solar Cell Scan 100 (SCS100)—Gilden Photonics Ltd.).

2.9. Time-Dependent Density Functional Theory. The Cl-1 dye was modelled using Discovery Studio Visualizer package (Accelrys, San Diego, CA). Ground-state structural optimisation and excited-state calculations for the isolated dye and the dye in complex with a titania nanoparticle were then performed using Gaussian'09 using the linear response approach [25]. The two hexyl groups in the side chains were replaced by the methyl groups to reduce computational time. This should not affect the results, as these side chains do not participate in the photoexcitation of the dye. The transitions of interest are predominantly charge-transfer in character. Such excitations are better described by conventional hybrid and range-separated *xc*-functionals [26, 27]. For the CL1 dye, two approximations were used: (a) Becke-3 Lee-Yang-Parr (B3LYP) hybrid functional [28–30] and (b) Coulomb-attenuated functional, CAM-B3LYP [31]. The standard 6−31G* basis set, which provides a very good compromise between the accuracy and computational time, was also used. The calculations were performed in two stages:

(1) Structural optimisation was carried out for the isolated dye using both functionals. After geometry optimisation, TD-DFT was used to obtain the UV/Vis absorption spectra of the dye within the two functionals. The theoretical results were then compared

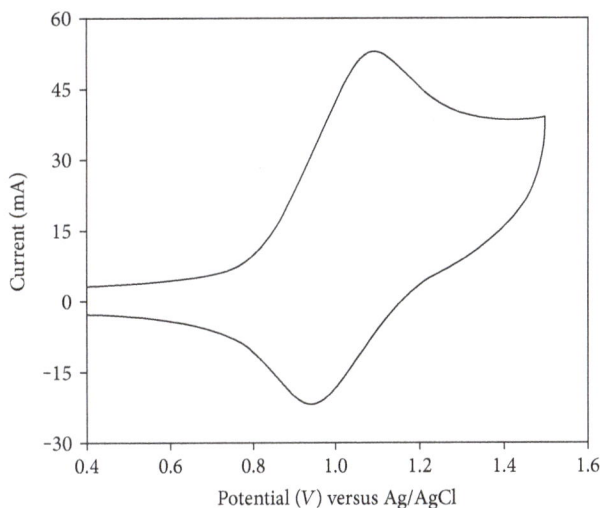

FIGURE 3: Cyclic voltammogram of adsorbed CL1 dye at a scan rate of 100 mV/s, in 0.1 M TBAPF$_6$ in acetonitrile showing quasi-reversible behaviour; the oxidation and reduction peaks can be attributed to the triphenylamine moieties and cyanoacrylic acid, respectively.

with experimental data to choose the most appropriate functional.

(2) The ground- and excited-state properties of the dye-titania complex were characterised using the optimal functional. To compare theoretical results with their experimental counterparts, solvent (ethanol) effects were added using the polarisable continuum solvation model (C-PCM) [32] in all calculations. The complex was created by attaching the dye to a 114 atom-containing nanoparticle TiO$_2$ cut from an anatase (101) surface. This cluster geometry has been employed in earlier studies and demonstrated that the lowest excitation energy was in agreement with the experimental semiconductor bandgap [33–37].

TABLE 1: Experimental data of electrochemical and spectroscopic properties of the CL1 dye.

$\lambda_{abs,max}$ (nm)	ε (M^{-1} cm^{-1})	$\lambda_{em,max}$ (nm)	E_{ox} (V) (versus NHE)	[a]E_{0-0} (eV)	E_{LUMO} (V) (versus NHE)
468	30,000	618	1.28	2.29	−1.01

[a]The LUMO (lowest unoccupied molecular orbital) level of the dye was calculated using equation $E_{ox} - E_{0-0}$ [47, 48], where E_{0-0} is the zeroth–zeroth transition energy of the dye estimated from the intersection between the absorption and emission spectra of the dye (Figure 2).

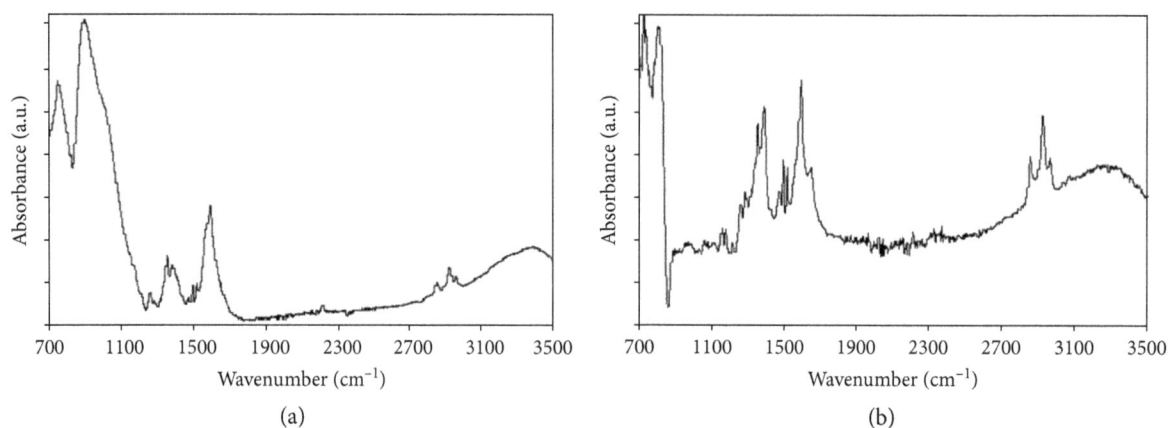

FIGURE 4: (a) ATR-FTIR spectrum of CL1 dye showing sharp phenyl and thiophene ring modes at 1517, 1491, 1375, and 1256 cm^{-1}, whilst the C–N stretching mode of triarylamine appeared at 1348 cm^{-1}. (b) ATR-FTIR spectrum of CL1 dye adsorbed on TiO$_2$ showing the v_{asym}(COO$^-$) and v_{sym}(COO$^-$) stretching modes of carboxylate linker groups at 1590 and 1382 cm^{-1}, respectively, whilst the C≡N stretch frequency was unchanged at 2209 cm^{-1}.

Geometrical optimisation of the isolated titania cluster was performed under CAM-B3LYP. The ground-state structure of the nanoparticle was then used to create the CL1 dye-titania complex. Once the complexed structure was also optimised, TD-DFT calculations were performed to obtain the UV/Vis absorption spectra for the dye-titania complex.

2.10. Molecular Dynamics. Classical MD simulations were used to investigate and quantify hydrogen-bond lifetimes with ethanol as a solvent. Given the small-molecule nature of the dye, the MMFF94 force field was utilized [38]. All MD calculations were performed using the MOE software package [39]. Nonbonded interactions were treated using a twin-range method [40], with short and long cut-off radii of 10 and 12 Å, respectively, with reaction field electrostatics [41] with a cut-off radius of 15 Å. The dielectric constant was set at the experimental value of 24.3 [42]. Following gas-phase geometry optimisation, the dye was placed in the centre of a rectangular periodic box surrounded by 845 ethanol molecules under periodic boundary conditions (PBC) [43], relaxed via MD in the liquid state at 298 K and 1 atm. Prior to MD under PBC, the heavy atoms in the simulation box were fixed and the system was relaxed by energy minimization. This was followed by "heating" of the system to 300 K in 25 K increments by MD in stages of 10 ps duration in the NVT ensemble, using velocity assignments from the Maxwell-Boltzmann distribution at the start of each step. A production simulation was then carried out in the NPT ensemble for 100 ps, and bond lengths were constrained with a relative tolerance of 10^{-8} [43]. A time step of 1 fs was used.

The period of the thermal and barostat reservoirs [43] was set to 1 and 5 ps, to allow for relatively weak coupling.

3. Results and Discussion

3.1. Absorption and Emission Spectra. The absorption spectrum of the CL1 dye in ethanol (Figure 1) showed a broad absorption band at 468 nm with a molar extinction coefficient (ε) of 30×10^3 M^{-1} cm^{-1}, which arises from $\pi - \pi^*$ charge-transfer transition [15, 44, 45]. The absorption spectra of adsorbed dyes on TiO$_2$ displayed a slight red shift in the absorption bands, indicative of interactions between the dye molecule and the semiconductor. The dye in solution exhibits considerable emission characteristics, when excited with light of suitable wavelengths. The fluorescence spectrum of the CL1 dye in ethanol shows an emission peak centred at 618 nm upon excitation at 468 nm. The absorption spectrum of CL1 dye in ethanol, shown in Figure 1(a), is presented together with its fluorescence spectrum in Figure 2, which shows a stoke shift of 150 nm.

3.2. Cyclic Voltammetry. Cyclic voltammetry was used to measure the ground-state oxidation potential (E_{ox}) of the dye. The cyclic voltammogram (Figure 3) of the dye showed quasi-reversible behaviour; the oxidation and reduction peaks can be attributed to the triphenylamine moieties and cyanoacrylic acid, respectively [15]. The value of E_{ox} of the dye, 1.28 V (versus NHE), is more positive (Table 1) than the redox potential (0.4 V versus NHE) [46] of the iodide/triiodide couple. Thus, the oxidised dye can be regenerated by I$^-$ in the electrolyte enabling efficient charge separation. The LUMO energy level of the dye, −1.01 V

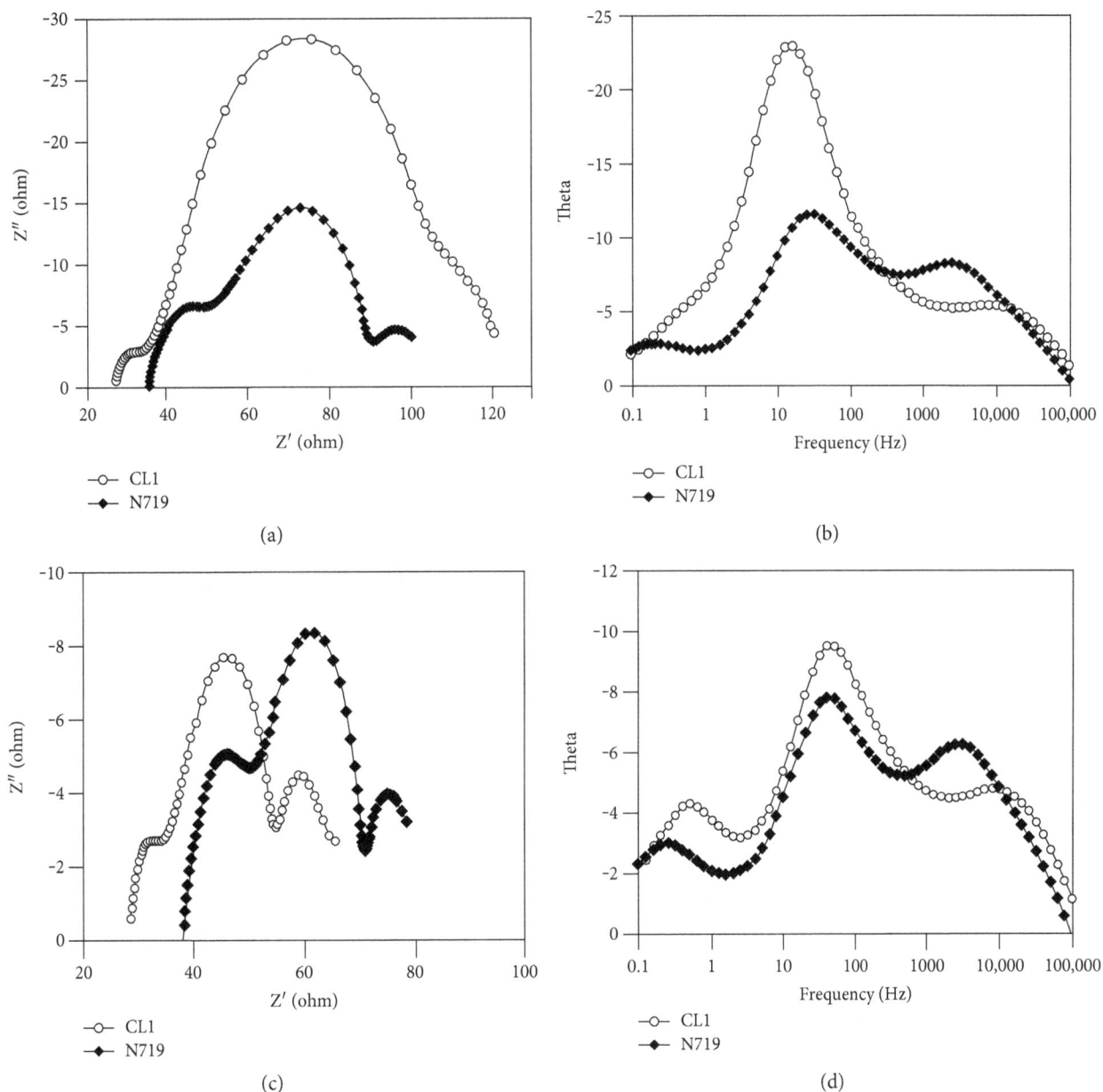

FIGURE 5: Nyquist ((a) and (c)) and Bode phase ((b) and (d)) plots of N719 and CL1 dye sensitised DSSCs in the dark ((a) and (b)) at −0.67 V and under 1 sun illumination ((c) and (d)) at open-circuit voltages.

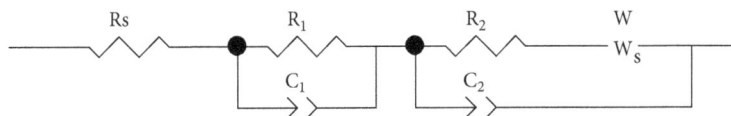

FIGURE 6: Equivalent circuit used to fit EIS data.

(versus NHE), is more negative than the conduction band edge (−0.5 V versus NHE) [46] of TiO_2 (Table 1), thus providing sufficient driving force to inject an electron from the excited dye to the conduction band of TiO_2.

3.3. ATR-FTIR Spectra. ATR-FTIR spectra of the dye (Figure 4(a)) showed sharp phenyl and thiophene ring modes [42, 49, 50] at 1517, 1491, 1375, and 1256 cm^{-1}, whilst the C–N stretching mode of triarylamine appeared at 1348 cm^{-1}. Stretching modes assigned to v (C=O), v (Si–C), and v (C–H) stretch were observed at 1588, 886, and 2957 cm^{-1}, respectively. The ATR-FTIR spectrum (Figure 4(b)) of adsorbed CL1 dye shows the $v_{asym}(COO^-)$ and $v_{sym}(COO^-)$ stretching modes of carboxylate linker

TABLE 2: Electronic device parameters obtained by fitting the EIS data to the equivalent circuit model (Figure 6).

DSC	Rs (Ω)	R_1 (Ω)	R_2 (Ω)	ω_{max} (Hz)
AM 1.5 sunlight at an intensity of 1 sun (1000 W/m^2)				
CL1	28.7	6.8	19.7	31.6
N719	38.1	10.7	22.6	39.8
In dark				
CL1	27.4	8.0	69.8	15.8
N719	35.7	12.9	41.3	19.9

TABLE 3: Photovoltaic parameters of DSSC cells fabricated using CL-1 and N719 dyes recorded under simulated AM 1.5 sunlight at an intensity of 1 sun (1000 W/m^2).

Dye	V_{oc} (V)	J_{sc} (mA/cm^2)	FF	η (%)
Si dye	0.742	14.4	65	6.90
N719	0.745	17.4	62	8.05

groups at 1590 and 1382 cm^{-1}, respectively [42, 49, 51]. The C≡N stretch frequency was unchanged at 2209 cm^{-1}. Possible binding modes between the carboxylate group and TiO$_2$ can be either bidentate bridging or chelation [52]. As the chelation mode is known to be unstable [53], the bridging bidentate mode is more likely to occur.

3.4. Impedance Analysis. The electrochemical impedance spectra (Figure 5) were fitted to an equivalent circuit model [54–56], containing a constant phase element (CPE) and resistance (R) (Figure 6). Rs is the resistance at the FTO/TiO$_2$ interface and R$_1$ and C$_1$ are the charge-transfer resistance and capacitance at the electrolyte/Pt-FTO interface. Variations in Rs arise from the electrical contacts and wiring of the device [57, 58]. R$_2$ and C$_2$ are the charge recombination resistance and capacitance at the TiO$_2$/dye/electrolyte interface. The parameters obtained upon fitting the spectra to the equivalent circuit (Figure 6) are shown in Table 2.

The impedance spectra of N719 and the CL1-DSSC measured at V_{oc} under illumination and in the dark at an applied bias voltage equivalent to V_{oc} of the cell are shown in Figure 6. Three semi-circles were observed in the Nyquist plots over the frequency range of 0.1–10^5 Hz. The smaller semicircle in the high-frequency region is associated with charge transfer at the electrolyte/Pt-FTO interface; the larger semicircle in the middle-frequency region, to the electron transport and recombination mechanism at the TiO$_2$/dye/electrolyte interface, whilst the low-frequency region semicircle may be attributed to diffusion of I$_3^-$ in the electrolyte. The impedance values of the TiO$_2$/dye/electrolyte interface, represented by the semicircles at the intermediate-frequency region in the Nyquist plots, are much smaller under illumination than in the dark. Under illumination, and subject to solubility conditions [59], I$_3^-$ is formed at the TiO$_2$/electrolyte interface by dye regeneration, whilst in the dark, I$_3^-$ is produced at the counter electrode. This indicates that recombination of CB electrons is accelerated under illumination hence decreasing the electron lifetime in the TiO$_2$ film. In the dark, the charge-transfer resistance of CL1 DSC (69.8 Ω), in the intermediate frequency region, was higher compared to N719 (41.3 Ω); under illumination, the charge-transfer resistances were similar (6.8 and 10.7 Ω, for CL1 and N719, respectively), indicative of similar recombination rates for both dyes.

EIS Bode phase plots also exhibited two characteristic peaks under 1 sun and in the dark; the peak at higher frequency can be attributed to charge transfer at the counter electrode, whilst the peak in the middle-frequency region is associated with electron transfer at the TiO$_2$/dye/electrolyte interface. The characteristic middle-frequency peak of the Bode plot can be used to provide a measure of the charge recombination rate [60]. For simple circuits, the reciprocal of this frequency peak is a direct measure of the electron lifetime in TiO$_2$; however, the relationship is more complex for the Randles circuit described in Figure 6. Under illumination, the midfrequency peak of the CL1-based DSC (Figure 5(d)) is slightly shifted to higher frequency compared to that of the N719 DSC, indicative of a shorter electron lifetime. This suggests a slightly lower rate of electron injection and lower charge collection efficiency for the CL1-based cell compared to the N719 cell, hence leading to a lower overall value of J_{sc}.

3.5. Cell Efficiency Analysis. A PCE of 6.90% was achieved for the CL1-containing DSSC. The fill factor and open-circuit voltages (V_{oc}) were similar to those of the N719 dye (cf. Table 3). The N719 dye achieved an efficiency of 8.05%, which is due to a higher cell current of 17.4 mA/cm^2 compared to the CL1 value of 14.4 mA/cm^2. The difference in cell current can be rationalised in terms of the spectral response of each dye, as exhibited in the IPCE spectra in Figure 7. The N719 dye displayed an increased absorbance over the spectral region (500–750 nm), resulting in an increase in the overall photocurrent yield. Note that the Si dye has a higher absorbance than N719 in the 400–470 nm region, which may be advantageous depending on the desired DSSC application and lighting conditions, for example, under indoor lighting conditions. It is equally useful for tandem cell configurations, where the light management between the top and bottom cell need to be optimised for obtaining devices with maximum open-circuit voltage and photo-current density.

The possibility of CL1 showing considerable molecular aggregation on the surface of TiO$_2$ cannot be discounted. In fact, the dye-sensitised TiO$_2$ electrode surface shows a strong and attractive red colour, suggesting the possibility of molecular aggregation in a qualitative manner. However, with the limited architectural details available on the packing of the new dye in an adsorbed state over TiO$_2$ at this stage, no conclusive evidence is available to determine the magnitude of molecular aggregation and its dependence on the TiO$_2$-dyeing process itself. Whether a neat monolayer could be obtained on TiO$_2$ surface without any molecular aggregation is also not clear as of now.

3.6. Comparison of CL-1 with Other Silicon-Based Dyes. Lin et al. synthesised a series of dithienosilole dyes (TPCADTS

(a)

(b)

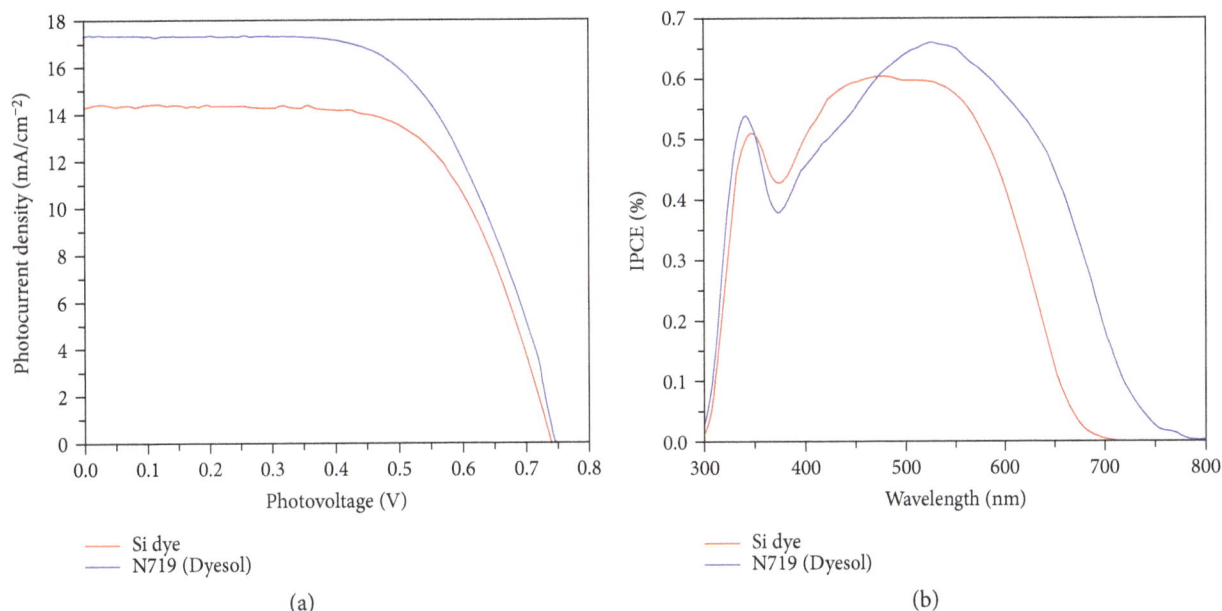

FIGURE 7: (a) J-V characteristics of N719 and Si dye (CL-1)-sensitised DSSC under standard AM 1.5 testing conditions. Si dye showed a V_{oc}, J_{sc}, FF, and η (%) of 0.742, 14.4, 65, and 6.90, respectively. N719 showed a V_{oc}, J_{sc}, FF, and η (%) of 0.745, 17.4, 62, and 8.05, respectively. (b) IPCE spectra of these cells displayed an increased absorbance over the spectral region (500–750 nm) for N719, but a relatively lower absorbance in the 400–470 nm region compared with Si dye.

and TP6CADTS), which are structurally similar to CL-1 with comparable PCE values [15]. The structures are shown in Figure 8. TPCADTS contained diphenyl-substituted dithienosilole and showed a very similar efficiency of 6.65%. When extra electron donors were added by addition of O-hexyl groups, the efficiency increased to 7.60%. When comparing the more similar TPCADTS with CL-1, it is evident where the improvements lie. The largest improvement lay in the value of J_{sc} for CL-1 which increased from 12.7 to 14.4 mA/cm^2 on addition of the dihexyl-substituted dithienosilole core. The values obtained for V_{oc} remained very similar. It has been reported that on addition of alkyl chains, V_{oc} can increase; but, both the diphenyl and dihexyl cores had similar effects on the cell performance.

Ko et al. synthesised a series of silole-spaced triarylamine derivatives with an efficiency ranging from 6.73% to 7.50% [14]. These dyes also had similar structural features to CL-1. The one major difference which leads to the increase in efficiency in comparison to the CL-1 was that the added bulky electron donation was present in the form of a 3-5′-N,N-bis (9,9 dimethylfluorene-2-yl)-phenyl unit (DTS). This dye was named 7b. Diphenyl- rather than dihexyl-substituted dithienosilole was used, and, when combined with the DTS electron-donating group, it had a significantly greater efficiency of 7.50%. Surprisingly enough, the J_{sc} value decreased to 13.9 mA/cm^2 compared to 14.4 mA/cm^2 for CL-1. No significant difference was observed in the V_{oc}. The increased efficiency of 7b likely arises from the superior FF which increased by 9% compared with CL-1. When a dimethyl group (7a) was used instead of a diphenyl (7b), the efficiency value decreased to 6.73% which is lower than that of CL1. This suggests that lower aggregation levels were found with CL-1.

3.7. Molecular Modelling

3.7.1. Hydrogen-Bonding Dynamics with Ethanol.
From the NPT production simulations of classical MD, the persistence times of hydrogen bonding events from hydrogen atoms in ethanol to the carboxylic acid's oxygen atom in the dye were measured. It was found that the hydrogen bonds were short lived and transient, occurring some 4-5% of the time overall on subpicosecond timescales; the average persistence time, between breakage and possible reformation, was 0.072 ± 0.023 ps. As ethanol molecules underwent self-diffusion in the solvation layer of the dye, and rotational motion therein, the identities of the donors to the carboxylic group oxygen atom change. This underlines the key role of the interactions of the dye with ethanol and also on dye-solvent hydrogen bonding. These frequent dye-solvent hydrogen-bond rearrangements serve to rationalise the potential chemical transformation observed experimentally when placed in ethanol for several days or more.

3.7.2. Spectra of Isolated Systems.
Turning to the use of TD-DFT to determine absorption spectra and the underlying transitions, the UV/Vis absorption spectra are provided in Figure 9 for the isolated dye (using the continuum solvation model for ethanol, as discussed earlier).

The experimental UV/Vis absorption spectrum has a maximum absorbance (λ_{max}) at 468.0 nm. Figure 9 shows the theoretical UV/Vis absorption spectra of the isolated dye as calculated within the B3LYP and the CAM-B3LYP approximations. Using B3LYP, a charge-transfer excitation energy of 618.4 nm was obtained. On the other hand, the CAM-B3LYP calculation yielded a more accurate result of

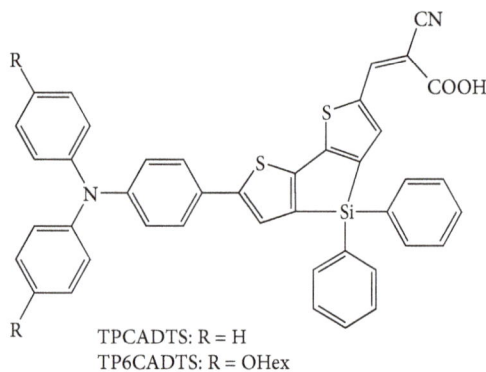

FIGURE 8: Molecular structures of TPCADTS and TP6CADTS.

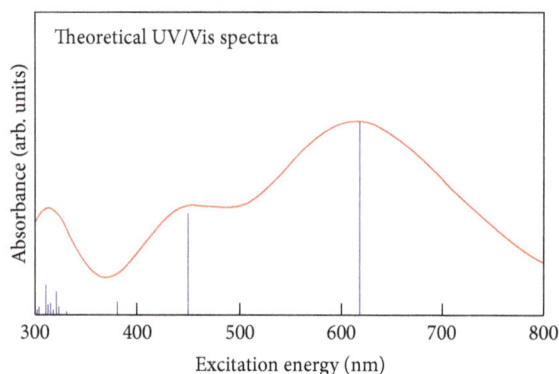

(a) B3LYP

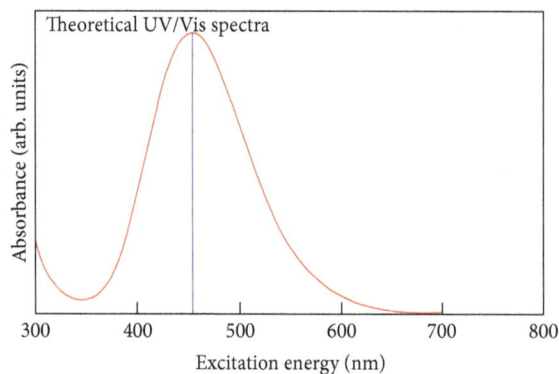

(b) CAM-B3LYP

FIGURE 9: Theoretical UV/Vis spectra of the CL1-dye calculated within the B3LYP approximations to xc-functional showing a charge-transfer excitation energy of 618.4 nm (a) and CAM-B3LYP approximations showing a more accurate result of 453.6 nm (b).

TABLE 4: Relevant calculation details for the CL1-dye: wavelength at maximum absorbance (λ_{max}^{Th}) and HOMO (H), LUMO (L) positions and H-L gap ($\Delta_{HL} = E_{LUMO} - E_{HOMO}$). For titania, we have reported the lowest transition valence band maximum (VBM) and the conduction band minimum (CBM) as well as the orbital-energy bandgap (E_g).

System	Approx.	λ_{max}^{Th} eV (nm)	H (eV)	L (eV)	Δ_{HL} (eV)
CL1	B3LYP	2.0 (618.4)	−5.0	−2.8	2.3
	CAM-B3LYP	2.7 (453.6)	−6.2	−1.7	4.6

System	Approx.	1st trans. eV (nm)	VBM (eV)	CBM (eV)	E_g (eV)
TiO$_2$	B3LYP	3.8 (326.3)	−7.5	−3.0	4.5
	CAM-B3LYP	4.4 (279.9)	−9.2	−1.7	7.6

TABLE 5: Transition corresponding to λ_{max}^{Th} for the CL1 dye within B3LYP and CAM-B3LYP: T.E.: transition energy ($=\lambda_{max}^{Th}$) in eV; O.S.: oscillator strength; and Coeff.: the magnitude of the configuration interaction singles coefficients. The most prominent transitions (indicated by the magnitude of the configuration interaction singles coefficients) are highlighted in bold.

Approx.	T.E. (O.S.) (eV)	Involved orbitals (Coeff.)
(1) B3LYP	2.0 (1.17)	**H → L (0.71)**
(2) CAM-B3LYP	2.7 (1.65)	**H → L (0.57)**, H-1 → L (0.37) H → L + 1 (0.14)

transition upon photoexcitation involves several promotions between the occupied and empty orbital pairs. However, the most prominent promotion (indicated by the magnitude of the configuration interaction singles coefficients) corresponds to the one from the highest occupied molecular orbital (HOMO) to the lowest unoccupied molecular orbitals (LUMO).

Figure 10 shows the isosurface plots (isovalue = 0.2e/a.u.^{3}) of the various MOs involved in the photoexcitation. The results show that the HOMO of the dye is delocalised over the entire molecule, with somewhat more localisation on the donor group compared to the acceptor group. In contrast, the LUMO of the dye is predominantly localised on the acceptor and the linker/spacer groups. In turn, this implies that there will be sufficient charge separation upon excitation, thereby, reducing the rate of charge recombination.

3.7.3. Absorption Spectra of the CL1-TiO$_2$ Complex. The excited-state properties of the dye-titania complex were modelled using the CAM-B3LYP approximation. CAM-B3LYP was utilised as it outperforms B3LYP in its description of the excited-state properties of the dye molecule, which is the photoactive subsystem in the complex; this superior performance of CAM-B3LYP was shown in the previous discussion of the isolated dye (cf. Tables 4 and 5 and Figures 10 and 11). To create a computer model of the dye-titania complex, the dye molecule was chemisorbed in silico onto the nanoparticle. This was accomplished by the removal of the hydrogen atom from the carboxylic acid anchor and

453.6 nm. Table 4 summarises these results, providing the values of λ_{max}^{Th}, the positions of HOMO (H), LUMO (L), and the HL gaps, $\Delta_{HL} = E_{LUMO} - E_{HOMO}$ for the dye. It also provides the values of valence band maximum (VBM), conduction band minimum (CBM), and the orbital energy bandgap (E_g) for the nanoparticle.

Table 5 provides further details for the transition corresponding to λ_{max}^{Th}. The B3LYP results displayed a large error in the excitation energy; therefore, we will focus mainly on CAM-B3LYP results. For CAM-B3LYP, the

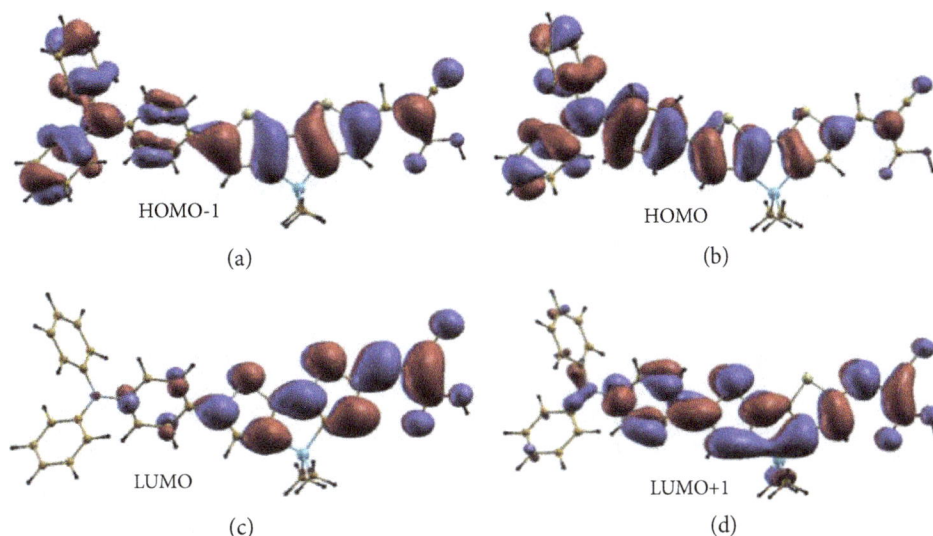

HOMO-1	HOMO
(a)	(b)
LUMO	LUMO+1
(c)	(d)

FIGURE 10: Isosurface plots (isovalue = 0.2e e/a.u.^{3}) for the molecular orbitals involved in the photoexcitation of the CL1-dye. The orbitals were calculated within CAM-B3LYP.

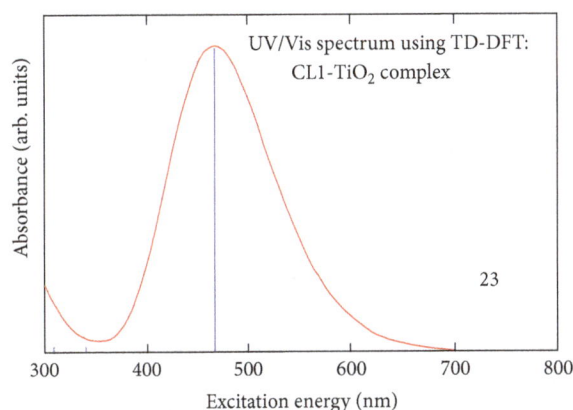

FIGURE 11: UV/Vis spectrum using TD-DFT of CL1-TiO$_2$ complex calculated within the CAM-B3LYP approximation, showing individual transitions as impulses with maximum absorbance values of 436.0 nm and 467.4 nm.

TABLE 6: Details of the most important transition ($=\lambda_{max}^{Th.}$) for the dye-titania complex. The calculations were done within the CAM-B3LYP approximation. Abbreviations and the use of bold font are as that of Table 5.

System	T.E. (eV) (O.S.)	Involved orbitals (Coeff.)
CL1-TiO$_2$	2.7	H-1 → L (0.38)
	(1.77)	**H → L (0.55)**

binding the two carboxylate oxygen atoms with two fivefold coordinated titanium atoms on the surface of the titania nanoparticle. The hydrogen atom removed was transferred to an undercoordinated oxygen atom on the titania surface to maintain the neutrality of the system, which was essential for the TD-DFT calculations. The resulting structure was relaxed within CAM-B3LYP. This optimised structure was then used to study the excited-state properties of the complex using TD-DFT.

Figure 11 shows the UV/Vis spectra for the complex as calculated using TD-DFT. The experimental and calculated values of the energy for maximum absorbance are in very good agreement at about 436.0 nm (2.8 eV) and 467.4 nm (2.7 eV), respectively.

Details of the most important transition corresponding to λ_{max}^{Th} and the involved orbital pairs are provided in Table 6. Figure 12 shows the molecular orbitals of the complex involved in the transition (HOMO-1 orbital is not

shown here; it is mostly derived from the dye's HOMO-1). From the isosurface plots, both HOMO and LUMO are predominantly localised on the dye molecule itself. However, the LUMO does show some hybridisation between the dyes LUMO and the d-orbitals of the titanium atoms at the surface. Such a result, where the only empty state involved is the LUMO of the complex, may indicate that only a direct transition is possible for the complex. However, it is difficult to pinpoint the precise mechanism in this case. This is due to the fact that although CAM-B3LYP improves the description of the photophysics of the *isolated* dye, it overestimates the bandgap of the titania nanoparticle. In turn, this results in a misalignment of dye states with regard to the valence and conduction bands of the titania nanoparticle; hence, there is an absence of a titania-derived conduction band manifold below the dye's LUMO. Clearly, it is difficult to obtain an accurate value for the charge-transfer excitation energy of the isolated dye and the dye-titania complex, whilst also getting a reasonable level alignment (at least within the *xc*-functionals explored here). The latter is important for the accurate description of the charge-transfer mechanism(s), and this is discussed further in [61].

4. Conclusions

The CL1 dye containing a silicon bridge showed a good light to electricity efficiency of 6.90%. The Si dye exhibited better

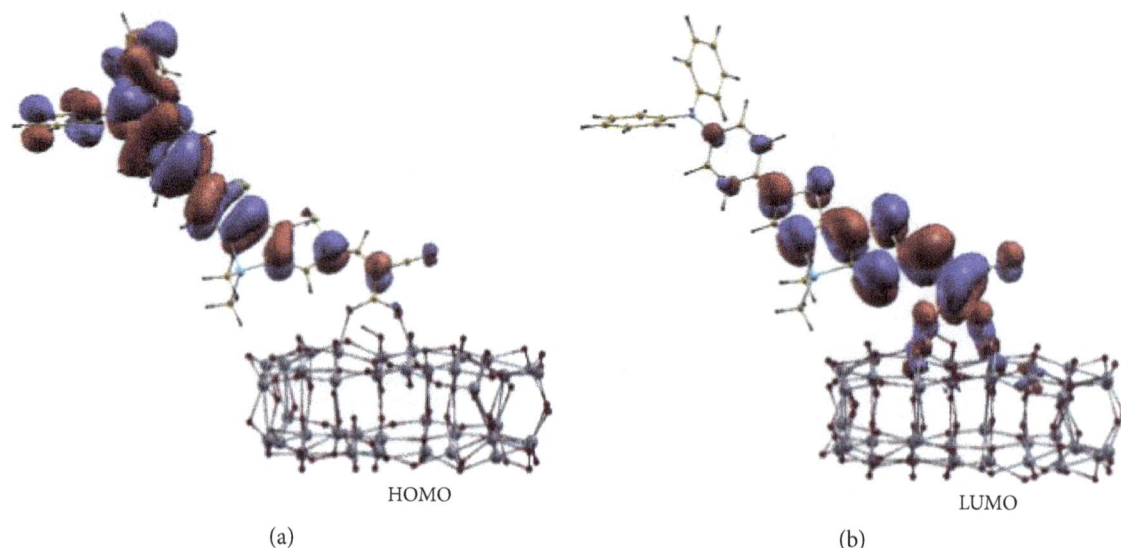

HOMO

(a)

LUMO

(b)

FIGURE 12: Molecular orbitals of CL1-TiO$_2$ complex involved in the most important transitions (HOMO and LUMO).

performance than N719 in the 400–470 nm region, which may be advantageous depending on the desired DSSC application and lighting conditions. CL-1 compares well with other Si-based dyes [9, 10], which achieve efficiencies of around 7.5%. Also, this is generally superior in performance than rival nanorod-based solar technologies [62] or DSSC approaches exploiting dye coverage with cobalt-based electrolytes [63]. From the view point of TD-DFT, the (conjectured) possibility that only a direct transition may take place due to some hybridisation between the LUMO of the dye and surface Ti atoms in the complexed state may limit the extent of photo-excited transition in the present dye—this may serve to rationalise why the observed overall energy conversion efficiency is not perhaps as high as the previously reported Si-based dyes. However, this conclusion is somewhat tentative, given that partial LUMO hybridisation does not necessarily preclude additional indirect transitions. These dyes are eventually useful for certain indoor applications where the spectral availability matches with the dye's absorption spectrum as well as for building tandem DSSC devices with optimal light management in order to obtain maximum short-circuit current densities. From classical MD simulations, it was evident that ethanol-formed hydrogen bonds and other dispersive and Coulombic interactions with the dye in dissolved state compromise its stability in solution form, whilst TD-DFT has provided a good agreement with the experimental data for prediction of optical absorption and identification of the underlying transitions responsible. Further work has been done on the addition of donor groups and will be presented in a later paper.

Disclosure

Pratibha Dev's current address is at Department of Physics and Astronomy, Howard University, Washington, DC 20059, USA, and Praveen K. Surolia's current address is at Department of Chemistry, Manipal University Jaipur, Rajasthan, India.

Conflicts of Interest

The authors declare that they have no conflicts of interest.

Acknowledgments

This material is based upon work supported by the Science Foundation Ireland under Grant no. [07/SRC/B1160]. K. Ravindranathan Thampi acknowledges the support received under SFI-Airtricity-Funded SFI-Stokes professorship grant and the SMARTOP project (Grant no. 265769) financed by the EC-FP7 programme. The authors acknowledge the computational support provided by the Irish Centre for High-End Computing (ICHEC). Neelima Rathi and Edmond Magner acknowledge support from the HEA-funded PRTLI4 programme, INSPIRE. Pratibha Dev and Praveen K. Surolia acknowledge the financial support for this work through EMPOWER fellowships granted by the Irish Research Council for Science and Engineering (IRCSET).

References

[1] B. O'Regan and M. Grätzel, "A low-cost, high-efficiency solar cell based on dye-sensitized," *Nature*, vol. 353, no. 6346, pp. 737–740, 1991.

[2] R. Komiya, A. Fukui, N. Murofushi, N. Koide, R. Yamanaka, and H. Katayama, *Technical Digest, 21st International Photovoltaic Science and Engineering Conference*, Fukuoka, November 2011, 2 C-5O-08.

[3] M. A. Green, K. Emery, Y. Hishikawa, W. Warta, and E. D. Dunlop, "Solar cell efficiency tables (version 45)," *Progress in Photovoltaics*, vol. 23, no. 1, pp. 1–9, 2015.

[4] H. Imahori, T. Umeyama, and S. Ito, "Large π-aromatic molecules as potential sensitizers for highly efficient dye-sensitized solar cells," *Accounts of Chemical Research*, vol. 42, no. 11, pp. 1809–1818, 2009.

[5] A. Yella, H. W. Lee, H. N. Tsao et al., "Porphyrin-sensitized solar cells with cobalt (II/III)–based redox electrolyte

exceed 12 percent efficiency," *Science*, vol. 334, no. 6056, pp. 629–634, 2011.

[6] S. Mathew, A. Yella, P. Gao et al., "Dye-sensitized solar cells with 13% efficiency achieved through the molecular engineering of porphyrin sensitizers," *Nature Chemistry*, vol. 6, no. 3, pp. 242–247, 2014.

[7] K. Kakiage, Y. Aoyama, T. Yano, K. Oya, J. Fujisawa, and M. Hanaya, "Highly-efficient dye-sensitized solar cells with collaborative sensitization by silyl-anchor and carboxy-anchor dyes," *Chemical Communications*, vol. 51, no. 88, pp. 15894–15897, 2015.

[8] J. N. Clifford, G. Yahioglu, L. R. Milgrom, and J. R. Durrant, "Molecular control of recombination dynamics in dye sensitised nanocrystalline TiO 2 films," *Chemical Communications*, no. 12, pp. 1260–1261, 2002.

[9] T. Bessho, S. M. Zakeeruddin, C. Y. Yeh, E. W. G. Diau, and M. Grätzel, "Highly efficient mesoscopic dye-sensitized solar cells based on donor-acceptor-substituted porphyrins," *Angewandte Chemie, International Edition*, vol. 49, no. 37, pp. 6646–6649, 2010.

[10] S. M. Feldt, E. A. Gibson, E. Gabrielsson, L. Sun, G. Boschloo, and A. Hagfeldt, "Design of organic dyes and cobalt polypyridine redox mediators for high-efficiency dye-sensitized solar cells," *Journal of the American Chemical Society*, vol. 132, no. 46, pp. 16714–16724, 2010.

[11] H. N. Tsao, C. Yi, T. Moehl et al., "Cyclopentadithiophene bridged donor-acceptor dyes achieve high power conversion efficiencies in dye-sensitized solar cells based on the tris-cobalt bipyridine redox couple," *ChemSusChem*, vol. 4, no. 5, pp. 591–594, 2011.

[12] L. Liao, A. Cirpan, Q. Chu, F. E. Karasz, and Y. Pang, "Synthesis and optical properties of light-emitting π-conjugated polymers containing biphenyl and dithienosilole," *Journal of Polymer Science Part A: Polymer Chemistry*, vol. 45, no. 10, pp. 2048–2058, 2007.

[13] T.-Y. Chu, J. Lu, S. Beaupré et al., "Effects of the molecular weight and the side-chain length on the photovoltaic performance of dithienosilole/thienopyrrolodione copolymers," *Advanced Functional Materials*, vol. 22, no. 11, pp. 2345–2351, 2012.

[14] S. Ko, H. Choi, M.-S. Kang et al., "Silole-spaced triarylamine derivatives as highly efficient organic sensitizers in dye-sensitized solar cells (DSSCs)," *Journal of Materials Chemistry*, vol. 20, no. 12, pp. 2391–2399, 2010.

[15] L.-Y. Lin, C.-H. Tsai, K.-T. Wong et al., "Organic dyes containing coplanar diphenyl-substituted dithienosilole core for efficient dye-sensitized solar cells," *The Journal of Organic Chemistry*, vol. 75, no. 14, pp. 4778–4785, 2010.

[16] M. J. Ross and K. R. William, *Impedance spectroscopy*, Wiley, New York, 1987.

[17] J. Bisquert, "Theory of the impedance of charge transfer via surface states in dye-sensitized solar cells," *Journal of Electroanalytical Chemistry*, vol. 646, no. 1-2, pp. 43–51, 2010.

[18] L. Andrade, R. Cruz, H. Ribeiro, and A. Mendes, "Impedance characterization of dye-sensitized solar cells in a tandem arrangement for hydrogen production by water splitting," *International Journal of Hydrogen Energy*, vol. 35, no. 17, pp. 8876–8883, 2010.

[19] F. Fabregat-Santiago, J. Bisquert, E. Palomares et al., "Correlation between photovoltaic performance and impedance spectroscopy of dye-sensitized solar cells based on ionic liquids," *Journal of Physical Chemistry C*, vol. 111, no. 17, pp. 6550–6560, 2007.

[20] R. Kern, R. Sastrawan, J. Ferber, R. Stangl, and J. Luther, "Modeling and interpretation of electrical impedance spectra of dye solar cells operated under open-circuit conditions," *Electrochimica Acta*, vol. 47, no. 26, pp. 4213–4225, 2002.

[21] S. Ito, T. N. Murakami, P. Comte et al., "Fabrication of thin film dye sensitized solar cells with solar to electric power conversion efficiency over 10%," *Thin Solid Films*, vol. 516, no. 14, pp. 4613–4619, 2008.

[22] S. Ito, P. Chen, P. Comte et al., "Fabrication of screen-printing pastes from TiO2 powders for dye-sensitised solar cells," *Progress in Photovoltaics Research and Applications*, vol. 15, no. 7, pp. 603–612, 2007.

[23] G. Lu, H. Usta, C. Risko et al., "Synthesis, characterization, and transistor response of semiconducting silole polymers with substantial hole mobility and air stability. Experiment and theory," *Journal of the American Chemical Society*, vol. 130, no. 24, pp. 7670–7685, 2008.

[24] J. M. Kroon, N. J. Bakker, H. J. P. Smit et al., "Nanocrystalline dye-sensitized solar cells having maximum performance," *Progress in Photovoltaics Research and Applications*, vol. 15, no. 1, pp. 1–18, 2007.

[25] M. Frisch, G. Trucks, H. Schlegel et al., "Semiempirical GGA-type density functional constructed with a long-range dispersion correction," *Journal of Computational Chemistry*, vol. 27, no. 15, pp. 1787–1799, 2006.

[26] M. J. G. Peach, P. Benfield, T. Helgaker, and D. J. Tozer, "Excitation energies in density functional theory: an evaluation and a diagnostic test," *The Journal of Chemical Physics*, vol. 128, no. 4, p. 044118, 2008.

[27] P. Dev, S. Agrawal, and N. J. English, "Determining the appropriate exchange-correlation functional for time-dependent density functional theory studies of charge-transfer excitations in organic dyes," *The Journal of Chemical Physics*, vol. 136, no. 22, p. 224301, 2012.

[28] A. D. Becke, "Density-functional thermochemistry. III. The role of exact exchange," *The Journal of Chemical Physics*, vol. 98, no. 7, pp. 5648–5652, 1993.

[29] R. Kavathekar, P. Dev, N. J. English, and J. M. D. MacElroy, "Molecular dynamics study of water in contact with TiO$_2$ rutile-110, 100, 101, 001 and anatase-101, 001 surfaces," *Molecular Physics*, vol. 109, no. 13, pp. 1649–1656, 2011.

[30] P. J. Stephens, F. J. Devlin, C. S. Ashvar, C. F. Chabalowski, and M. J. Frisch, "Theoretical calculation of vibrational circular dichroism spectra," *Faraday Discussions*, vol. 99, pp. 103–119, 1994.

[31] T. Yanai, D. P. Tew, and N. C. Handy, "A new hybrid exchange–correlation functional using the Coulomb-attenuating method (CAM-B3LYP)," *Chemical Physics Letters*, vol. 393, no. 1–3, pp. 51–57, 2004.

[32] M. Cossi, N. Rega, G. Scalmani, and V. Barone, "Energies, structures, and electronic properties of molecules in solution with the C-PCM solvation model," *Journal of Computational Chemistry*, vol. 24, no. 6, pp. 669–681, 2003.

[33] P. Persson, R. Bergström, and S. Lunell, "Quantum chemical study of photoinjection processes in dye-sensitized TiO2 nanoparticles," *The Journal of Physical Chemistry. B*, vol. 104, no. 44, pp. 10348–10351, 2000.

[34] F. De Angelis, A. Tilocca, and A. Selloni, "Time-dependent DFT study of [Fe (CN) 6] 4-sensitization of TiO2

nanoparticles," *Journal of the American Chemical Society*, vol. 126, no. 46, pp. 15024–15025, 2004.

[35] F. De Angelis, "Direct vs. indirect injection mechanisms in perylene dye-sensitized solar cells: a DFT/TDDFT investigation," *Chemical Physics Letters*, vol. 493, no. 4–6, pp. 323–327, 2010.

[36] S. Agrawal, P. Dev, N. J. English, K. R. Thampi, and J. MacElroy, "First-principles study of the excited-state properties of coumarin-derived dyes in dye-sensitized solar cells," *Journal of Materials Chemistry*, vol. 21, no. 30, pp. 11101–11108, 2011.

[37] S. Agrawal, P. Dev, N. J. English, K. R. Thampi, and J. MacElroy, "A TD-DFT study of the effects of structural variations on the photochemistry of polyene dyes," *Chemical Science*, vol. 3, no. 2, pp. 416–424, 2012.

[38] T. A. Halgren, "Merck molecular force field. I. Basis, form, scope, parameterization, and performance of MMFF94," *Journal of Computational Chemistry*, vol. 17, no. 5-6, pp. 490–519, 1996.

[39] MOE, *The Molecular Operating Environment from Chemical Computing Group Inc.*.

[40] N. J. English and J. M. D. MacElroy, "Atomistic simulations of liquid water using Lekner electrostatics," *Molecular Physics*, vol. 100, no. 23, pp. 3753–3769, 2002.

[41] N. J. English, "Effect of electrostatics techniques on the estimation of thermal conductivity via equilibrium molecular dynamics simulation: application to methane hydrate," *Molecular Physics*, vol. 106, no. 15, pp. 1887–1898, 2008.

[42] M. Xu, S. Wenger, H. Bala et al., "Tuning the energy level of organic sensitizers for high-performance dye-sensitized solar cells," *Journal of Physical Chemistry C*, vol. 113, no. 7, pp. 2966–2973, 2009.

[43] M. P. Allen and D. J. Tildesley, *Molecular Simulation of Liquids*, Clarendon, Oxford.

[44] S. Roquet, A. Cravino, P. Leriche, O. Alévêque, P. Frère, and J. Roncali, "Triphenylamine–thienylenevinylene hybrid systems with internal charge transfer as donor materials for heterojunction solar cells," *Journal of the American Chemical Society*, vol. 128, no. 10, pp. 3459–3466, 2006.

[45] J. H. Yum, D. P. Hagberg, S. J. Moon et al., "A light-resistant organic sensitizer for solar-cell applications," *Angewandte Chemie (International Edition in English)*, vol. 48, no. 9, pp. 1576–1580, 2009.

[46] A. Hagfeldt and M. Gratzel, "Light-induced redox reactions in nanocrystalline systems," *Chemical Reviews*, vol. 95, no. 1, pp. 49–68, 1995.

[47] C. Klein, M. K. Nazeeruddin, P. Liska et al., "Engineering of a novel ruthenium sensitizer and its application in dye-sensitized solar cells for conversion of sunlight into electricity," *Inorganic Chemistry*, vol. 44, no. 2, pp. 178–180, 2004.

[48] C. Teng, X. Yang, C. Yang et al., "Influence of triple bonds as π-spacer units in metal-free organic dyes for dye-sensitized solar cells," *Journal of Physical Chemistry C*, vol. 114, no. 25, pp. 11305–11313.

[49] D. Shi, Y. Cao, N. Pootrakulchote et al., "New organic sensitizer for stable dye-sensitized solar cells with solvent-free ionic liquid electrolytes," *Journal of Physical Chemistry C*, vol. 112, no. 44, pp. 17478–17485, 2008.

[50] R. M. Silverstein, G. C. Bassler, and T. C. Morrill, *Spectrometric identification of organic compounds*, 1974.

[51] M. K. Nazeeruddin, R. Humphry-Baker, D. L. Officer, W. M. Campbell, A. K. Burrell, and M. Gratzel, "Application of metalloporphyrins in nanocrystalline dye-sensitized solar cells for conversion of sunlight into electricity," *Langmuir*, vol. 20, no. 15, pp. 6514–6517, 2004.

[52] G. B. Deacon and R. J. Phillips, "Relationships between the carbon-oxygen stretching frequencies of carboxylato complexes and the type of carboxylate coordination," *Coordination Chemistry Reviews*, vol. 33, no. 3, pp. 227–250, 1980.

[53] A. Vittadini, A. Selloni, F. P. Rotzinger, and M. Gratzel, "Formic acid adsorption on dry and hydrated TiO2 anatase (101) surfaces by DFT calculations," *The Journal of Physical Chemistry. B*, vol. 104, no. 6, pp. 1300–1306, 2000.

[54] J. Bisquert, F. Fabregat-Santiago, I. Mora-Sero, G. Garcia-Belmonte, and S. Giménez, "Electron lifetime in dye-sensitized solar cells: theory and interpretation of measurements," *Journal of Physical Chemistry C*, vol. 113, no. 40, pp. 17278–17290, 2009.

[55] Q. Wang, J.-E. Moser, and M. Grätzel, "Electrochemical impedance spectroscopic analysis of dye-sensitized solar cells," *The Journal of Physical Chemistry B*, vol. 109, no. 31, pp. 14945–14953, 2005.

[56] L. Han, N. Koide, Y. Chiba, and T. Mitate, "Modeling of an equivalent circuit for dye-sensitized solar cells," *Applied Physics Letters*, vol. 84, no. 13, pp. 2433–2435, 2004.

[57] J. Halme, P. Vahermaa, K. Miettunen, and P. Lund, "Device physics of dye solar cells," *Advanced Materials*, vol. 22, no. 35, pp. 22E210–E234, 2010.

[58] L.-L. Li, Y.-C. Chang, H.-P. Wu, and E. W.-G. Diau, "Characterisation of electron transport and charge recombination using temporally resolved and frequency-domain techniques for dye-sensitised solar cells," *International Reviews in Physical Chemistry*, vol. 31, no. 3, pp. 420–467, 2012.

[59] N. J. English and D. G. Carroll, "Prediction of Henry's Law constants by a Quantitative Structure Property Relationship and neural networks," *Journal of Chemical Information and Computer Sciences*, vol. 41, no. 5, pp. 1150–1161, 2001.

[60] M. Adachi, M. Sakamoto, J. Jiu, Y. Ogata, and S. Isoda, "Determination of parameters of electron transport in dye-sensitized solar cells using electrochemical impedance spectroscopy," *The Journal of Physical Chemistry. B*, vol. 110, no. 28, pp. 13872–13880, 2006.

[61] P. Dev, S. Agrawal, and N. J. English, "Functional assessment for predicting charge-transfer excitations of dyes in complexed state: a study of triphenylamine-donor dyes on titania for dye-sensitized solar cells," *The Journal of Physical Chemistry. A*, vol. 117, no. 10, pp. 2114–2124, 2013.

[62] A. Singh, N. J. English, and K. M. Ryan, "Highly ordered nanorod assemblies extending over device scale areas and in controlled multilayers by electrophoretic deposition," *The Journal of Physical Chemistry. B*, vol. 117, no. 6, pp. 1608–1615, 2013.

[63] M. Pazoki, P. W. Lohse, N. Taghavinia, A. Hagfeldt, and G. Boschloo, "The effect of dye coverage on the performance of dye-sensitized solar cells with a cobalt-based electrolyte," *Physical Chemistry Chemical Physics*, vol. 16, no. 18, pp. 8503–8508, 2014.

13

A Density Peak-Based Clustering Approach for Fault Diagnosis of Photovoltaic Arrays

Peijie Lin,[1] **Yaohai Lin,**[2] **Zhicong Chen,**[1] **Lijun Wu,**[1] **Lingchen Chen,**[1] **and Shuying Cheng**[1]

[1]*Institute of Micro/Nano Devices and Solar Cells, College of Physics and Information Engineering, Fuzhou University, Fuzhou 350116, China*
[2]*College of Computer and Information Sciences, Fujian Agriculture and Forestry University, Fuzhou 350002, China*

Correspondence should be addressed to Shuying Cheng; sycheng@fzu.edu.cn

Academic Editor: Cheuk-Lam Ho

Fault diagnosis of photovoltaic (PV) arrays plays a significant role in safe and reliable operation of PV systems. In this paper, the distribution of the PV systems' daily operating data under different operating conditions is analyzed. The results show that the data distribution features significant nonspherical clustering, the cluster center has a relatively large distance from any points with a higher local density, and the cluster number cannot be predetermined. Based on these features, a density peak-based clustering approach is then proposed to automatically cluster the PV data. And then, a set of labeled data with various conditions are employed to compute the minimum distance vector between each cluster and the reference data. According to the distance vector, the clusters can be identified and categorized into various conditions and/or faults. Simulation results demonstrate the feasibility of the proposed method in the diagnosis of certain faults occurring in a PV array. Moreover, a 1.8 kW grid-connected PV system with 6×3 PV array is established and experimentally tested to investigate the performance of the developed method.

1. Introduction

The rapid increase in the amount of grid-connected photovoltaic (PV) systems has put forward a significant research topic, that is, operating condition analysis and fault diagnosis of PV systems. As one of the most important components, the performance of PV arrays (DC side) usually affects the operation of the entire system. However, due to complex outdoor working environments, the PV array is susceptible to thermal cycling, humidity, ultraviolet light, hard shadows, and other environmental factors that cause various faults such as cracking, hot spots, modules' short circuit, and PV strings' open circuit. As a result, these will lead to power losses and even fire hazards [1]. The overcurrent protection devices (OCPDs) and ground fault detection interrupters (GFDIs) are usually installed as the traditional fault detection and protection for the PV arrays [2]. However, due to the nonlinear output characteristics of the PV array, various faults remain and cannot be eliminated by the protection devices [3, 4].

To address these problems, various fault diagnosis approaches for PV arrays have been studied, including thermal imaging [5–7], earth capacitance measurement (ECM), time-domain reflectometry (TDR) [8, 9], power loss analysis [10–12], current and voltage indicators evaluation [13–16], and machine learning [17–23]. The infrared thermal imaging method is applied to detect and identify the hot spot and degradation fault in PV modules according to the temperature characteristics of the PV module. The ECM is presented to detect the location of open-circuit faults in PV strings, and the TDR is applied to identify the degradation of a PV array. Power loss analysis method is proposed to detect various types of faults occurring in solar PV systems by comparing the measured and theoretical output power of the PV array. The automatic supervision and fault detection procedure that based on evaluation of current and voltage indicators in grid-connected PV systems is proposed to identify the short circuits and open circuits in PV arrays [13] as well as inverter disconnection and partial shading conditions [14]. Moreover, the procedure

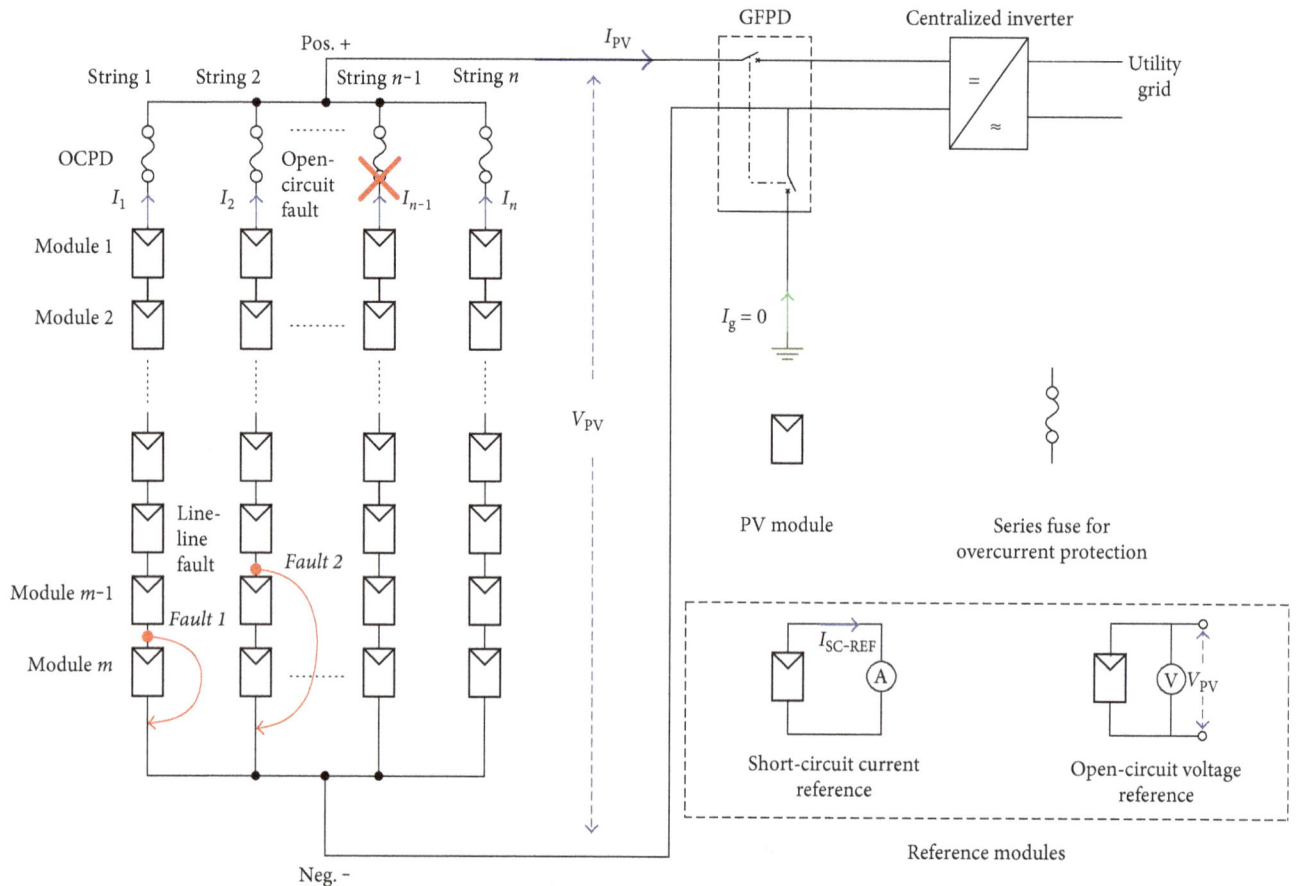

FIGURE 1: Schematic diagram of series-parallel grid-connected PV system.

is combined with an OLE (Object Linking and Embedding) for Process Control (OPC) monitoring for remote supervision and diagnosis of grid-connected PV systems [15]. Furthermore, the analysis of current and voltage indicators is applied to detect, in real time, the faults related to bypassed PV modules, open-circuit strings and partial shading for a PV plant connected to a single-phase grid [16].

Furthermore, to better detect and classify PV faults, machine learning algorithms are widely carried out. A fault detection and classification model based on decision tree is presented to deal with the line-line, open-circuit, and partial shade faults in PV arrays [17]. Artificial neural network technique is applied to monitor the health status, measure degradation, and indicate maintenance schedules of a PV system [18]. The study in [19] proposed a method to identifying the short-circuit location of PV modules in one string by using three-layered feed-forward neural network. An online PV modules' fault diagnosis model is established based on back propagation neural network [20]. The Bayesian neural network and polynomial regression models are researched for the evaluation of soiling effects on PV plants [21]. A new artificial neural network approach is implemented in a field-programmable gate array (FPGA) and has the ability to identify eight types of fault occurring in a PV array [22]. A semisupervised learning model is employed for line-line and open fault detection and classification in PV arrays [23].

In practice, daily operational data from various PV systems are stored in the monitoring systems, enabling the working condition estimation of PV arrays and fault diagnosis based on the data [24–26]. According to the distribution characteristics of PV data analyzed in this paper, a density peak-based clustering approach for fault diagnosis in PV arrays is proposed. The approach diagnoses the PV faults by clustering and classifying the daily operational data. The advantage of the proposed approach is that a larger amount of training data and tedious training process are not needed and only few labeled reference data obtained from a simulated PV system is required to identify clusters.

The rest of this paper is organized as follows: Section 2 depicts the distribution characteristics of PV data and the process of the proposed method. The simulation results are presented in Section 3, and several working conditions of PV array are studied. In Section 4, experiments and result analysis are carried out. Finally, some conclusions are drawn in Section 5.

2. Proposed Models

In this section, the features of PV data are analyzed, such as data distribution, cluster shape, and cluster number. Then, the procedure of the proposed approach is described in detail.

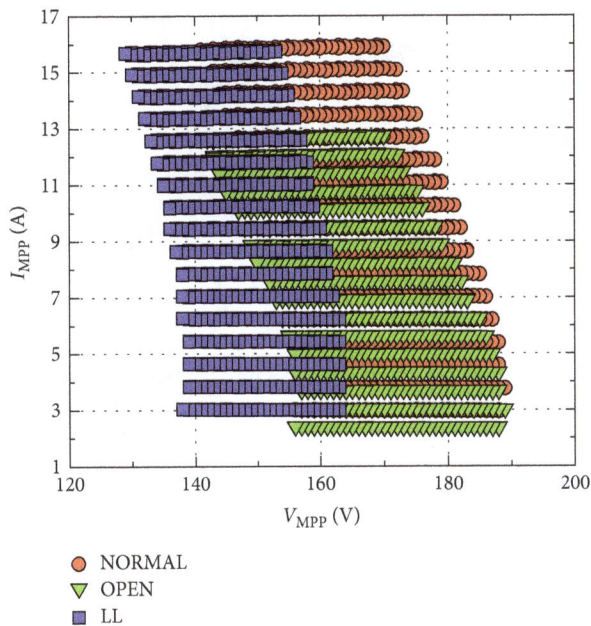

NORMAL ●
OPEN ▽
LL ■

FIGURE 2: The V_{MPP} versus I_{MPP} of PV array over a range of irradiance and temperature.

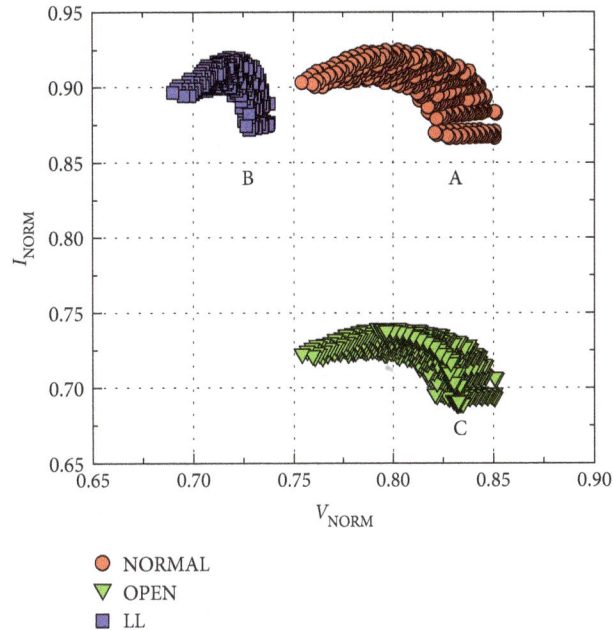

NORMAL ●
OPEN ▽
LL ■

FIGURE 3: The distribution of PV data over a range of irradiance and temperature.

2.1. Photovoltaic Data Distribution. The schematic diagram of a typical series-parallel grid-connected PV system is shown in Figure 1. The system generally is comprised of $m \times n$ PV array, a centralized inverter, protection devices (such as OCPD and GFPD), and connection wires [27]. Usually, the PV array can output maximum power under variable environment due to the maximum power point tracking (MPPT) technology of inverters. When faults occur, however, the MPPT is possible to keep the optimal power output if the PV array can reach the inverter's working voltage. As a result, the current of the PV array may be significantly reduced, leading to the failure of the OCPD to clear the fault [3].

In every daily operation cycle, the voltage (V_{MPP}) and current of PV array at MPPs change due to the variations of solar irradiance and atmospheric temperature. In order to investigate the changes of V_{MPP} and I_{MPP} under different conditions, a normal (NORMAL) and two common faults of a specific PV array are considered. As shown in Figure 1, the two faults are line-line (LL) fault and open-circuit (OPEN) fault, which may be difficult to be cleared by conventional OCPD. The simulated V_{MPP} versus I_{MPP} over a range of irradiance and temperature is shown in Figure 2. Obviously, part of the V_{MPP} and I_{MPP} overlaps, causing difficulties for the PV fault diagnosis.

To make better visualization and identification of PV faults, the approach proposed in [13, 23] is applied to normalize the V_{MPP} and I_{MPP}. The normalization formula can be expressed as follows:

$$V_{\mathrm{NORM}} = \frac{V_{\mathrm{MPP}}}{m \times V_{\mathrm{OC\text{-}REF}}},$$
$$I_{\mathrm{NORM}} = \frac{I_{\mathrm{MPP}}}{n \times I_{\mathrm{SC\text{-}REF}}}, \tag{1}$$

where V_{NORM} and I_{NORM} are the normalized PV voltage and PV current, respectively; $V_{\mathrm{OC\text{-}REF}}$ is the open-circuit voltage of reference PV module; $I_{\mathrm{SC\text{-}REF}}$ is the short-circuit current of reference PV module (as shown in Figure 1); m is the number of modules in series in each PV string; and n is the number of strings in parallel in the array. Hereafter, the data set of V_{NORM} and I_{NORM} is simply referred to as PV data, which is the input data of the proposed model. The PV data distribution of a PV array over a range of irradiance and temperature is shown in Figure 3.

It is clearly demonstrated that the PV data have good data clustering and the clusters are nonspherical in shape. In each cluster, data from the bottom to the upper-left indicate the data from low irradiance to high irradiance. In daily operation, the PV system generally runs under NORMAL condition and the corresponding PV data are distributed in only a cluster, that is, cluster A in Figure 3. When fault occurs, such as LL fault, the data distribution is changed from cluster A to cluster B. Furthermore, the data may vary from cluster B to cluster C if another fault happens, such as OPEN fault. Hence, the number of clusters cannot be predefined. Moreover, the center of each cluster has a relatively large distance from any points with a higher local density. Therefore, the PV data can be clustered by using an appropriate clustering algorithm and then further analyzed for PV array faults.

2.2. Procedure of the Proposed Approach. There are two phases in our proposed approach. Firstly, the daily PV operation data are recorded and assigned into several clusters by using a clustering algorithm. Each cluster represents a kind of work conditions of the PV array. Secondly, with the aid of the labeled reference data, each cluster will be identified, respectively. Thus, the recorded PV data can be divided into

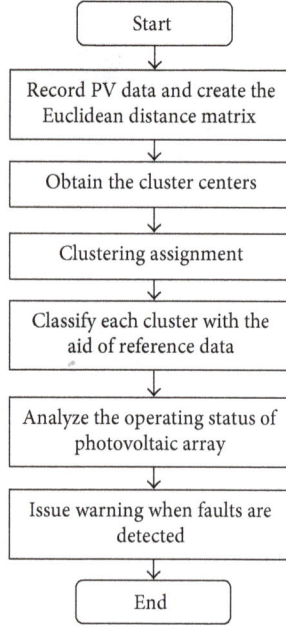

FIGURE 4: Flowchart of the proposed approach.

TABLE 1: Main parameters of SM55.

Parameters	Values
Maximum power P_{MPP} (W)	55
Maximum power current I_{MPP} (A)	3.15
Maximum power voltage V_{MPP} (V)	17.4
Short-circuit current I_{SC} (A)	3.45
Open-circuit voltage V_{OC} (V)	21.7

the aforementioned work conditions, that is, NORMAL, LL, OPEN, or their combinations.

2.2.1. Phase 1 PV Data Clustering. Recently, an algorithm implementing clustering by fast search and find of density peaks (CFSFDP) published on *Science* is proposed by Rodriguez and Laio [28]. This method is based on two assumptions: the cluster centers must have the highest local density and they have relatively large distance to the points with higher density. It has an excellent ability to analyze arbitrary shape clusters as well as different dimensional cases and to find cluster centers. As discussed in Section 2.1, PV data have some features, such as nonspherical, cluster centers have a relatively large distance from any points with a higher local density, and cluster number cannot be predefined. Therefore, the CFSFDP algorithm is very suitable for the analysis of the PV data.

In CFSFDP, two important indicators are defined and computed: ρ_i and δ_i, which represent the local density of a data point and the distance from data points of higher density, respectively. In the proposed approach, for each PV data point i, the procedure for calculating its ρ_i and δ_i is as follows:

Firstly, the PV data are recorded and organized as $X = [x_1, x_2, \ldots, x_N]$, where $x_i = [V_{NORMi}\ I_{NORMi}]^T$ and N is

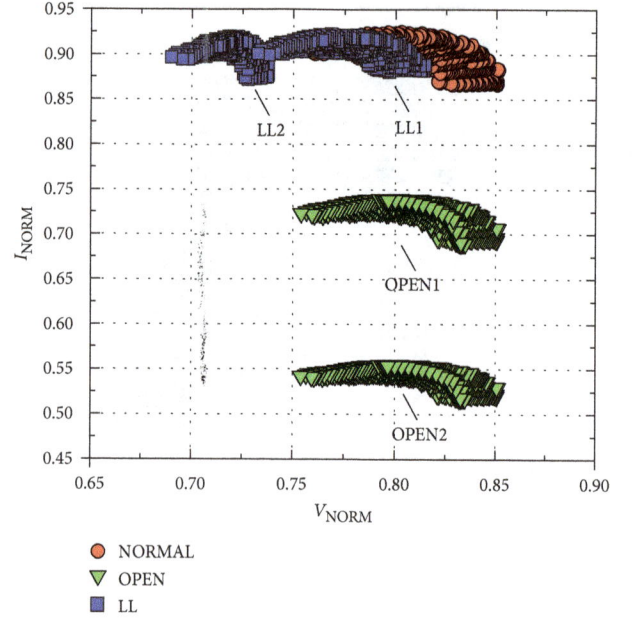

FIGURE 5: The distribution of PV data.

the number of PV data points. The distance matrix of data points should be calculated. Let d_{ij} represent the Euclidean distance between x_i and x_j; then

$$d_{ij} = \left\| x_i - x_j \right\|_2,\qquad(2)$$

where $\|\cdot\|$ denotes the 2-norm operator.

Then ρ_i is calculated by using the Gaussian kernel function, as follows:

$$\rho_i = \sum_{i=1}^{N} \exp\left(-\frac{d_{ij}}{2d_c^2}\right),\qquad(3)$$

where d_c is the cutoff distance, which represents the neighborhood range of data point i. The CFSFDP algorithm suggests that one can choose d_c so that the average number of neighbors is around 1% to 2% of the total number of points in the PV data set and 2% is applied in this study. And δ_i is computed as follows:

$$\delta_i = \min_{j:\rho_j > \rho_i} \left(d_{ij}\right).\qquad(4)$$

For the point with the highest density, the δ_i is defined as $\max_j(d_{ij})$. It is obvious that points with local or global maxima density have large δ_i. According to ρ_i and δ_i, there are some characteristics that can be obtained as follows: a point has high ρ and low δ, which means that the point i is close to the clustering center; a point has low ρ and low δ, which indicates that the point is located in the boundary of the clustering; a point has low ρ and high δ, which implies that the point is far away from each clustering and can be noise or outliers. So only the points with both high ρ and high δ are the clustering centers. Therefore, the product of ρ_i and δ_i is applied to measure the probability of cluster centers, which is denoted as γ_i [28].

(a)

(b)

(c)

(d)

(e)

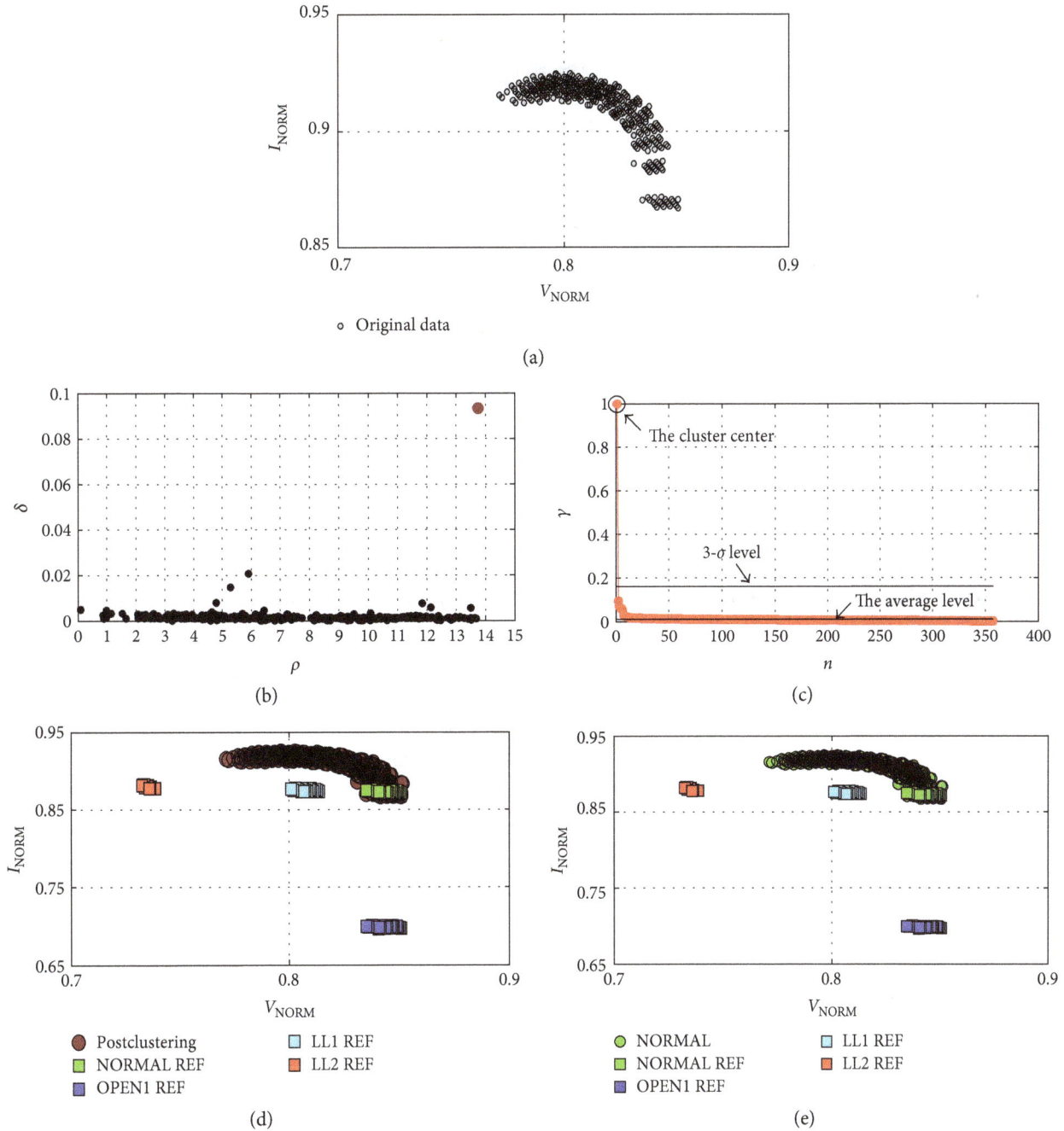

FIGURE 6: Analysis for the NORMAL case: (a) original data, (b) decision graph, (c) the value of γ in decreasing order, (d) data after clustering, and (e) cluster after identifying.

In this study, ρ_i and δ_i are normalized and employed to calculate γ_i as follows:

$$\gamma_i = \frac{\rho_i}{\rho_{max}} \cdot \frac{\delta_i}{\delta_{max}}. \tag{5}$$

Thus, only the data points with large γ can be selected as cluster centers. In our study, each cluster corresponds to an operational condition of the PV systems and the number of daily conditions is much smaller than the total amount of data. Therefore, the 3-sigma (3-σ) rule is applied as the criterion to automatically select the large γ and then determine the cluster centers [29].

Finally, after the cluster centers have been found, the CFSFDP algorithm constructs clusters by assigning other points to the same cluster as its nearest neighbor of higher density. The cluster assignment is performed in a single

(a)

(b)

(c)

(d)

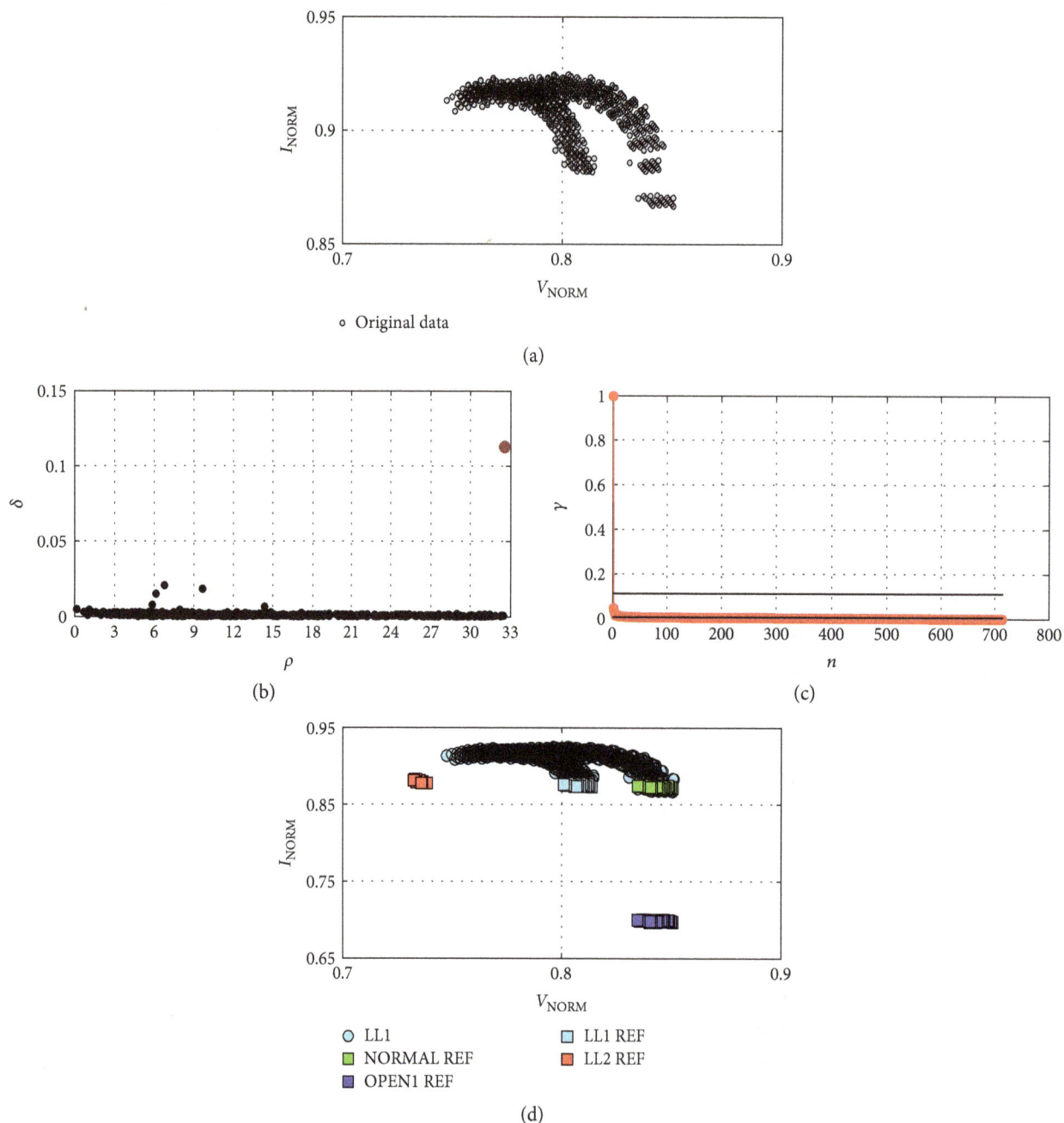

FIGURE 7: Analysis for the combination of NORMAL and OPEN1: (a) original data, (b) decision graph, (c) the value of γ in decreasing order, and (d) data after clustering and identifying.

step and does not require optimizing any objective function iteratively.

2.2.2. Phase 2 Cluster Classification. To identify the class of each cluster, a set of labeled reference data should be created first. From Section 2.1, PV data have a relatively great distance among different work conditions at low irradiation. Therefore, the labeled PV data obtained under low irradiation is adopted as the reference data. In addition, the reference data are obtained based on PV simulation models to avoid shortcomings that may be caused by experimental method, such as the potential safety issue and additional labor cost.

Subsequently, the minimum distance between the labeled reference data and the clusters is applied to define their correlation. Let N_R represent the number of the reference data categories and $r \in [1, N_R]$ the id of the reference data categories. Let N_C represent the number of clusters and $c \in [1, N_C]$ the id of cluster. For cluster c, the minimum distance between it and each reference data category can be expressed as a row vector:

$$D_M^c = \left[d_{c,1}, \ldots, d_{c,r}, \ldots, d_{c,N_R} \right]. \tag{6}$$

Then each element in the vector is compared with the cutoff distance d_c, respectively. If $d_{c,r} < d_c$, this illustrates that

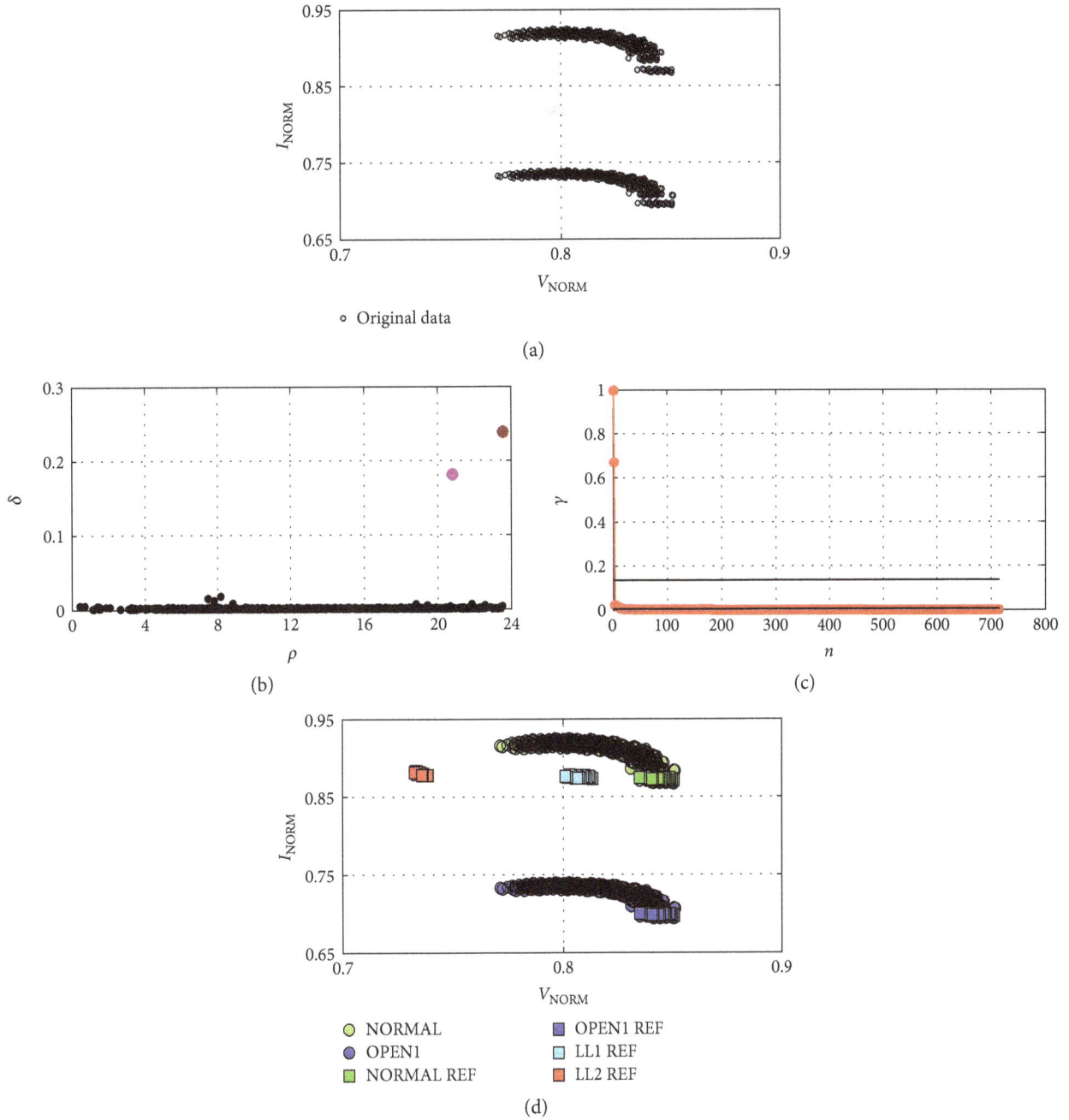

FIGURE 8: Analysis for the combination of NORMAL and LL1: (a) original data, (b) decision graph, (c) the value of γ in decreasing order, and (d) data after clustering and identifying.

the reference data of r category can be assigned to cluster c. In other words, cluster c can be labeled as r category. If all the elements are bigger than d_c, then the category of the smallest elements will be found and used to label cluster c.

Consequently, the flowchart of the proposed approach for PV array analysis is shown in Figure 4. First, the daily PV running data, that is, $X_i = [V_{\text{NORM1}}, V_{\text{NORM2}}, V_{\text{NORM3}}, \ldots, V_{\text{NORM}N}]$ and $Y_i = [I_{\text{NORM1}}, I_{\text{NORM2}}, I_{\text{NORM3}}, \ldots, I_{\text{NORM}N}]$, are recorded, and the Euclidean distance matrix is created. Subsequently, the neighborhood range of data

points is selected to calculate the local density and the minimum distance between a point and any other point with higher density, namely, ρ_i and δ_i, respectively. Cluster centers are obtained based on the product of ρ_i and δ_i and then followed by the cluster assignment of all data points. Finally, clusters are classified by investigating the minimum distance between the data of each reference category and that of each cluster. According to the labeled cluster, the operating status of PV array can be identified. When a fault is detected, the alarm will be sent out if necessary.

(a)

(b)

(c)

NORMAL   NORMAL REF   LL2 REF
OPEN1   OPEN1 REF
OPEN   LL1 REF

(d)

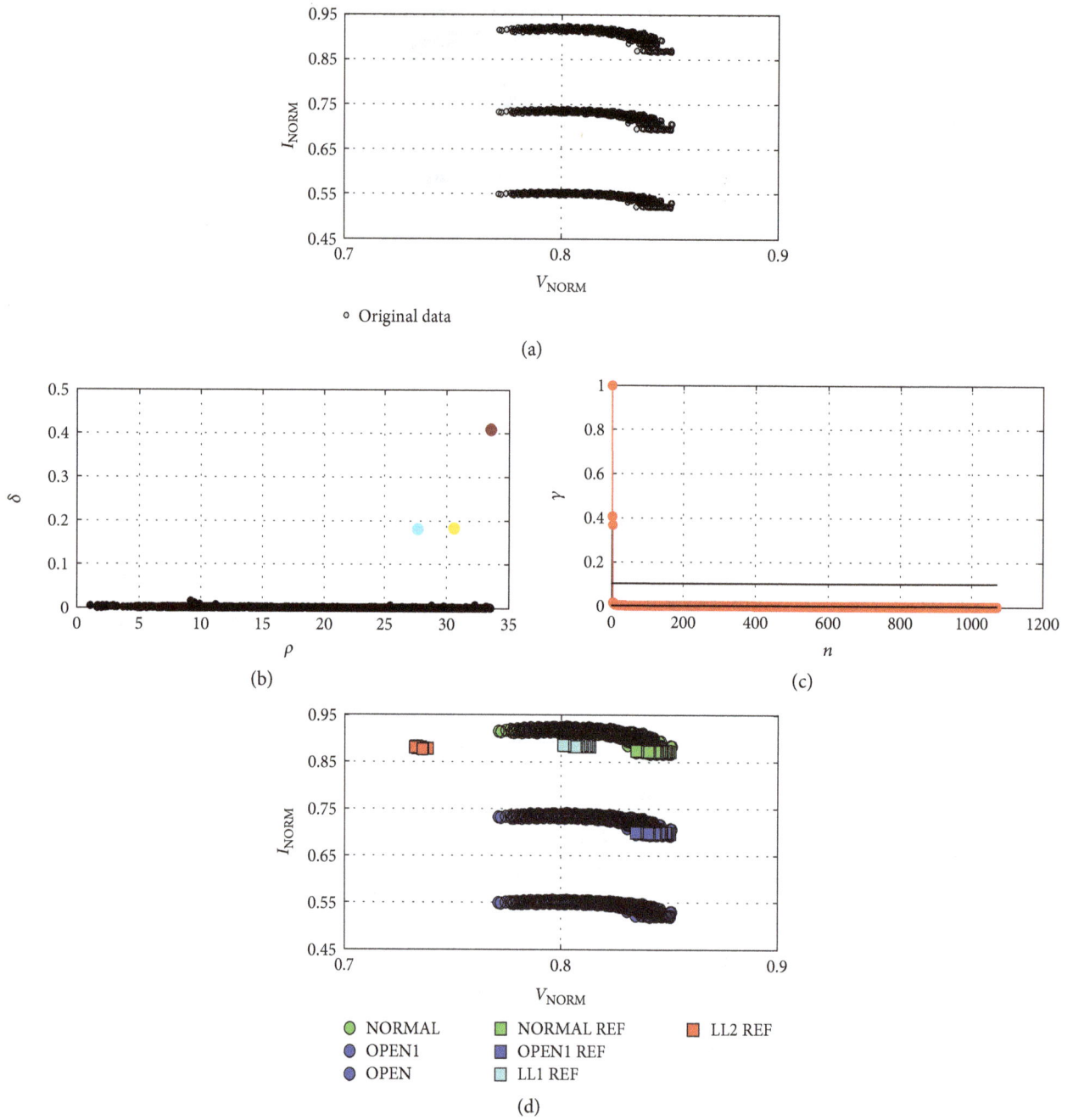

FIGURE 9: Analysis for the combination of NORMAL, OPEN1, and OPEN2 case: (a) original data, (b) decision graph, (c) the value of γ in decreasing order, and (d) data after clustering and identifying.

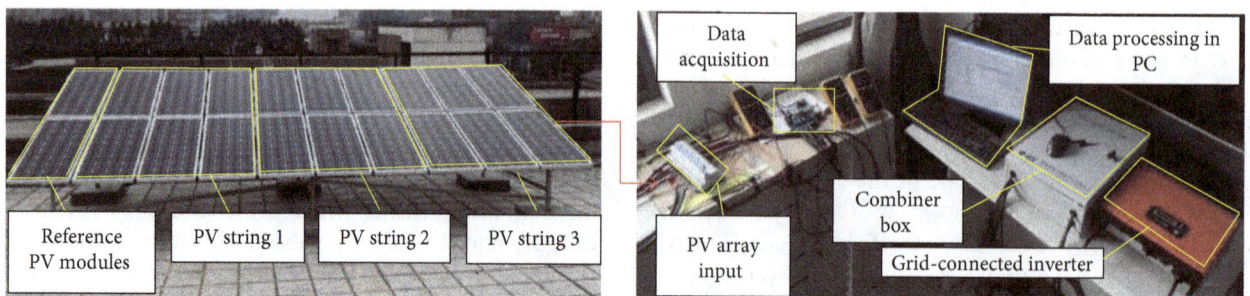

FIGURE 10: The experimental platform.

TABLE 2: Parameters of the PV components.

Components	Type	Parameters
PV modules and reference modules	Monocrystalline	At STC: P_{MPP}: 100 W, I_{MPP}: 5.71 A, V_{MPP}: 17.5 V, I_{SC}: 6.03 A, V_{OC}: 21.5 V
PV array	6 × 3 modules	At STC: P_{MPP}: 1.8 kW, I_{MPP}: 17.13 A, V_{MPP}: 105 V
Grid-connected inverter	Goodwe GW2500-NS	Max. output power: 2500 W; PV voltage range: 80~500 V; MPPT voltage range: 80~450 V

TABLE 3: Experimental environment and data.

Case	Solar irradiances	Ambient temperature	Amount of data
NORMAL	~170–1020 W/m^2	26–35°C	1380
Combination of NORMAL and LL1	~180–930 W/m^2	9–19°C	NORMAL: 710; LL1: 530
Combination of NORMAL and OPEN1	~180–920 W/m^2	5–16°C	NORMAL: 760; LL1: 500

3. Simulation and Results

In this section, several data sets are constructed to investigate the performance of the proposed method. First, the settings of simulation system are introduced. Furthermore, the test data under different conditions are simulated and briefly described. Finally, simulation results are presented.

3.1. Simulated PV System. In this study, we adopt one-diode model for PV module and apply the monocrystalline PV module SM55 to build a simulation PV system in MATLAB/Simulink [30]. The schematic diagram of the system is shown in Figure 1. The system consists of 10×5 PV modules, that is, $m = 10$ and $n = 5$. The main parameters of each PV module at standard test conditions (STC) are shown in Table 1 [31].

The module-plane solar irradiance (G_T) and ambient air temperature (T_{amb}) can be used for finding the operating solar cell temperature (T_{cell}) with the following equation [32]:

$$T_{cell} = T_{amb} + \frac{NOCT - 20°C}{800 \text{ W/m}^2} \cdot G_T, \tag{7}$$

where NOCT is the nominal operating cell temperature of the PV module SM55 and is chosen as 45°C [31].

3.2. Simulation Data under Different Conditions. As shown in Figure 5, there are three categories in operating conditions of the PV system, that is, normal condition, line-line (LL) fault, and open-circuit (OPEN) fault. The test data are obtained by simulating a whole daily running status of the PV system. The input ambient parameters for the simulation system are as follows: the solar irradiance (G_T) widely varying from 100 to 1000 W/m^2 with step change of 50 W/m^2 and the ambient temperature (T_{amb}) changes from 0°C to 40°C with step by 1°C. The PV data (V_{NORM} versus I_{NORM}) under the three conditions are plotted in Figure 5 and analyzed as follows:

(1) Normal condition: Under the changing of solar irradiance and temperature, the PV data usually have the following operating range: V_{NORM} (0.77,0.86) and $I_{NORM} \in$ (0.86,0.92).

(2) Line-line fault: The LL fault category contains two types of faults: LL1 and LL2. The LL1 fault presents that there is one-module mismatch between the fault point "Fault1" and negative conductor (Fault1-Neg) in the faulted string. Similarly, the LL2 fault is defined as two-module mismatch in the fault string. Compared with NORMAL, I_{NORM} of LL is slightly reduced, whereas V_{NORM} is observably decreased. Besides, the data of NORMAL and LL1 overlap at high solar irradiance.

(3) Open-circuit fault: the OPEN fault category consists of two kinds of faults: OPEN1 and OPEN2. They are defined as open-circuit faults on one string and two strings, respectively. It is obvious that the OPEN fault has the same V_{NORM} as the one of NORMAL condition. However, I_{NORM} is reduced in proportion according to the number of open strings.

3.3. Simulation Results. Although the daily operating temperature range of a PV system is changing, the daily normalized data of the PV system has similar data distribution. Therefore, to simulate daily running condition of the PV system, only the data obtained under a low temperature range (0°C to 20°C) is selected as the test data for analysis in this paper. The reference data are simulated under the solar irradiance of 210 W/m^2 to distinguish them from the test data. The reference data consist of four categories and are arranged in accordance with the following order: NORMAL, OPEN1, LL1, and LL2; thus $N_r = 4$ and $r \in [1, 4]$.

As discussed in Section 2.1, there may be a variety of conditions in the daily operating of the PV system. Therefore, three cases are researched, including one condition, the combination of two conditions, and the combination of three conditions. Simulation results of all cases are shown in Figures 6–9 and are discussed as follows.

(1) Case Study I: One Condition. The NORMAL condition is studied in this case, and the original test data are plotted in

○ Original data

(a)

(b)

(c)

○ NORMAL   ■ OPEN1 REF

■ NORMAL REF   □ LL1 REF

(d)

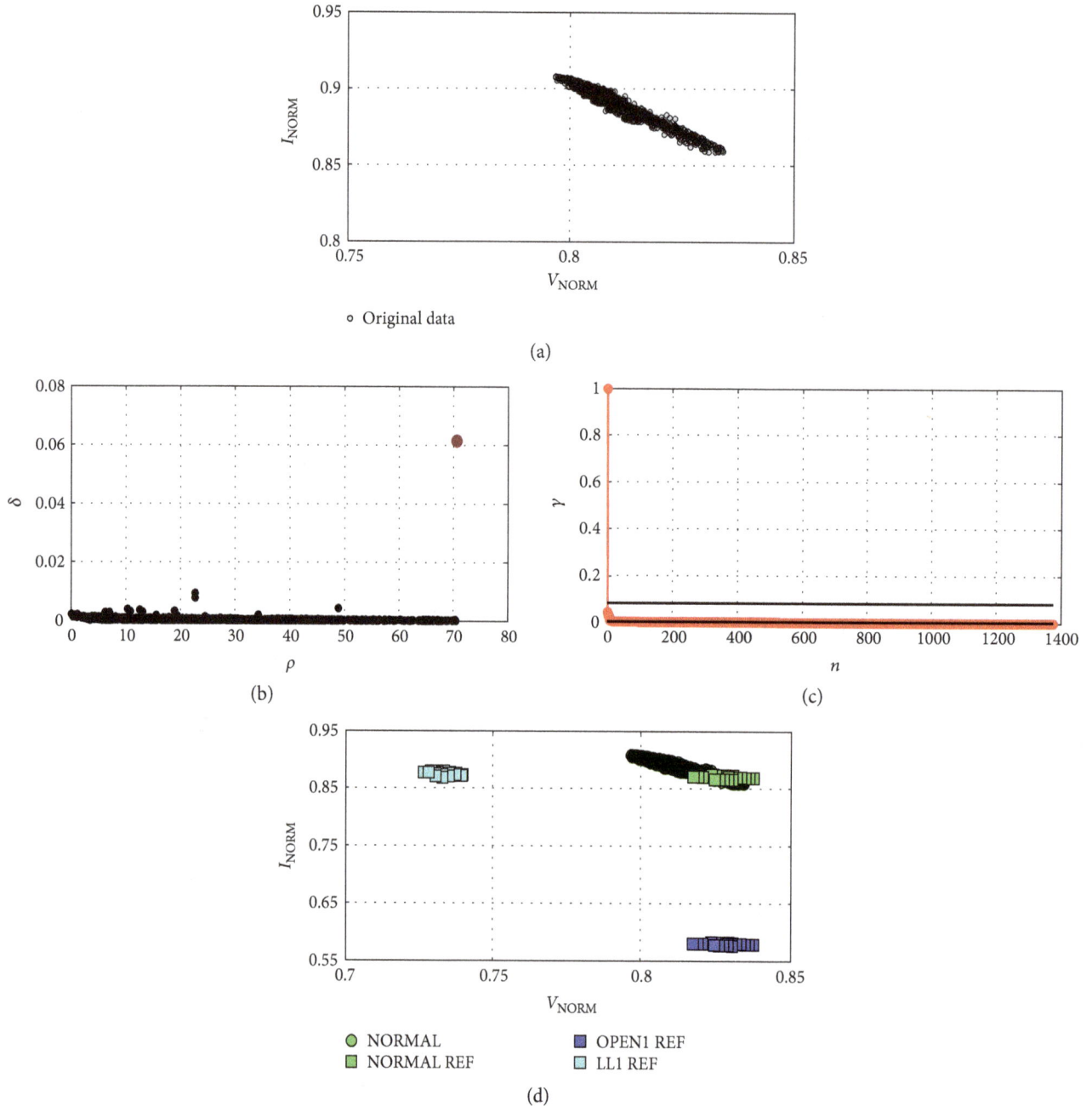

FIGURE 11: Experimental result of the NORMAL case: (a) original data, (b) decision graph, (c) the value of γ in decreasing order, and (d) data after clustering and identifying.

Figure 6(a) and are represented with black. According to the CFSFDP algorithm, the ρ_i and δ_i of all data points are calculated, respectively. Figure 6(b) shows the graph of δ_i as a function of ρ_i for each data point, which is called the decision graph. The γ_i in decreasing order is plotted in Figure 6(c). Compared to the 3-σ level, it is clear that only the top one can be chosen as the cluster center, indicating that there exists one cluster. Then, other points are assigned to the cluster as its nearest neighbor of higher density, as shown in Figure 6(d). The data points are colored when they belong to the cluster. It is obviously that all the test data are correctly clustered.

After the completion of the data clustering, the cluster is characterized by using the four types of reference data which are shown in Figure 6(d) with different colors. The d_c is chosen to be 0.00289 so that the average number of neighbors is around 2% of the total number of data. And the minimum distance vector D_M^1 is calculated to be [0.00013, 0.16708, 0.01753, 0.04974]. It can be concluded that the first element of D_M^1 is smaller than the d_c, so the cluster can be characterized as NORMAL and is painted with the same color of the NORMAL REF, as shown in Figure 6(e). Therefore, the test data of NORMAL condition can be accurately clustered and characterized.

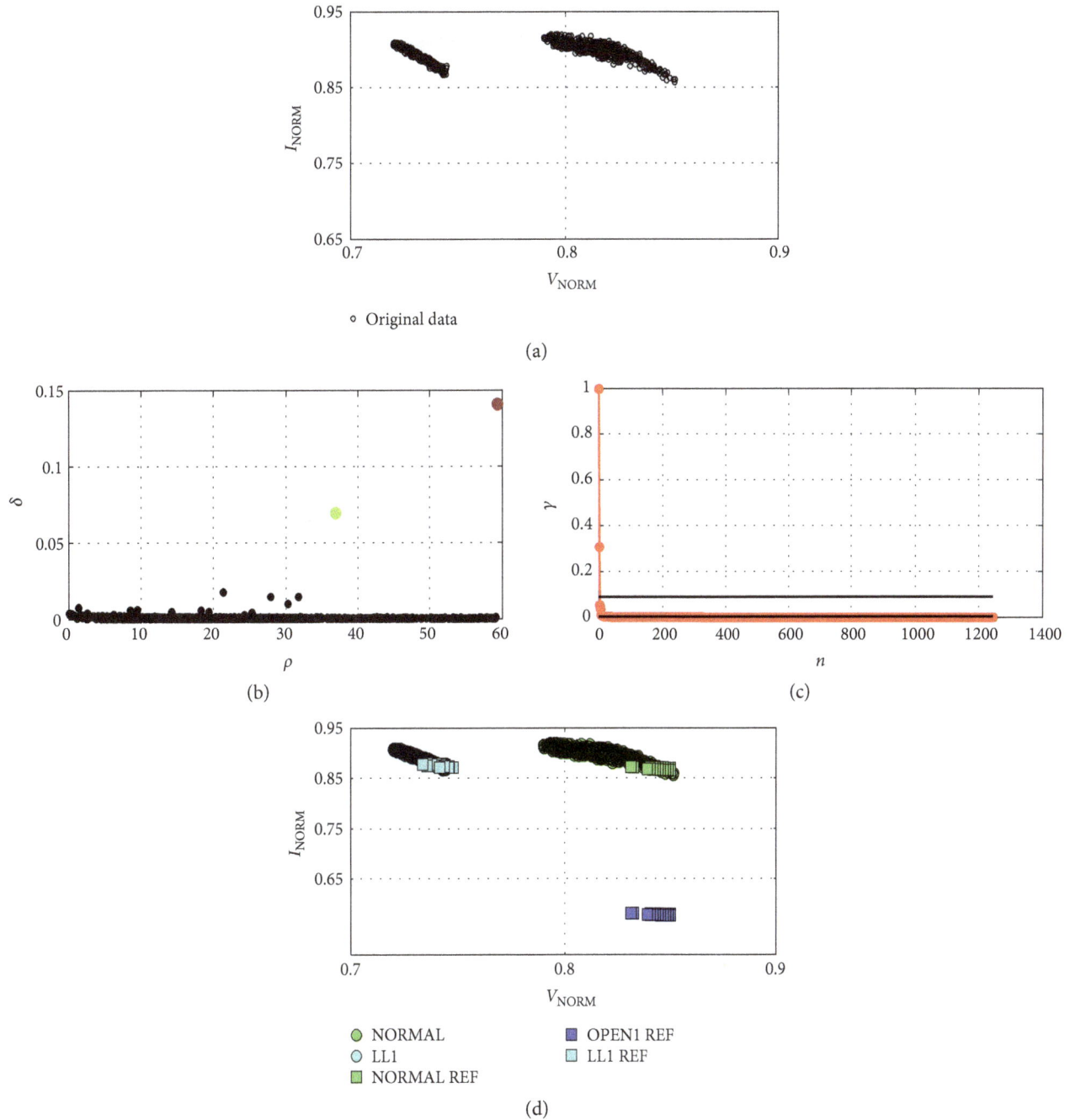

FIGURE 12: Experimental result of the combination of NORMAL and LL1 case: (a) original data, (b) decision graph, (c) the value of γ in decreasing order, and (d) data after clustering and identifying.

(2) Case Study II: Combination of Two Conditions. In this case, the combination of NORMAL and LL1 and the combination of NORMAL and OPEN1 are studied, respectively. For the first one, as can be seen from Figures 7(b) and 7(c), only one cluster center is found due to the NORMAL and LL1 with many data overlapping. Thus, the test data are grouped into the cluster. The minimum distance vector D_M^1 equals [0.00013, 0.16709, 0.00052, 0.03143] and the d_c equals 0.00321, which illustrates that the first and third elements in the vector are smaller than d_c. However, there is only one

cluster to be identified, and the proposed approach tends to classify the cluster as LL1 fault (shown in Figure 7(d)) since the condition of PV array has changed from normal to fault.

For the second combination, as shown in Figures 8(b) and 8(c), two cluster centers are obtained. The d_c is 0.00368. For the two clusters, the D_M^1 is [0.00013, 0.16708, 0.01753, 0.04974] and D_M^2 is [0.13573, 0.00039, 0.14365, 0.14686]. Accordingly, it is clear that the first element of D_M^1 and second element of D_M^2 are smaller than d_c. Thus, the other data points are assigned to two clusters based on

(a)

(b)

(c)

(d)

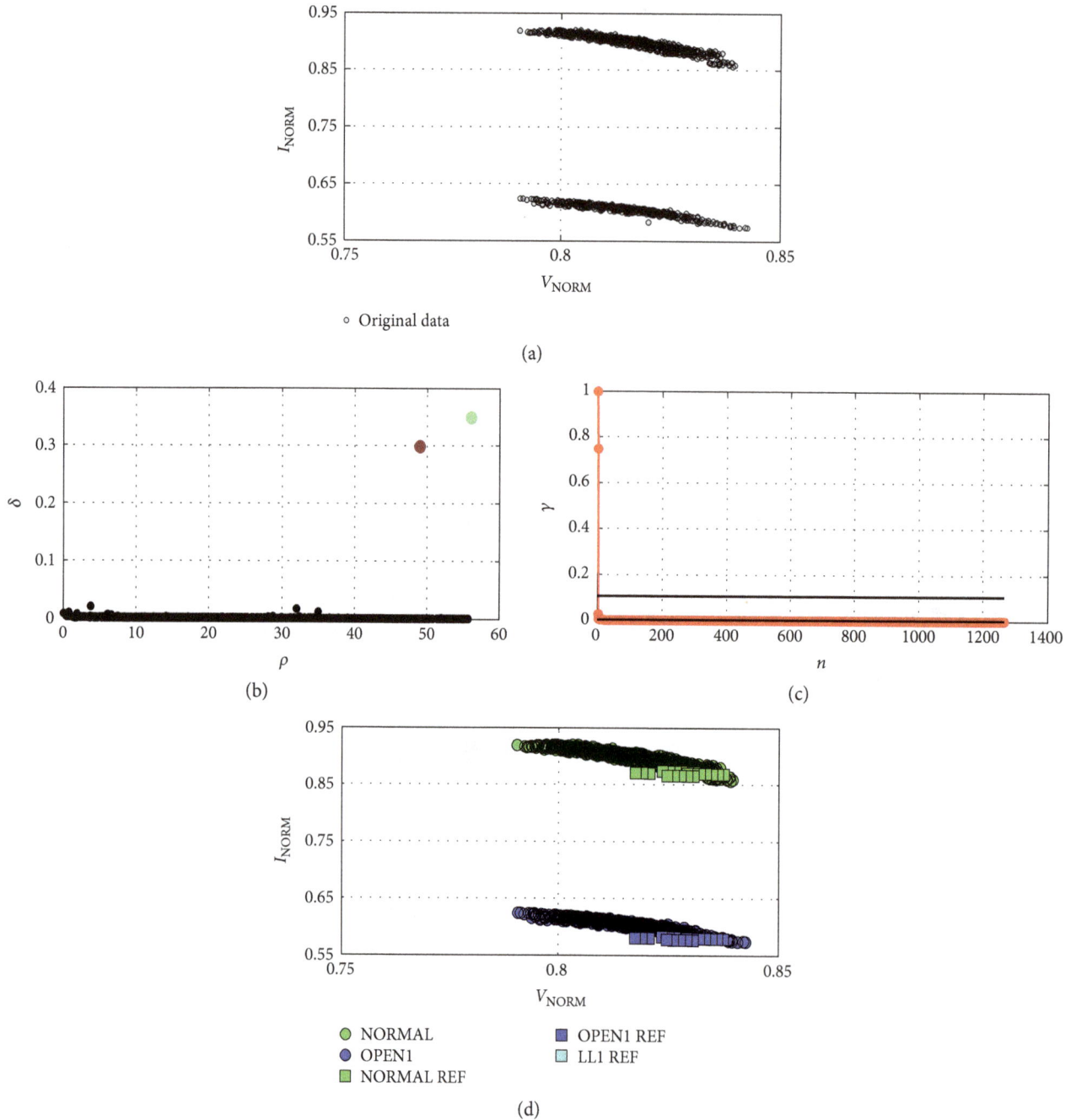

FIGURE 13: Experimental result of the combination of NORMAL and OPEN1 cases: (a) original data, (b) decision graph, (c) the value of γ in decreasing order, and (d) data after clustering and identifying.

the two cluster centers and recognized as NORMAL and OPEN1, respectively, as shown in Figure 8(d). Consequently, the test data of this combination can be accurately clustered and characterized.

(3) Case Study III: Combination of Three Conditions. The combination of three conditions, that is, NORMAL, OPEN1, and OPEN2, is investigated. The case represents the three conditions which successively occur in one day. Hence, there should be three data clusters. The original data is shown in

Figure 9(a). From Figures 9(b) and 9(c), it is clear that three cluster centers are properly chosen, that is, $N_c = 3$. For the three clusters, the minimum distance vectors are as follows: $D_M^1 = [0.00013, 0.16708, 0.01753, 0.04974]$, $D_M^2 = [0.13573, 0.00039, 0.14365, 0.14686]$, and $D_M^3 = [0.31827, 0.14515, 0.32846, 0.32709]$. The d_c equals 0.00454; thus, it can be illustrated that the first element of D_M^1 and the second elements of D_M^2 are smaller than σ. Therefore, clusters one and two can be classified as NORMAL and OPEN1, respectively. For the third cluster, it can be found that all the elements in D_M^3 are larger than d_c, while the second elements are the smallest. Thus, the cluster can be identified as the category of OPEN, as shown in Figure 9(d).

Consequently, the proposed approach has the ability to accurately cluster the PV data in various simulated cases and diagnoses the faults in PV arrays.

4. Experimental Results

In this section, the presented approach is tested with an experimental PV system, and the experimental platform as well as the experimental results is presented.

4.1. Experimental Platform. A 1.8 kW grid-connected photovoltaic system is applied to test the performance of the proposed algorithm under the real working conditions, as shown in Figure 10. The PV array consists of three PV strings in parallel, and each string has six modules in series. The reference PV modules have the same electrical parameters with the PV array. Moreover, it can be assumed that the PV array and the reference PV modules have the identical working environment since they are installed together. Therefore, the reference PV modules are applied real time normalizing the PV data online. The overview for parameters of components in the PV system is given in Table 2.

Three instances are implemented and studied, including NORMAL, the combination of NORMAL and LL1, and the combination of NORMAL and OPEN1. The first case is carried out in summer with a high running temperature range, and the other two cases are operated in spring with a relatively low temperature range. The detailed description about these conditions has been presented in Section 3.2. The experimental environment for the PV array and the amount of data recorded during the experiments are given in Table 3.

Besides, the reference data are obtained by using a PV simulation based on the parameters from Table 2. The reference data include three categories and are arranged in such a sequence: NORMAL, OPEN1, and LL1. The solar irradiance for the PV simulation is fixed at 200 W/m². According to the operating temperature range of the three cases, the ambient temperature range for the PV simulation is 21–40°C for the first case and 0–20°C for the others, respectively.

4.2. Experimental Results. Figures 11–13 illustrate the experimental results of the aforementioned three cases. It is obvious that the distribution of experimental data has remarkable clustering, which is similar to the simulated ones. For the NORMAL condition, as shown in Figure 11, only a cluster is found by the proposed approach. And D_M^1 equals [0.00011, 0.27664, 0.06441] and d_c equals 0.00132, which indicates that the cluster can be accurately categorized as NORMAL.

Second, for the second case, as can be seen from Figure 12, the data are exactly clustered into two groups. And d_c is 0.00238, D_M^1 equals [0.00018, 0.27546, 0.05649], and D_M^2 equals [0.08573, 0.29803, 0.00016]. Accordingly, it is clear that the two clusters can be recognized as NORMAL and LL1, respectively.

Finally, for the third instance, as shown in Figure 13, two clusters are exactly obtained. The d_c is 0.00242. For the two clusters, D_M^1 and D_M^2 equal [0.00055, 0.27494, 0.06779] and [0.24378, 0.00052, 0.25088], respectively. Therefore, the test data of this instance can be characterized as NORMAL and OPEN1, respectively.

Consequently, according to the experimental results, the proposed approach has the ability to cluster and classify the daily data of the PV array.

5. Conclusions

According to the distribution features of the daily operating data from a PV system, a clustering approach has been presented to identify the working conditions of the PV system and further diagnose the faults in the PV array. The proposed method has the ability to cluster the PV data and identify the clusters based on the minimum distance vector between the reference data and the clusters. Three kinds of daily work cases are simulated to validate the effectiveness of the approach, that is, the normal condition, the combination of normal condition with one fault, and the combination of normal condition with two faults. The simulated results indicate that the method can accurately cluster the PV data and identify the faults in each case. Furthermore, a grid-connected PV system is built to test the experimental performance of the developed approach. Under different temperatures and irradiation ranges, three daily operating status of the PV system are implemented and the experimental results also demonstrate the usefulness of the algorithm in a practical system.

Conflicts of Interest

The authors declare that they have no conflicts of interest.

Acknowledgments

The authors would like to thank Dr. Ye Zhao from the Power Electronics Research Group at Northeastern University for the generous offer of valuable suggestions about photovoltaic modeling and fault analysis. This work was supported by the National Natural Science Foundation of China (Grant nos. 61574038, 31300473, 61601127, and 51508105), the Science Foundation of Fujian Education Department of China (Grant no. JAT160073), the Science Foundation of Fujian Science & Technology Department of China (Grant nos. 2015H0021, 2015J05124, and 2016H6012), the Fujian Provincial Economic and Information Technology Commission of China (Grant nos. 830020 and 83016006), and the Scientific Research Foundation for the Returned Overseas Chinese Scholars, State Education Ministry (Grant no. LXKQ201504).

References

[1] V. Sharma and S. S. Chandel, "Performance and degradation analysis for long term reliability of solar photovoltaic systems: a review," *Renewable and Sustainable Energy Reviews*, vol. 27, pp. 753–767, 2013.

[2] *Article 690—Solar Photovoltaic Systems, NFPA70, National Electrical Code*, 2014.

[3] Y. Zhao, J. F. de Palma, J. Mosesian, R. Lyons, and B. Lehman, "Line–line fault analysis and protection challenges in solar photovoltaic arrays," *IEEE Transactions on Industrial Electronics*, vol. 60, no. 9, pp. 3784–3795, 2013.

[4] J. Flicker and J. Johnson, "Analysis of fuses for blind spot ground fault detection in photovoltaic power systems," *Sandia National Laboratories Report*, Tech. Rep., NM, USA, 2013.

[5] Y. Hu, W. Cao, J. Wu, B. Ji, and D. Holliday, "Thermography-based virtual MPPT scheme for improving PV energy efficiency under partial shading conditions," *IEEE Transactions on Power Electronics*, vol. 29, no. 11, pp. 5667–5672, 2014.

[6] Z. Zou, Y. Hu, B. Gao, W. L. Woo, and X. Zhao, "Study of the gradual change phenomenon in the infrared image when monitoring photovoltaic array," *Journal of Applied Physics*, vol. 115, no. 4, pp. 1–11, 2014.

[7] C. Buerhop, D. Schlegel, M. Niess, C. Vodermayer, R. Weißmann, and C. J. Brabec, "Reliability of IR-imaging of PV-plants under operating conditions," *Solar Energy Materials and Solar Cells*, vol. 107, pp. 154–164, 2012.

[8] T. Takashima, J. Yamaguchi, K. Otani, T. Oozeki, K. Kato, and M. Ishida, "Experimental studies of fault location in PV module strings," *Solar Energy Materials and Solar Cells*, vol. 93, no. 6, pp. 1079–1082, 2009.

[9] T. Takashima, J. Yamaguchi, and M. Ishida, "Fault detection by signal response in PV module strings," in *33rd IEEE Photovoltaic Specialists Conference, 2008 (PVSC'08), IEEE*, pp. 1–5, California, USA, 2008.

[10] S. Silvestre, A. Chouder, and E. Karatepe, "Automatic fault detection in grid connected PV systems," *Solar Energy*, vol. 94, pp. 119–127, 2013.

[11] A. Chouder and S. Silvestre, "Automatic supervision and fault detection of PV systems based on power losses analysis," *Energy Conversion and Management*, vol. 51, no. 10, pp. 1929–1937, 2010.

[12] W. Chine, A. Mellit, A. M. Pavan, and S. A. Kalogirou, "Fault detection method for grid-connected photovoltaic plants," *Renewable Energy*, vol. 66, pp. 99–110, 2014.

[13] S. Silvestre, M. A. da Silva, A. Chouder, D. Guasch, and E. Karatepe, "New procedure for fault detection in grid connected PV systems based on the evaluation of current and voltage indicators," *Energy Conversion and Management*, vol. 86, pp. 241–249, 2014.

[14] S. Silvestre, S. Kichou, A. Chouder, G. Nofuentes, and E. Karatepe, "Analysis of current and voltage indicators in grid connected PV (photovoltaic) systems working in faulty and partial shading conditions," *Energy*, vol. 86, pp. 42–50, 2015.

[15] S. Silvestre, L. Mora-López, S. Kichou, F. Sánchez-Pacheco, and M. Dominguez-Pumar, "Remote supervision and fault detection on OPC monitored PV systems," *Solar Energy*, vol. 137, pp. 424–433, 2016.

[16] I. Yahyaoui and M. E. V. Segatto, "A practical technique for on-line monitoring of a photovoltaic plant connected to a single-phase grid," *Energy Conversion and Management*, vol. 132, pp. 198–206, 2017.

[17] Y. Zhao, L. Yang, B. Lehman, J. F. de Palma, J. Mosesian, and R. Lyons, "Decision tree-based fault detection and classification in solar photovoltaic arrays," in *Twenty-Seventh Annual IEEE Applied Power Electronics Conference and Exposition (APEC), 2012, IEEE*, pp. 93–99, Florida, USA, 2012.

[18] D. Riley and J. Johnson, "Photovoltaic prognostics and health management using learning algorithms," in *38th IEEE Photovoltaic Specialists Conference (PVSC), 2012, IEEE*, pp. 001535–001539, Texas, USA, 2012.

[19] S. Syafaruddin, E. Karatepe, and T. Hiyama, "Controlling of artificial neural network for fault diagnosis of photovoltaic array," in *16th International Conference on Intelligent System Application to Power Systems (ISAP), 2011, IEEE*, pp. 1–6, Hersonissos, Greece, 2011.

[20] Y. Wang, Z. Li, C. Wu, D. Q. Zhou, and L. Fu, "A survey of online fault diagnosis for PV module based on BP neural network," *Power System Technology*, vol. 37, no. 8, pp. 2094–2100, 2013.

[21] A. M. Pavan, A. Mellit, D. De Pieri, and S. A. Kalogirou, "A comparison between BNN and regression polynomial methods for the evaluation of the effect of soiling in large scale photovoltaic plants," *Applied Energy*, vol. 108, pp. 392–401, 2013.

[22] W. Chine, A. Mellit, V. Lughi, A. Malek, G. Sulligoi, and A. M. Pavan, "A novel fault diagnosis technique for photovoltaic systems based on artificial neural networks," *Renewable Energy*, vol. 90, pp. 501–512, 2016.

[23] Y. Zhao, R. Ball, J. Mosesian, J. F. de Palma, and B. Lehman, "Graph-based semi-supervised learning for fault detection and classification in solar photovoltaic arrays," *IEEE Transactions on Power Electronics*, vol. 30, no. 5, pp. 2848–2858, 2015.

[24] B. Fang, X. Yin, Y. Tan et al., "The contributions of cloud technologies to smart grid," *Renewable and Sustainable Energy Reviews*, vol. 59, pp. 1326–1331, 2016.

[25] T. Hu, M. Zheng, J. Tan, L. Zhu, and W. Miao, "Intelligent photovoltaic monitoring based on solar irradiance big data and wireless sensor networks," *Ad Hoc Networks*, vol. 35, pp. 127–136, 2015.

[26] K. H. Tseng, H. J. Wu, G. H. Lin, and P. T. Cheng, "Establishment and case analysis of a photovoltaic cloud management system," in *IEEE 11th Conference on Industrial Electronics and Applications (ICIEA), 2016, IEEE*, pp. 831–836, Hefei, China, 2016.

[27] Y. Zhao, B. Lehman, J. F. de Palma, J. Mosesian, and R. Lyons, "Fault analysis in solar PV arrays under: low irradiance conditions and reverse connections," in *37th IEEE Photovoltaic Specialists Conference (PVSC), 2011, IEEE*, pp. 002000–002005, Washington, USA, 2011.

[28] A. Rodriguez and A. Laio, "Clustering by fast search and find of density peaks," *Science*, vol. 344, no. 6191, pp. 1492–1496, 2014.

[29] Z. A. Bakar, R. Mohemad, A. Ahmad, and M. M. Deris, "A comparative study for outlier detection techniques in data mining," in *2006 IEEE Conference on Cybernetics and Intelligent Systems, IEEE*, pp. 1–6, Bangkok, Thailand, 2006.

[30] S. M. MacAlpine, R. W. Erickson, and M. J. Brandemuehl, "Characterization of power optimizer potential to increase energy capture in photovoltaic systems operating under non-uniform conditions," *IEEE Transactions on Power Electronics*, vol. 28, no. 6, pp. 2936–2945, 2013.

[31] SHELL, *Shell SM55 Photovoltaic Solar Module*, http://www.solarquest.com/microsolar/suppliers/siemens/sm55.pdf.

[32] E. Skoplaki and J. A. Palyvos, "Operating temperature of photovoltaic modules: a survey of pertinent correlations," *Renewable Energy*, vol. 34, no. 1, pp. 23–29, 2009.

14

Mismatch Based Diagnosis of PV Fields Relying on Monitored String Currents

Pierluigi Guerriero,[1] **Luigi Piegari,**[2] **Renato Rizzo,**[1] **and Santolo Daliento**[1]

[1]*Department of Electrical Engineering and Information Technology (DIETI), University of Naples Federico II, Via Claudio 21, 80125 Naples, Italy*

[2]*Department of Electronics, Information and Bioengineering (DEIB), Politecnico di Milano, Piazza Leonardo da Vinci 32, 20133 Milan, Italy*

Correspondence should be addressed to Pierluigi Guerriero; pierluigi.guerriero@unina.it

Academic Editor: Jürgen Hüpkes

This paper presents a DC side oriented diagnostic method for photovoltaic fields which operates on string currents previously supplied by an appropriate monitoring system. The relevance of the work relies on the definition of an effective and reliable day-by-day target for the power that every string of the field should have produced. The procedure is carried out by comparing the instantaneous power produced by all solar strings having the same orientation and by attributing, as producible power for all of them, the maximum value. As figure of merit, the difference between the maximum allowed energy production (evaluated as the integral of the power during a defined time interval) and the energy actually produced by the strings is defined. Such a definition accounts for both weather and irradiance conditions, without needing additional sensors. The reliability of the approach was experimentally verified by analyzing the performance of two medium size solar fields that were monitored over a period of four years. Results allowed quantifying energy losses attributable to underperforming solar strings and precisely locating their position in the field.

1. Introduction

It is commonly known that photovoltaic (PV) fields can have a lower energy yield than expected. This occurrence can be due to many factors, the main ones being the adoption of low-quality materials (in order to reduce costs), careless assembly (because of poorly skilled manpower), and wrong design. Often the bad performance is only recognized after that the degradation has become so impressive that revenues fall well behind the nominal targets. The chance to detect malfunction events early depends on the adoption of reliable monitoring/diagnostic (M&D) systems (a complete literature survey about M&D methods can be found in [1]).

A rough classification of M&D techniques can be based on the "level of granularity" (LoG). The lowest LoG corresponds to treating of the solar field as a whole, thus monitoring the instantaneous output power, at either the DC side or the AC side, while the highest LoG corresponds to the monitoring of each individual solar panel embedded in the solar field. Independently of the LoG, the yield of a photovoltaic system can be evaluated by means of the performance ratio (PR), which, according to the IEC 61724 [2], is defined as the ratio between the measured instantaneous power, P_i, and the nominal power of the system, P_{nom}, corrected by taking into account the instantaneous irradiance G_i with respect to the irradiance at STC.

$$PR = \frac{P_i}{P_{nom}} \frac{G_{STC}}{G_i}. \tag{1}$$

In order to take into account thermal effects, the improved version was proposed in [3]:

$$PR(T) = \frac{P_i}{P_{nom} + \beta \cdot \Delta T} \frac{G_{STC}}{G_i}, \tag{2}$$

where β is the temperature coefficient and ΔT is the temperature increment with respect to 25°C.

In [4], P_{nom} was replaced with the expected ac power P_{ac}, evaluated as

$$P_{\text{ac}} = G_i \left(a_1 + a_2 G_i + a_3 \log\left(G_i\right)\right)\left(1 + a_4\left(\Delta T\right)\right), \quad (3)$$

where $a_1, a_2, a_3,$ and a_4 are fitting parameters, while, in [5, 6], (1) was modified as follows:

$$\text{PR}\left(T\right) = \frac{P_{\text{nom}}\left(G_i/G_{\text{STC}}\right) - L}{P_{\text{nom}}\left(G_i/G_{\text{STC}}\right)}, \quad (4)$$

where L is a loss term, taking into account both temperature and mismatches effects.

Methods based on the monitoring at low LoG level have the drawback of not being suitable for locating faulted components. Such a feature can be achieved by moving towards a higher LoG, which implies the increasing of the number of monitored parameters. This approach is illustrated in [7], where string currents are compared with a previously defined nominal reference.

A slight different approach can be found in [8], where an inferential algorithm is adopted to individuate, after an initial training, one or more reference strings whose yields are used in place of the nominal power for the definition of the performance ratio. In [9], a restricted dataset of observed string currents and voltages are analyzed for the determination of possible faults that could have caused that dataset.

M&D methods based on the PR require the definition of a range where PR is considered as normal (even though less than 1); the definition of the range is usually made on statistical bases by considering the standard deviation σ. The optimal width of the confidence interval in terms of σ is still debated; as an example, in [10] the ineffectiveness of the 3σ rule, usually adopted to recognize outliers strings, was evidenced. Alternative methods were proposed in [11, 12], where the plot of the whole I-V curve of individual strings was used to recognize six categories of faults, including shadow effects and bypass diode fault.

Improved fault location capability can be attained by pushing the monitoring at the individual solar panel level. This solution is relatively simple to implement in distributed conversion systems [13], where each solar panel has its own dc/dc converter that can be properly controlled to plot or estimate the I-V characteristic [14]. Otherwise a restricted set of parameters can be measured. In [15] both the instantaneous solar panel operating voltage and the operating string current (which coincides with the panel current) are measured, so that the instantaneous power delivered by the solar panel can be calculated. In [16–18] the insertion of a series connected switch is exploited to temporarily disconnect individual solar panels from the string, thus allowing the measurement of both open circuit voltage and short circuit current. Unfortunately, single panel monitoring systems are too expensive and their adoption is very limited.

On the other hand, almost all large solar fields have some form of current monitoring. This information is usually provided by current sensors embedded into the parallel boxes (string-boxes), which logs operating currents into a devised database. Querying of the database is usually allowed and data can be organized in a large variety of plots and tables;

however, interpretation is left to the customer. The main issue is that data might be not significant by itself; in fact, there is no way to interpret, for example, the power delivered by a PV string in a given day without information about the expected power in that day. Unfortunately, as evidenced by (1)–(4), the latter information depends on the specific weather conditions. This problem has been tackled (e.g., [3, 8]) by enriching the monitoring system with additional sensors (irradiance, temperature, wind speed, and so on), allowing to convert weather information into suitable yield targets. The main drawback of this approach (in addition to the need for a more pervasive sensor network) is that it requires accurate models and reliable parameters.

In this paper, an automated procedure for analyzing string current is proposed. Differently from other methods, the proposed approach does not require irradiance and/or temperature sensors. Moreover, since no provisional models are adopted, calibration and parameter extraction are not required as well. The method, indeed, relies on the generation of a site dependent target, which is built by exploiting only the string currents, with no need for weather information. As will be detailed in Section 2 the target is built by evaluating the instantaneous power produced by all solar strings and by attributing, as producible power for all of them, the maximum value. Thanks to this approach real-time warnings can be given, along with detailed information about production losses. Moreover, malfunctioning strings can be located and, in selected cases, the origin of the performance limitation can be suggested.

Since the procedure is based on DC side measurements its main limit is that losses coming from AC faults cannot be quantified. However, it should be considered that catastrophic faults occurring in the AC side (e.g., undervoltage/overvoltage and underfrequency/overfrequency) result in the disconnection of the inverter from the utility grid. This event is recognized thanks to the simultaneous (and sudden) zeroing of all strings currents.

The paper is organized as follows. Section 2 reports the method for generating the power versus time (P-T) reference curves. Section 3 illustrates the results of a wide experimental campaign, performed on two solar fields (280 kWp and 420 kWp). Discussion and comments about the results are also provided. Conclusions are drawn in Section 4.

2. Target Generation

As mentioned above, string current monitoring systems (string-boxes) are largely available and widely adopted in large PV fields. In this paper the commercial system [19] was exploited. This system populates a devised database with string currents measured every 5 min; subsequently, currents were converted into power by reading the string voltages (since the performance of strings belonging to the same subarray was compared, the voltage was just a proportionality constant between current and power). It must be remarked that both analyzed solar fields were not equipped with irradiance/temperature sensors; hence, data analysis with approaches presented elsewhere (e.g., [3, 8]) was not allowed.

Power profiles available for each string had the form shown in Figure 1. This figure reports the power produced, for example, on March 28, 2015, by PV string #3, connected to inverter #1 (notation (n, m) in the legend means that the curve refers to string #m which is connected to inverter #n).

The analysis of the curve poses several issues. First, a medium size solar field consists of tens of strings (the two fields analyzed in this paper had 61 and 85 PV strings, resp.). Thus, a large number of curves are produced every day. Second, the curves do not give information by themselves because the power produced in a given instant of time is the result of the specific irradiance conditions. As a consequence, the curves must be compared with an *expected* reference; references might be constructed by adopting historical weather series (as done in commercial tools for PV system design). In such a case, only a statistical comparison over a long observation period could be performed. Otherwise, references could be achieved by exploiting real-time data on weather conditions. In such a case, very reliable parameters and accurate modeling of the solar field structure would be needed [20]; special cases like architectural shadows [21, 22] could not be recognized.

In this paper, the target P-T curve was built by exploiting the string currents (converted into power) provided by the monitoring system. More specifically, it was assumed that, in a large solar field, it is likely that, for each instant of time, at least one PV string is properly working. On this basis, a virtual target curve is created, according to (5) by considering, for each instant of time, the power produced by the best performing string.

$$P_{n,m}(t) = \max\left([P_{n,1}(t) \cdots P_{n,M}(t)]\right). \tag{5}$$

In (5) the subscripts n and m have the meaning defined above (string #m connected to inverter #n), and M is the number of strings connected to inverter #n. In other terms, the target P-T curve (for a group of strings which are parallel-connected to the same inverter) consists of several pieces, and each piece is extracted from the P-T curve of the string that, in the specified interval of time, is producing more than all the other strings. The sense of this definition is that if in the array there is a string that can actually produce that power (as defined in (5)), all the strings could do that, based on the assumption that weather conditions, irradiance, and ambient temperature are the same.

As an example, Figure 2 shows, along with the curve already depicted in Figure 1, the target P-T curve evaluated according to (5). The analysis of Figure 2 clearly points out a reliability issue affecting string #3, while Figure 1, without the comparison with the target, could have been interpreted as a reduction of irradiance (clouds) occurring at 15:00.

It is worth nothing that the definition of the reference P-T curve allows determining of the daily energy target as well; indeed, energy is defined as the integral of the P-T curve and is graphically illustrated in Figure 3(b).

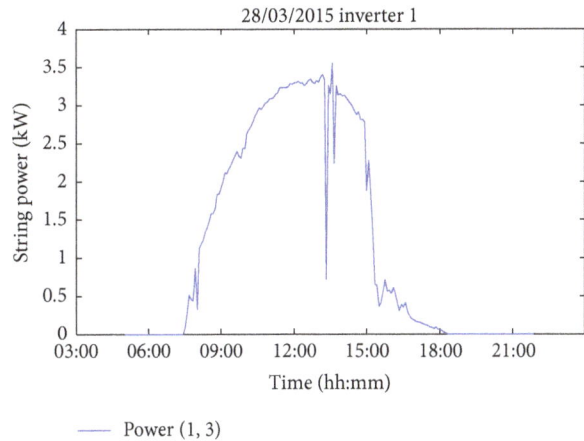

FIGURE 1: Power versus time (P-T) curve measured in March 2015 for string #3 connected to inverter #1.

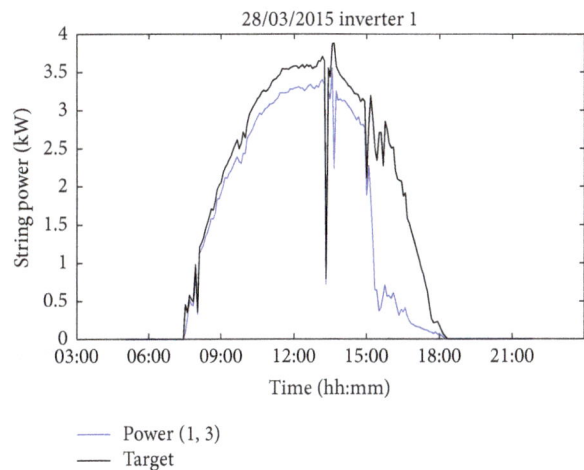

FIGURE 2: P-T curve already reported in Figure 1 compared with P-T target curve evaluated for specified day.

3. Experiments and Discussion

The daily production target defined above was exploited for the diagnosis of malfunctions occurring in two solar fields, both performing unexpected low energy production.

Two kinds of analysis were performed: a horizontal analysis, where the performance of each PV string was compared with other strings and with the target, over one year of observation, and a vertical analysis, where the performance of each string is compared with itself over four years of observation. Comparisons were made at the subarray level, where the subarray is defined as a group of parallel-connected strings with same orientation. Since different subarrays experienced slightly different tilt and azimuth orientations, devised P-T targets were defined for each of them. With the aim of avoiding the presentation of almost duplicate data, the results of the horizontal analysis will be shown with reference to the 280 kWp solar field, while the results of the vertical analysis refer to the 420 kWp solar field.

(a)

(b)

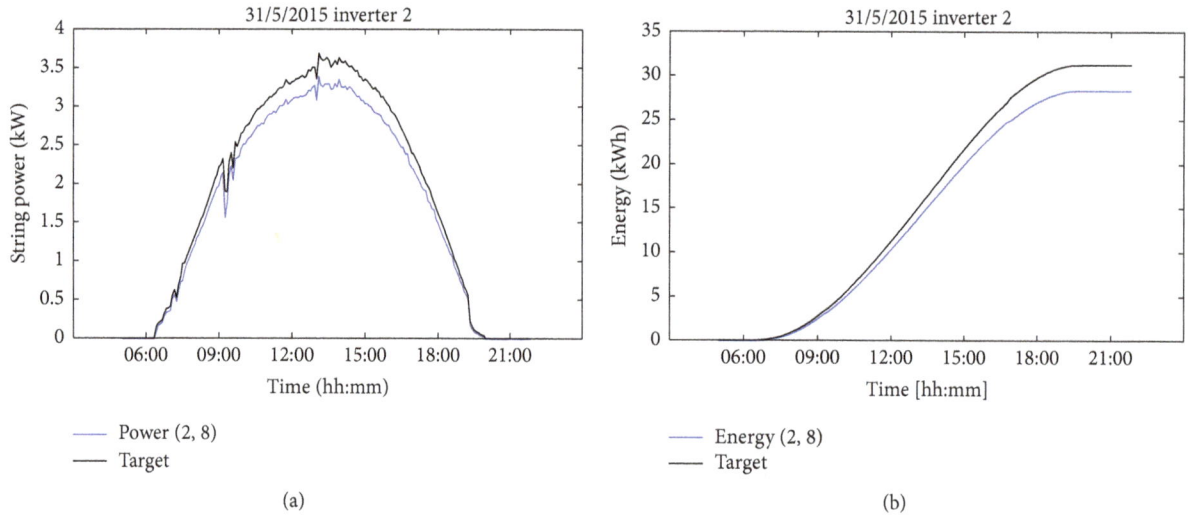

FIGURE 3: (a) Actual and target *P-T* curves in Figure 1 and (b) corresponding energy versus time (*E-T*) curves.

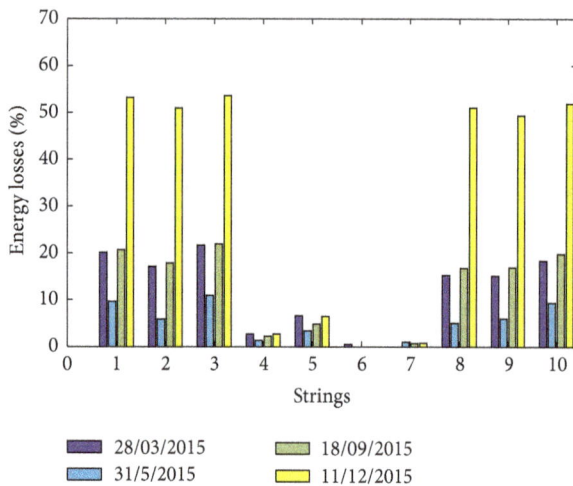

FIGURE 4: Percentage of energy losses with respect to target of ten strings connected to inverter #1.

3.1. Horizontal Analysis.

In the horizontal analysis, the energy produced every day by each PV string of a subarray was monitored for a one-year period (2015) and compared with the energy target defined by integrating the corresponding *P-T* curve, the target being defined with reference to the same subarray. The comparison between the energy actually produced and the target made it possible to quantify the energy losses accumulated by the specified string during the given day.

In Figure 4, an example of the results is shown. For the sake of simplicity, only the energy losses referring to four days of the year are reported. These four days are representative of the four seasons; in order for the data to be representative, four bright days were chosen. The figure shows the performance of the ten strings connected to the inverter denoted as #1.

From the figure, it can be argued that two strings, #6 and #7, perform almost like the target (which means that

according to (5) the target was built mainly by taking pieces of the *P-T* curves from string #6 and string #7); some of the others show relevant energy losses, with peaks of about 50% in December (yellow bars). The latter circumstance suggested the eventuality that losses could have been attributed to architectural shadowing. In order to point out the effects of architectural shading the relative energy losses were also evaluated by restricting the integration window of the *P-T* curve to only two hours, centered at noon (from 11:00 to 13:00), so as to have the sun as high as possible in the sky. Figure 5 shows the results.

The figure reveals a strong reduction of the percentage losses, thus confirming that when the sun appears lower in the sky the solar field is strongly subject to shadowing. However, significant gaps can be still seen around noon as well. In the proposed approach those strings exhibiting more than a given percentage (set by the customer) of energy losses, during a given interval of time (set by the customer as well), are classified as malfunctioning. In other terms, the customer can decide to receive a real-time alarm if a string produces (for example) less than 50% of the target during one hour of observation or a daily alarm if the daily production is less than (say) 20%. The analysis shown in Figures 4 and 5 was repeated for all eight inverters of the field. The results are collected in Figure 6.

It can be observed that there are large variations, both among the subarrays and through each subarray. In Table 1, the strings experiencing energy losses greater than 5% are reported. A cumulative loss of about 5 MWh/year can be attributed to them. The cumulative energy losses evaluated by considering all the strings (61) were about 10 MWh/year.

A further summary of the results is shown in Figure 7, which reports the average energy loss (with respect to the target) cumulated by each subarray; the standard deviation is also reported.

It is important to point out that each subarray had its own target; therefore percentage losses, shown in the figure, are reliable estimators of the energy that a proper maintenance

TABLE 1: Yearly losses of the worst strings.

Inverter	String	Energy loss [kWh/year]
1	1	510
1	2	575
1	8	380
2	8	510
3	2	380
3	3	510
3	4	450
4	5	510
6	1	380
6	6	510

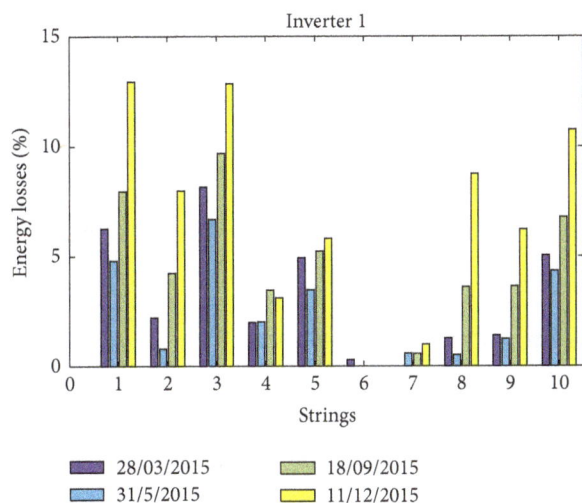

FIGURE 5: Percentage of energy losses with respect to target accumulated from 11:00 to 13:00.

could recover. The width of the standard deviation can be used to individuate those subarrays embedding outliers string. For example, the large standard deviation found for inverter #7 is a clear warning about the presence of some strongly underperforming string, as confirmed by Figure 6(g) (inverter #7), which shows large losses attributable to string #5; the fact that losses strongly increase in December suggests that the string is subject to architectural shadowing.

The above results are not trivial. It should be pointed out, indeed, that the analysis was fully automated. The tool can operate in blind mode on every kind of database containing string currents. Moreover, as reported in Table 1, strongly underperforming strings are precisely located, and the amount of energy loss can be used as a criterion for deciding the priority of maintenance.

3.2. Vertical Analysis. In the vertical analysis the performance of each solar string was monitored over a period of four years. This kind of analysis made it possible to quantify deterioration over time. Examples of vertical analyses are reported in Figures 8(a) and 8(b).

The comparison of the measurements performed in October 2012 and October 2015 shows a dramatic deterioration of string #9. It is important to note that the target curves were almost identical; this observation allowed to surely classify the string as faulty. In the current version, the procedure automatically classifies as "probably faulty" those strings exhibiting an increment of energy loss (with respect to the target) greater than 5% from one year to another.

For the sake of completeness, Figure 9 illustrates the cumulated energies corresponding to the *P-T* curves of Figure 8.

As can be seen the daily energy lost by string #9 was about 40%.

Summary results of the vertical analysis for all the subarrays are reported in Figure 10 (notice that results referring to 2013 were omitted because in that year the solar field was subject to dramatic failure on the AC side).

As can be seen, in some cases (see, e.g., Figure 10(b), string #9) degradation occurs suddenly from one year to another. This fact indicates some disruptive phenomenon occurring in solar panels rather than the gradual worsening of the parameters. Another observation is that, in many cases, those strings which show a large deterioration from one year to another (see, e.g., Figure 10(c), string #1) were characterized by high losses since the first year of operation, thus suggesting the presence of defective solar panels, already prone to reliability issues. The above considerations can be better appreciated by analyzing degradation on a statistical base. Let us consider Figure 11, which represents the percentage energy yield (with respect to the maximum allowed energy production given by the target) over time for the strings connected, for example, to the inverter #4. The bars refer, respectively, to the best performing string, the worst performing string, and the average yield, the latter being evaluated by considering all the strings connected to inverter #4. From the figure it can be argued that the decreasing of the average value is almost entirely caused by the strong degradation of the worst string (as confirmed by Figure 10(d), string #9), thus confirming, again, that degradation is not a uniform phenomenon. As mentioned above, it is also interesting to note that the worst string was already underperforming since the first year; this observation suggests the hypothesis that disruptive phenomena take place because of poor components which might cause damage propagation.

A summary, referring to the entire filed, is reported in Figure 12. The figure shows the mean percentage energy yield (with respect to the target) evaluated by considering the whole solar field; the standard deviation is also reported (results referring to inverter #4 are shown for comparison). The analysis of the figure confirms that degradation is mainly a spotted phenomenon. Indeed, the standard deviation increases with time as a result of largely spread individual performance.

3.3. How to Set the Alarm Threshold. From the above discussion it comes that the threshold set for the alarms affects the number of generated alarms.

At first glance the minimum applicable value of the threshold should be adopted so as to achieve the maximum fault detection sensitivity. Such a minimum value depends on the accuracy of instrumentation needed to perform electrical

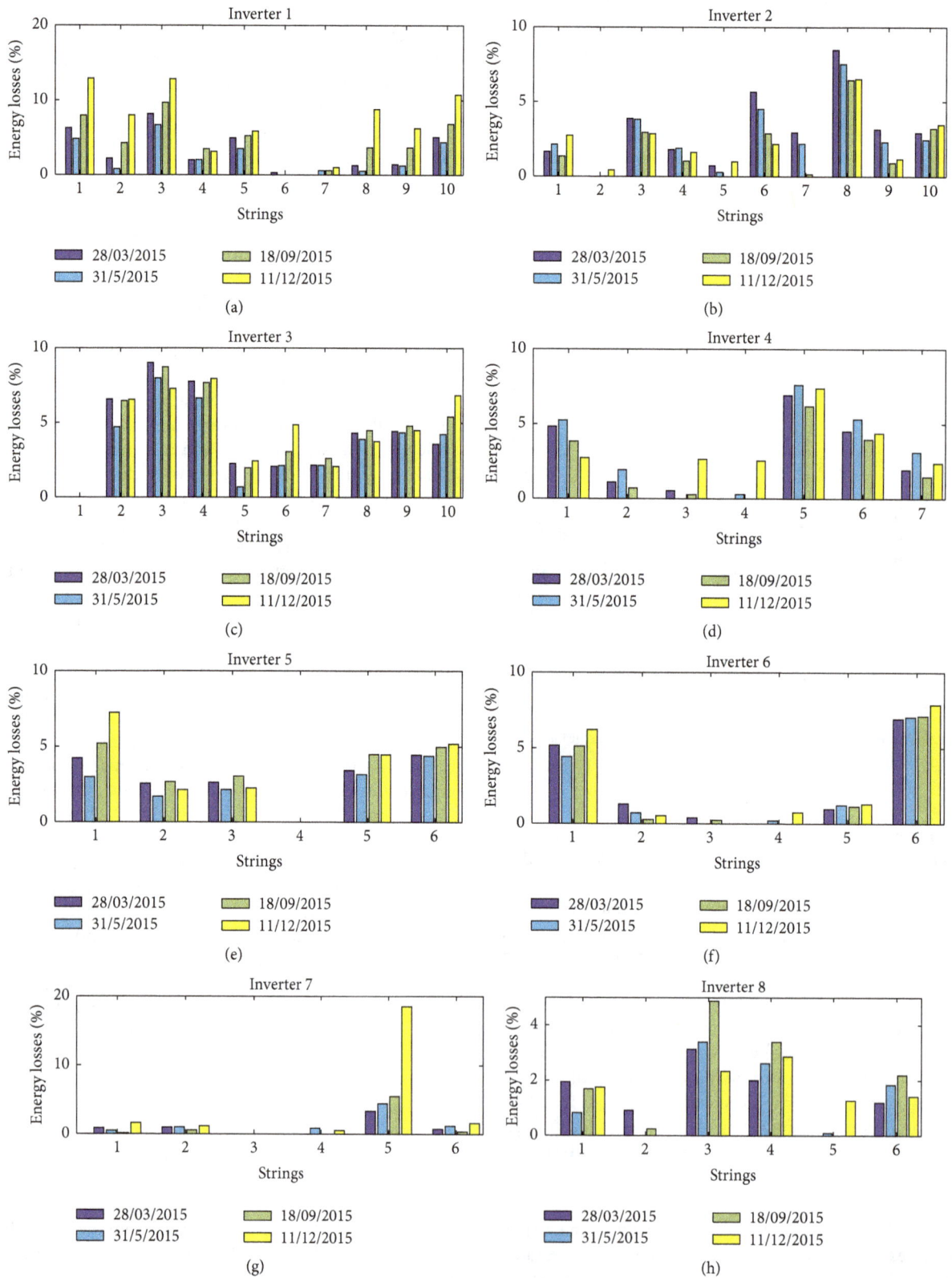

FIGURE 6: Percentage of energy losses for all subarrays under investigation.

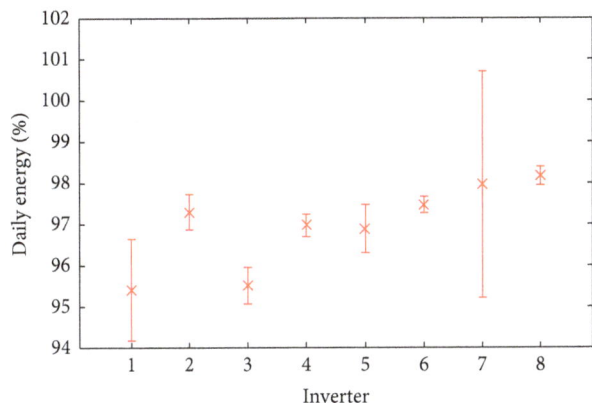

FIGURE 7: Relative energy yield performed by the subarrays.

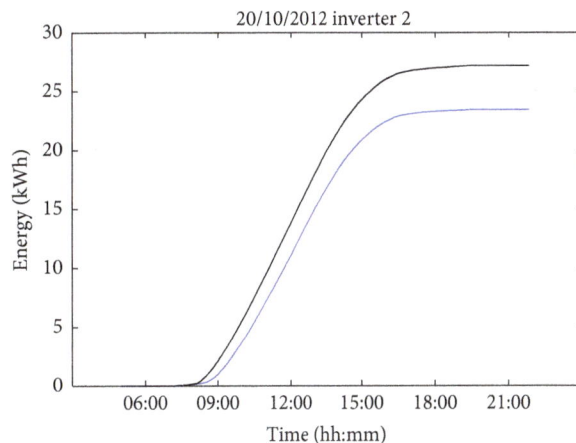

(a)

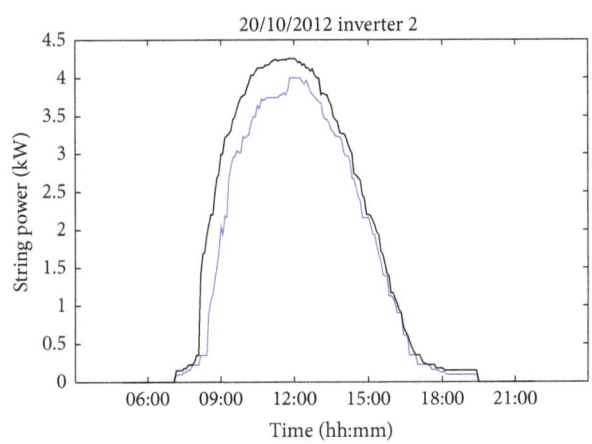

(b)

FIGURE 8: *P-T* curves of string #9 in first year of operation (a) and after 3 years (b).

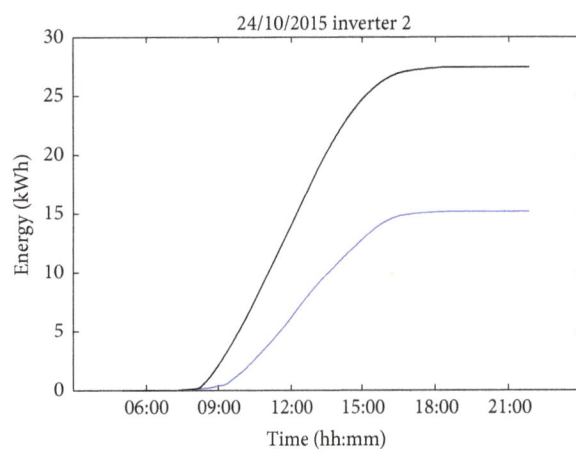

(a)

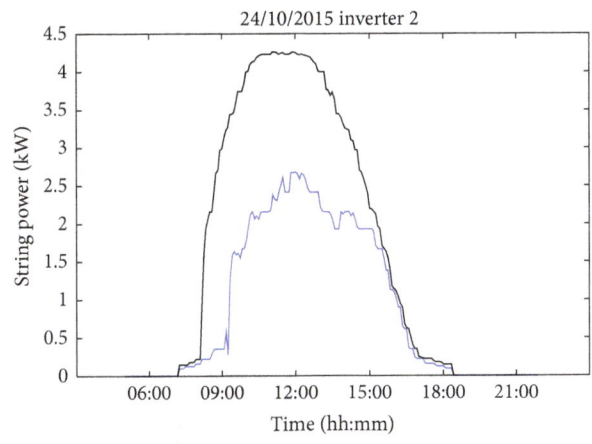

(b)

FIGURE 9: Vertical analysis of accumulated energies. Data refer to the *P-T* curves reported in Figure 8.

lower the threshold, the higher the possible occurrence of false positive alarms, specifically if the value of the threshold is close to the measurement error. Second, a low threshold might identify some events associated with *negligible* energy losses as significant, where the concept of negligible should be related to the economic value of the losses and the expected cost for fixing the fault. Therefore the threshold should be set by trying to fulfill two constraints: it should be sufficiently higher than the measurements errors (about 1% for the systems under investigation) and its value should select only energetically significant losses.

The value of 5% adopted in the above discussed experiments takes into account both the minimum amount of energy loss justifying a maintenance activity and the accuracy provided by the monitoring system.

The effect of a different choice is shown in Figure 13, where the number of alarms is shown as function of the threshold

measurements (voltage and current sensing). However, two drawbacks occur when the threshold is too low. First, the

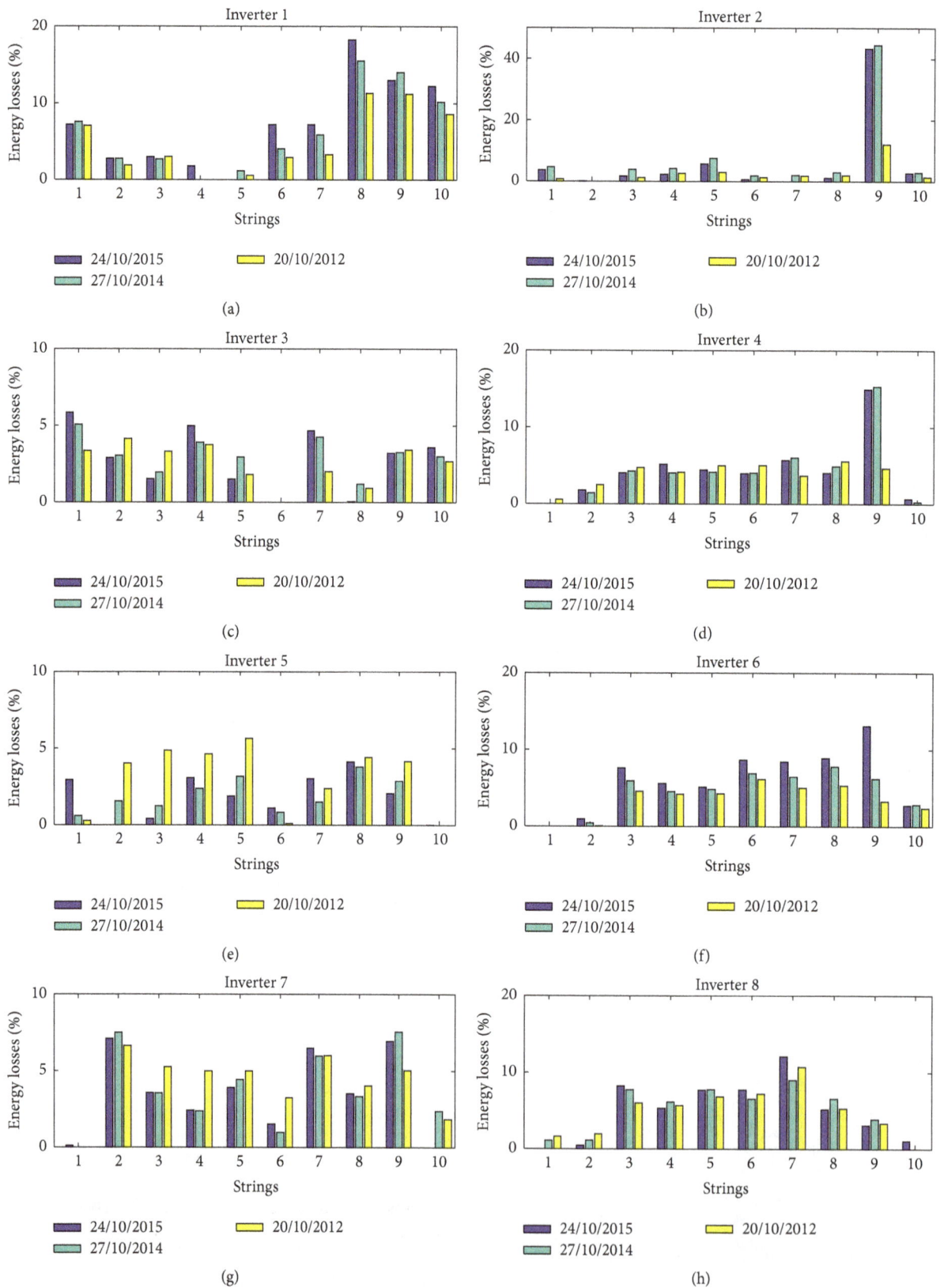

FIGURE 10: Vertical analysis of energy losses for four subarrays of the solar field.

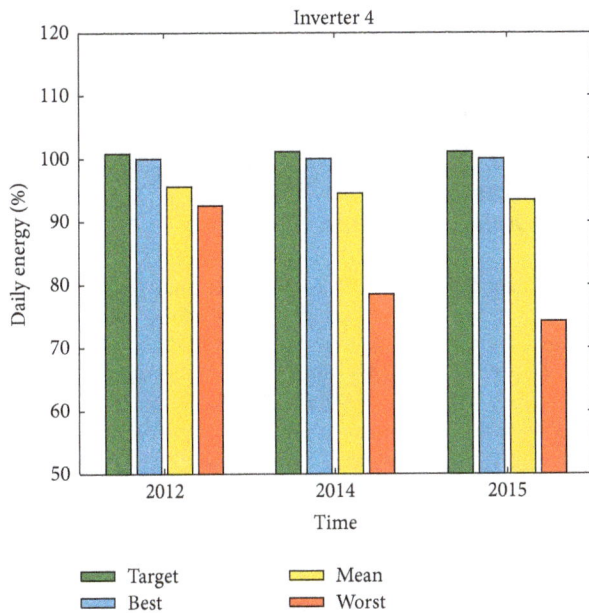

FIGURE 11: Relative energy yield over time. The performance of both the best and worst strings is compared with the mean yield of the subarray and with the target.

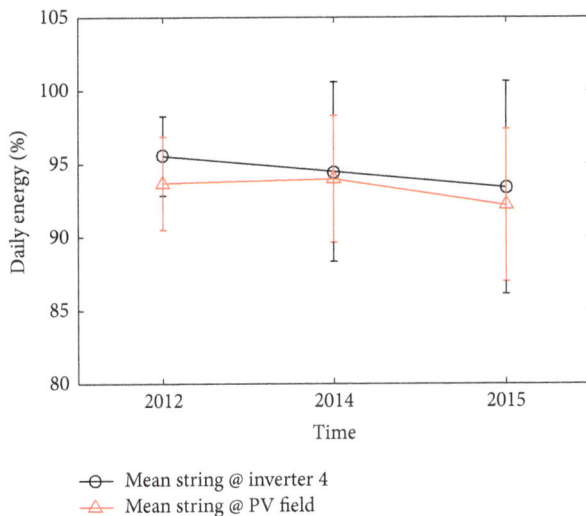

FIGURE 12: Mean value of the relative energy yield performed by the whole solar field (red curve) and by string #4 (black curve).

FIGURE 13: Number of alarms versus threshold: (a) horizontal analysis and (b) vertical analysis.

for both aforementioned horizontal and vertical analyses. As expected, the algorithm identifies about one fault per string if the threshold is 1%. while the number of alarms falls down as the threshold value increases.

It is interesting to note from Figure 13(b) (referring to the vertical analysis performed on the 420 kWp field) that the number of alarms increases with time. This fact means that some strings deteriorate faster than others, with an acceleration from 2014 to 2015 as already observed with reference to Figure 10.

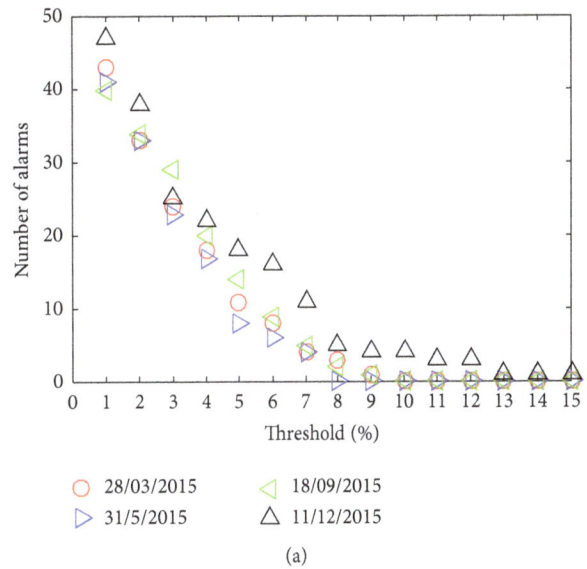

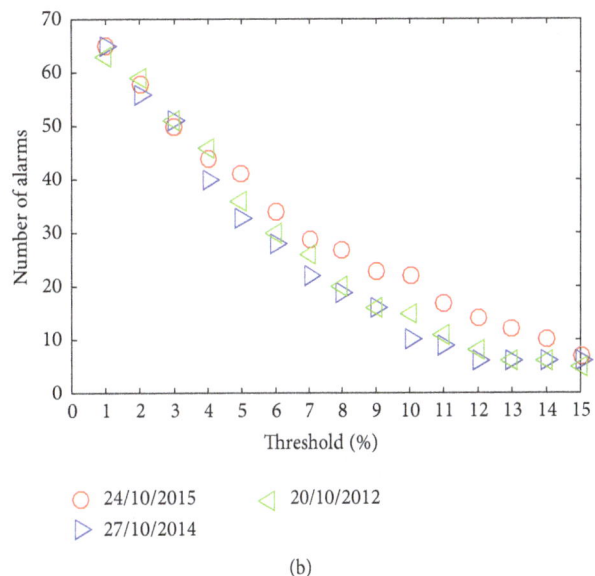

4. Conclusions

In this paper, a method for analyzing the data provided by PV field monitoring systems, working at the single string level, was presented. The method takes advantage of a site dependent target definition for the power that each PV string could produce. The target inherently accounts for weather conditions and makes it possible to quantify the energy losses attributable to each underperforming PV string. As case study, the data provided by the monitoring systems of two medium size solar fields were analyzed. Analysis was conducted among the strings, during a given year (horizontal analysis) and over time for a given string (vertical analysis). The horizontal analysis showed up to 50% of energy losses of some strings with respect to the reference and cumulate losses of about 10 MWh in a year, 5 MWh being attributable to only 10 strings. The vertical analysis allowed individuating some strings exhibiting huge performance deterioration from an year to another, not justifiable with normal ageing. It was interesting to note that those strings experiencing such strong lowering of performance were already selected as underperforming strings after the first period of operation, thus suggesting that an early warning, followed by effective maintenance, could have produced significant revenues.

Competing Interests

The authors declare that they have no competing interests.

References

[1] S. Daliento, A. Chouder, P. Guerriero et al., "Monitoring, diagnosis, and power forecasting for photovoltaic fields: a review," *International Journal of Photoenergy*, vol. 2017, Article ID 1356851, 13 pages, 2017.

[2] IEC 61724, https://webstore.iec.ch/preview/info_iec61724ed1.0en.pdf.

[3] F. Bizzarri, A. Brambilla, L. Caretta, and C. Guardiani, "Monitoring performance and efficiency of photovoltaic parks," *Renewable Energy*, vol. 78, pp. 314–321, 2015.

[4] R. Platon, J. Martel, N. Woodruff, and T. Y. Chau, "Online fault detection in PV systems," *IEEE Transactions on Sustainable Energy*, vol. 6, no. 4, pp. 1200–1207, 2015.

[5] A. Tahri, T. Oozeki, and A. Draou, "Monitoring and evaluation of photovoltaic system," *Energy Procedia*, vol. 42, pp. 456–464, 2013.

[6] R. Hariharan, M. Chakkarapani, G. Saravana Ilango, and C. Nagamani, "A method to detect photovoltaic array faults and partial shading in PV systems," *IEEE Journal of Photovoltaics*, vol. 6, no. 5, pp. 1278–1285, 2016.

[7] M. Baba, T. Shimakage, and N. Takeuchi, "Examination of fault detection technique in PV systems," in *Proceedings of the 35th International Telecommunications Energy Conference 'Smart Power and Efficiency' (INTELEC '13)*, pp. 431–434, October 2013.

[8] L. Cristaldi, M. Faifer, G. Leone, and S. Vergura, "Reference strings for statistical monitoring of the energy performance of photovoltaic fields," in *Proceedings of the 5th International Conference on Clean Electrical Power (ICCEP '15)*, pp. 591–596, June 2015.

[9] S. Ben-Menahem and S. C. Yang, "Online photovoltaic array hot-spot Bayesian diagnostics from streaming string-level electric data," in *Proceedings of the 38th IEEE Photovoltaic Specialists Conference (PVSC '12)*, pp. 2432–2437, IEEE, Austin, Tex, USA, June 2012.

[10] Y. Zhao, F. Balboni, T. Arnaud, J. Mosesian, R. Ball, and B. Lehman, "Fault experiments in a commercial-scale PV laboratory and fault detection using local outlier factor," in *Proceedings of the 40th IEEE Photovoltaic Specialist Conference (PVSC '14)*, pp. 3398–3403, Denver, Colo, USA, June 2014.

[11] M. Davarifar, A. Rabhi, A. Hajjaji, E. Kamal, and Z. Daneshifar, "Partial shading fault diagnosis in PV system with discrete wavelet transform (DWT)," in *Proceedings of the 3rd International Conference on Renewable Energy Research and Applications (ICRERA '14)*, pp. 810–814, Milwakuee, Wis, USA, October 2014.

[12] W. Chine, A. Mellit, A. Massi Pavan, and V. Lughi, "Fault diagnosis in photovoltaic arrays," in *Proceedings of the 5th International Conference on Clean Electrical Power: Renewable Energy Resources Impact (ICCEP '15)*, vol. 7, pp. 62–72, Taormina, Italy, 2015.

[13] M. Coppola, P. Guerriero, F. Di Napoli, S. Daliento, D. Lauria, and A. Del Pizzo, "A PV AC-module based on coupled-inductors boost DC/AC converter," in *Proceedings of the 2014 International Symposium on Power Electronics, Electrical Drives, Automation and Motion (SPEEDAM '14)*, pp. 1015–1020, June 2014.

[14] B. Andò, S. Baglio, A. Pistorio, G. M. Tina, and C. Ventura, "Sentinella: smart monitoring of photovoltaic systems at panel level," *IEEE Transactions on Instrumentation and Measurement*, vol. 64, no. 8, pp. 2188–2199, 2015.

[15] L. Ciani, L. Cristaldi, M. Faifer, M. Lazzaroni, and M. Rossi, "Design and implementation of a on-board device for photovoltaic panels monitoring," in *Proceedings of the IEEE International Instrumentation and Measurement Technology Conference: Instrumentation and Measurement for Life (I2MTC '13)*, pp. 1599–1604, Minneapolis, Minn, USA, May 2013.

[16] P. Guerriero, F. Di Napoli, G. Vallone, V. d'Alessandro, and S. Daliento, "Monitoring and diagnostics of PV plants by a wireless self-powered sensor for individual panels," *IEEE Journal of Photovoltaics*, vol. 6, no. 1, pp. 286–294, 2016.

[17] P. Guerriero, V. d'Alessandro, L. Petrazzuoli, G. Vallone, and S. Daliento, "Effective real-time performance monitoring and diagnostics of individual panels in PV plants," in *Proceedings of the 4th International Conference on Clean Electrical Power: Renewable Energy Resources Impact (ICCEP '13)*, pp. 14–19, June 2013.

[18] P. Guerriero, V. d'Alessandro, L. Petrazzuoli, G. Vallone, and S. Daliento, "Effective real-time performance monitoring and diagnostics of individual panels in PV plants," in *Proceedings of the International Conference on Clean Electrical Power (ICCEP '13)*, pp. 14–19, June 2013.

[19] http://www.energie-rinnovabili.net/archivio/files/GECO-RETAIL-versione2.pdf.

[20] R. Ayaz, I. Nakir, and M. Tanrioven, "An improved matlab-simulink model of PV module considering ambient conditions," *International Journal of Photoenergy*, vol. 2014, Article ID 315893, 6 pages, 2014.

[21] W. He, F. Liu, J. Ji, S. Zhang, and H. Chen, "Safety analysis of solar module under partial shading," *International Journal of Photoenergy*, vol. 2015, Article ID 907282, 8 pages, 2015.

[22] A. J. Hanson, C. A. Deline, S. M. MacAlpine, J. T. Stauth, and C. R. Sullivan, "Partial-shading assessment of photovoltaic installations via module-level monitoring," *IEEE Journal of Photovoltaics*, vol. 4, no. 6, pp. 1618–1624, 2014.

15

Perovskite Thin Film Solar Cells Based on Inorganic Hole Conducting Materials

Pan-Pan Zhang,[1,2] **Zheng-Ji Zhou,**[1,2] **Dong-Xing Kou,**[1,2] **and Si-Xin Wu**[1,2]

[1]*Key Laboratory for Special Functional Materials of Ministry of Education, Henan University, Kaifeng, Henan Province 475004, China*
[2]*Collaborative Innovation Center of Nano Functional Materials and Applications, Henan University, Kaifeng, Henan Province 475004, China*

Correspondence should be addressed to Zheng-Ji Zhou; zzj@henu.edu.cn

Academic Editor: Wilfried G.J.H.M. Van Sark

Organic-inorganic metal halide perovskites have recently shown great potential for application, due to their advantages of low-cost, excellent photoelectric properties and high power conversion efficiency. Perovskite-based thin film solar cells have achieved a power conversion efficiency (PCE) of up to 20%. Hole transport materials (HTMs) are one of the most important components of perovskite solar cells (PSCs), having functions of optimizing interface, adjusting the energy match, and helping to obtain higher PCE. Inorganic p-type semiconductors are alternative HTMs due to their chemical stability, higher mobility, high transparency in the visible region, and applicable valence band (VB) energy level. This review analyzed the advantages, disadvantages, and development prospects of several popular inorganic HTMs in PSCs.

1. Introduction

Perovskite solar cells (PSCs) based on organic-inorganic metal halide perovskites have recently attracted considerable attention as the power conversion efficiency (PCE) has increased dramatically from the initial 3.9% in 2009 to current 22.1% in a short span of several years [1–8]. Previous results demonstrated that PSCs may be the first in the history of photovoltaics (PV) combining high efficiency with low cost. The configurations of PSCs were evolved from dye-sensitized solar cells (DSSCs), and the key materials for the perovskite are compounds with the chemical formula ABX_3 (A = CH_3NH_3, B = Pb or Sn, and X = Cl, Br, or I), which have received extensive attention due to their favorable photovoltaic properties [9, 10]. Generally, the device structure of PSCs can be categorized into mesoporous structure (Figure 1(a)) and planar junction structure (Figure 1(b)) [11]. The main function of perovskite is absorbing light to generate and transmit electron-hole pairs under continuous illumination. Then, the electrons and holes pass through the perovskite and shift to electron-transporting layer (ETL) and hole transport layer (HTL), respectively, to generate current, as schematic presented in Figure 1(c).

While although a mass of efforts such as solvent additives, molecular dipoles, or interface modification has been devoted to optimize device efficiency, none of the present PSCs provides high PCE with long-term stability. Undoubtedly, a lack of confirmed stability may become the biggest barrier on the path of PSCs towards commercialization. HTL is one of the most important components of PSCs, having functions of optimizing interface, adjusting the energy-match, and helping obtain higher PCE, which has a great effect on device performance and stability [12–14]. Poly(3,4-ethylenedioxythiophene)-polystyrenesulfonate (PEDOT:PSS) and 2,2′,7,7′-tetrakis (N,N-di-p-methoxyphenylamine)-9,9,-spirobifluorene (spiro-OMeTAD) have been widely employed as HTL in the field of PSCs. Although PSCs that applied PEDOT:PSS and spiro-OMeTAD as HTL have obtained high PCE, none of them can provide with long-term stability partly because of the problematic acidic and hygroscopic characteristics of the organic HTMs applied in PSCs [15, 16].

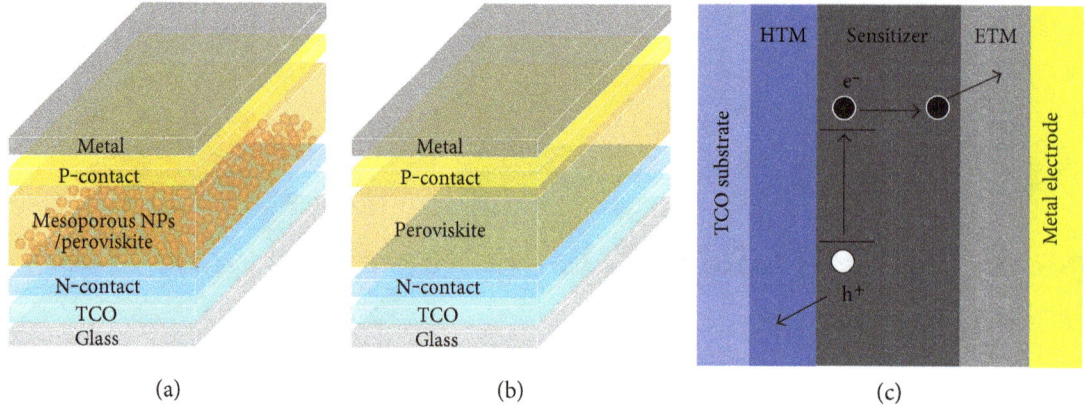

FIGURE 1: Device structures of (a) mesoporous structure, (b) n–i–p planar, planar junction structure, and (c) the scheme of carrier transport in PSCs.

TABLE 1: Device parameters for high-performance organometallic lead halide perovskite solar cells (PSCs) based on different inorganic hole transport materials (HTMs), the band gap, and deposition method of these HTMs.

HTM	Band gap (eV)	Deposition method	V_{oc} (V)	J_{sc} (mA/cm^2)	FF (%)	PCE (%)	Reference
NiO	3.5	ALD	1.04	21.87	72	16.40	[37]
Cu:NiO	$x > 3.5$	Solution-processed	1.12	19.17	73	15.40	[24]
Mg:Li:NiO	$x > 3.5$	Solution-processed	1.07	20.62	75	18.40	[36]
CuI	3.1	Solution-processed	1.04	21.06	62	13.58	[40]
CuSCN	3.8	Electrodeposition	1.00	21.90	76	16.60	[46]
CuO	1.3	Solution-processed	1.06	15.82	72	12.16	[52]
Cu$_2$O	2.1	Solution-processed	1.07	16.52	75	13.35	[52]
MoO$_3$	3.4	Thermal decomposition	1.00	21.49	69	14.87	[53]
VO$_x$	2.42	Solution-processed	0.90	22.29	71	14.23	[55]

Recently, to improve the stability and reduce the cost of PSCs, various inorganic hole transport materials have been discovered and applied. In this review, PSCs employing different inorganic HTMs as hole transport layer (HTL) have been discussed and summarized. To date, a series of p-type inorganic metal compounds have been employed in PSCs, such as CuI [17], CuSCN [18–20], NiO [21–26], CuO [27], Cu$_2$O [27, 28], MoO$_3$ [29–31], and VO$_x$ [32]. Compared to organic HTMs, inorganic p-type semiconductor materials have the advantages of high hole mobility, wide band gap, low cost, and solution-processed availability, which show promising prospects as hole-selective contacts in perovskite solar cells.

2. Inorganic HTMs for PSCs

There are some general requirements for inorganic HTMs used in PSCs, such as high transparency in the visible region, well chemical stability, higher mobility, and applicable valence band (VB) energy level. At present, inorganic HTMs such as CuI, CuSCN, NiO, CuO, Cu$_2$O, MoO$_3$, and VO$_x$ have already been used in PSCs and the solar cells employing inorganic p-type semiconductors as HTLs exhibited improvement in device performance and stability.

Table 1 presents the band gap and deposition method of various HTMs reported in the literatures, as well as their photovoltaic parameters of top-performing PSCs based on these HTMs. In the following, we will analyze and discuss these frequently used inorganic HTMs, respectively.

2.1. NiO. NiO is a well-known p-type semiconductor widely used as a p-sensitization electrode for DSSCs and a hole-selective contact for organic bulk heterojunction solar cells [33, 34].

Initially, NiO was widely applied in DSSCs. In recent years, nickel oxide (NiO) as a promising HTM has been studied by several groups. In 2014, the first announced PCE of PSCs using nickel oxide as HTL was up to 7.8% [35]. Then, NiO nanocrystals (NCs) were obtained by a simple sol-gel process adopted as the hole transport layer in an inverted PSC, which observed a high PCE of 9.11% [25]. In 2015, copper- (Cu-) doped NiO (Cu:NiO) as HTL of planar heterojunction PSCs achieved a PCE of 15.4% [24]. At present, the highest recorded PCE of PSCs using Li, Mg-codoped NiO (Li$_{0.05}$Mg$_{0.15}$Ni$_{0.8}$O) as HTL was 18.4% [36]. This study developed heavily p-doped (p$^+$) Ni$_x$Mg$_{1-x}$O to extract photogenerated hole from perovskite layer, and large size (>1 cm^2) PSCs with an efficiency of up to 16.2% were

(a)

(b)

(c)

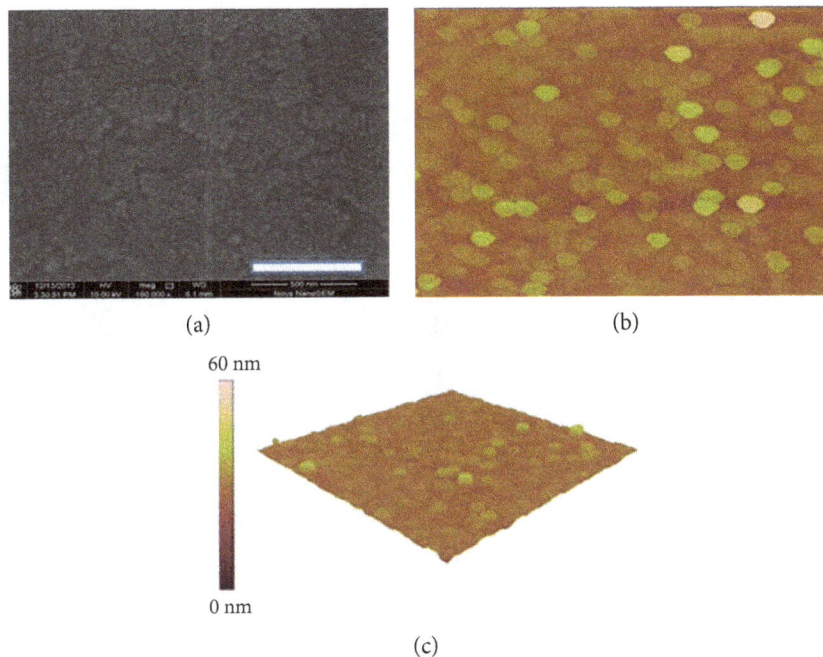

FIGURE 2: (a) SEM NiO film prepared by spin coating, (b) and (c) AFM of NiO film prepared by spin coating. Reprinted with permission from [21].

successfully fabricated. The latest reported PCE of PSCs employing pure NiO as HTL is up to 16.4%, and the NiO film was gotten by the atomic layer deposition (ALD) method [37].

NiO become a potential candidate for HTL, as its p-type characteristics of high optical transmittance, wide band gap (Eg > 3.50 eV), chemical stability, and an applicable valence band match with common light photoactive layers [33, 38]. The synthetic methods of NiO can be categorized into solution-processed and sol-gel, and NiO films can be prepared by various methods, such as pulsed laser deposition, electrodeposition, spray pyrolysis, spin coating, sputtering, and ALD method [37]. A frequently used method for preparing NiO is solution-processed technique by adding monoethanolamine and nickel acetate tetrahydrate in methoxyethanol then stirring for 10 hours to form NiO precursor solutions and via spin coating to achieve NiO film. The NiO film obtained by this way was composed of NiO nanoparticles (NCs), and the whole film was crack-free and smooth, as schematic presented in SEM and AFM (Figures 2(a), 2(b), and 2(c)). The photoluminescence (PL) quenching of perovskite based on NiO showed favorable charge transfer compatibility, presented in Figure 3. Perovskite excitation is at around 760 nm. Finally, PSC with NiO prepared by the above method as HTL obtained a 7.6% device performance [21].

Another common approach to achieve NiO is a simple sol-gel process. NiO film was fabricated by spin coating the sol-gel solution. This kind of NiO nanocrystal film with a flat and smooth surface guarantees the formation of a continuous and close-knit HTLs of PSCs as schematic presented in Figures 4(a) and 4(b). Hole extraction and transport properties of this film interfaced with the perovskite film were

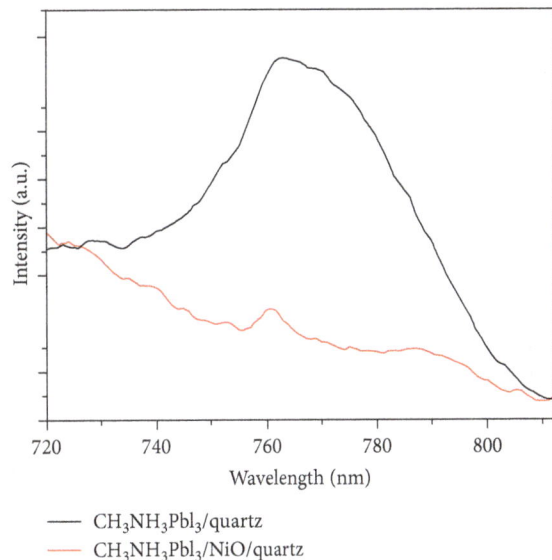

FIGURE 3: Photoluminescence spectra (excited by 600 nm laser) of CH3NH3PbI3 film deposited on top of NiO film or quartz substrate. Reprinted with permission from [21].

higher than those of organic HTLs, such as PEDOT:PSS, and perovskite deposited on NiO NC films is with homogeneous phase. The PSCs with HTL of NiO NC film at a thickness of around 35 nm exhibited the best PCE of 9.11%, as showed in Figure 4(c) [25].

Recently, ALD method has been used to fabricate ultrathin pure un-doped NiO films. And PSCs employing this kind of NiO film as HTL achieved a high PCE of 16.4%. We can fabricate highly sequential and dense ultrathin films at nanometer size following the ALD method.

(a)

(b)

(c)

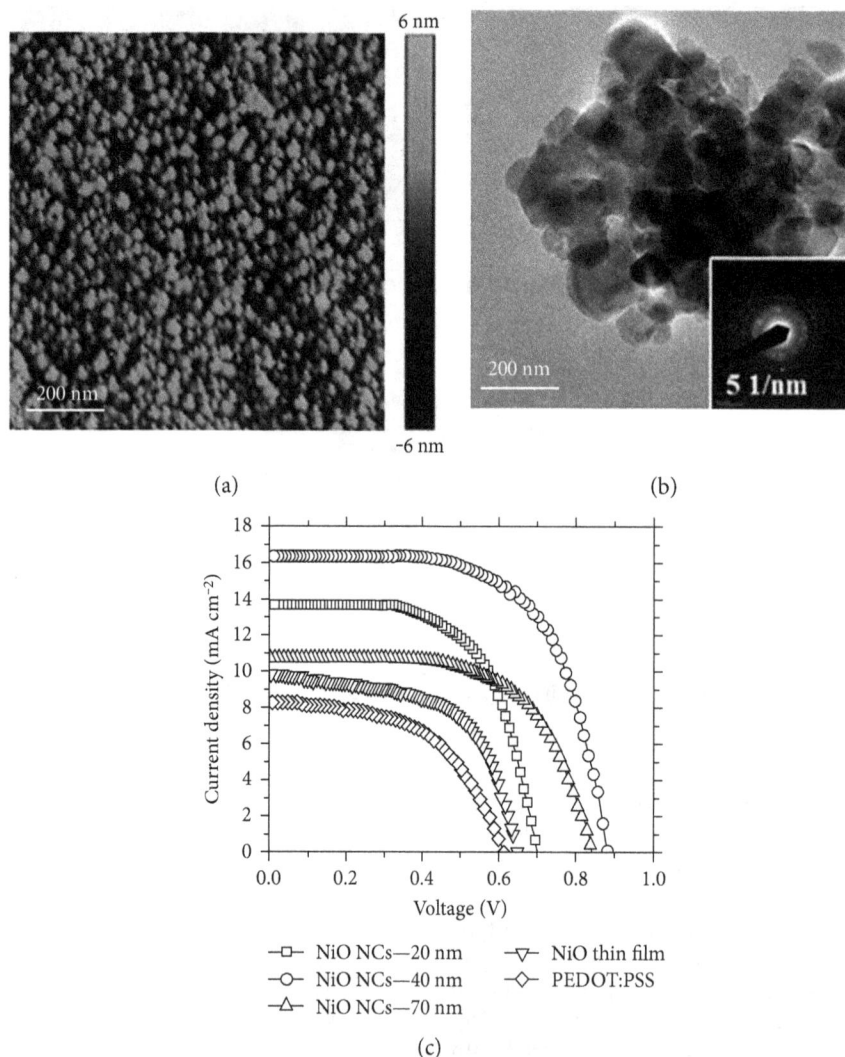

FIGURE 4: (a) AFM roughness image of a NiO nanocrystal film on FTO. (b) Bright-field TEM image of a sample of NiO NCs with the corresponding diffraction pattern (inset). (c) Typical J–V curves of the perovskite solar cells with different NiO hole transport layers and PEDOT:PSS. Reprinted with permission from [25].

The effective work function (WF) of ultrathin NiO apparently increased, which enormously promoted the hole extraction performance. In addition, ultrathin NiO films have a higher transparency which highly contribute to the photovoltaic devices. The freshly ultrathin pure undoped NiO films deposited as HTL of PSCs exhibited a high PCE of 16.4% with $J_{sc} = 21.9\,\text{mA cm}^{-2}$, $V_{oc} = 1.04\,\text{V}$, and $FF = 0.72$ [37].

A salient weakness with PSCs' use of NiO as HTL is that it is hard for NiO to support an ultrathin perovskite film (<60 nm) [22, 23], which has still limited the development of PCE of PSCs using NiO as HTLs. Another shortage with PSCs employing NiO as HTL is that FF and V_{oc} are lower than the common organic HTMs, in particular when the NiO was achieved by the solution-processed method. These parameters greatly affect the device performance [37].

Briefly speaking, an ideal p-NiO film for high PV performance should (1) have high transparency, (2) have favorable hole extraction and transport performance properties,

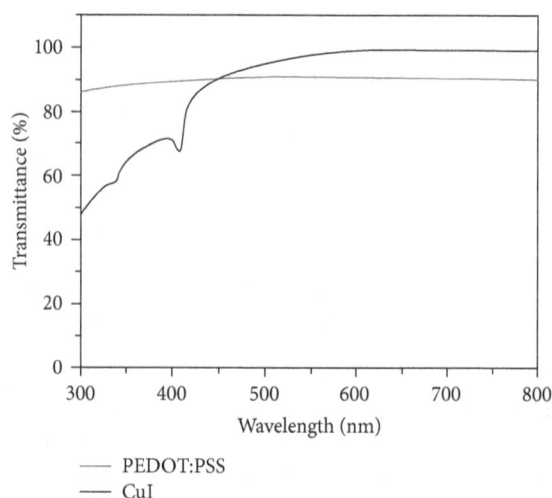

FIGURE 5: Optical transmission spectra of the PEDOT:PSS film and CuI film. Reprinted with permission from [40].

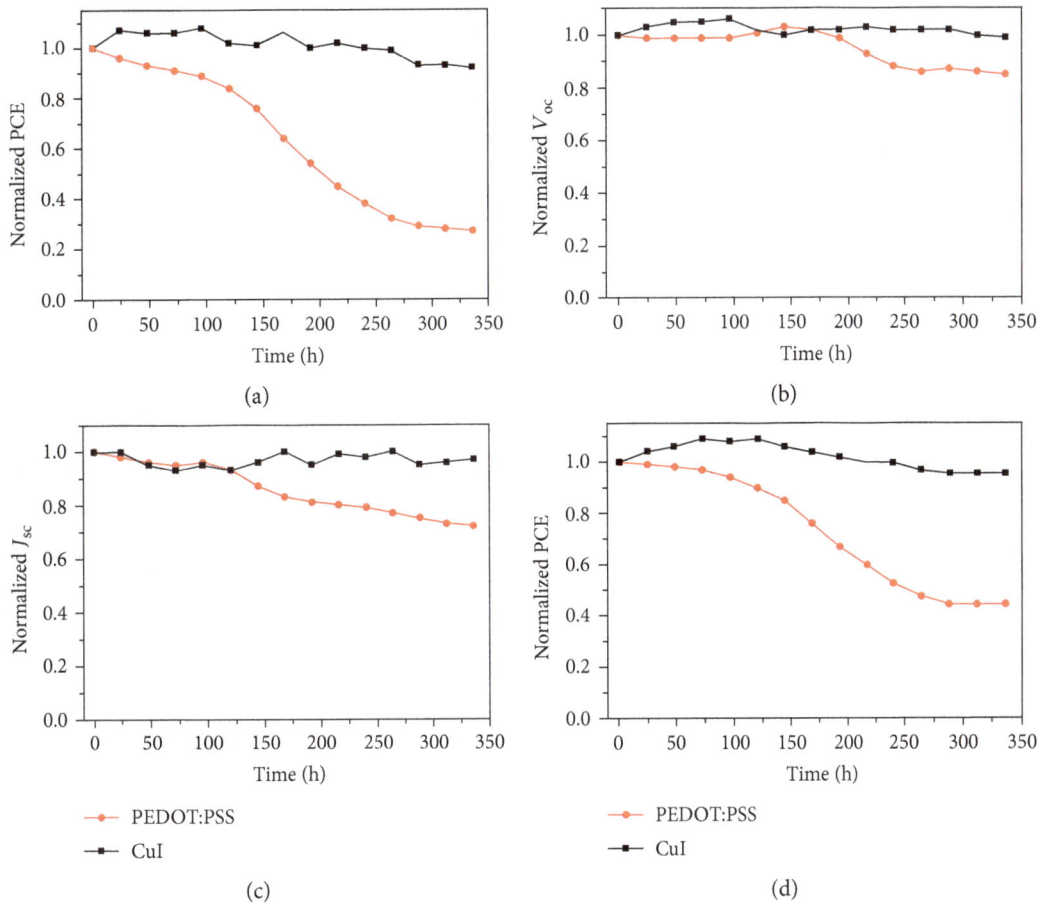

FIGURE 6: Normalized PCE (a), V_{oc} (b), J_{sc} (c), and FF (d) of perovskite solar cells employing CuI and PEDOT:PSS HTLs as a function of storage time in air. Reprinted with permission from [40].

(3) have applicable energy levels, and (4) have low cost and easy access [33].

2.2. CuI.

Inorganic p-type semiconductor copper iodide (CuI) becomes one of the promising HTLs for PSCs, due to its admirable properties such as wide band gap, high conductivity, low cost, and solution processable. At the beginning, CuI as hole conductors was employed in DSSCs and quantum dot-sensitized solar cells. In recent years, several groups have used CuI as HTLs for PSCs, and the first application of CuI for PSCs has successfully achieved a potential PCE of 6.0% [17]. Then, the PCE of PSCs using CuI as HTL increased to 7.5% [39]. In 2015, the PCE of PSCs applying CuI as HTL in inverted planar heterojunction perovskite solar cells reach up to 13.8% [40]. At present, the highest announced PCE of PSCs using CuI as HTL is 16.8% [41].

Solution-processed method is the frequently used approach to acquire CuI. The CuI films were prepared by spin coating CuI precursor solution in inert gas. Compared to PEDOT:PSS, CuI films exhibit higher transmittance in visible light from 450 to 800 nm (Figure 5), which make it potential to be used as a HTL. In addition, high transparency can allow more photo flux reach perovskite active layer to generate intense photocurrent. However, the surface morphology of CuI films is rough when compared to PEDOT:PSS

films, which may be because of the existence of large CuI grains [40].

At present, PSCs employing CuI as HTLs have achieved a relatively high PCE of 13.58% [40]. Above all, PSCs using CuI as the HTL exhibited improved air stability when compared to PSCs employing PEDOT:PSS as the HTL (Figure 6). In this respect, CuI is an excellent choice, because long-term stability of PSCs is of vital significance for practical applications [40].

Although PSCs using CuI as HTLs can replace conventional organic HTMs, the device open-circuit voltage (V_{oc}) is relatively low mainly because of a high recombination rate as determined by impedance spectroscopy. There are still numerous challenges in the optimization of PSC-employed CuI as HTLs. The primary problems include (1) how to control the surface morphology of CuI, (2) how to achieve favorable contacts between CuI layer and perovskite layer, and (3) how to dissolve CuI and eliminate large CuI grains.

2.3. CuSCN.

As mentioned above, copper thiocyanate (CuSCN) appears to be a good candidate of inorganic HTLs for its internal p-type characteristics of a wide band gap and high optical transparency [42–44].

CuSCN was first applied in mesoporous PSCs by Ito et al. with a device structure of $FTO/TiO_2/TiO_2/CH_3NH_3PbI_3/$

(a)

(b)

(c)

(d)

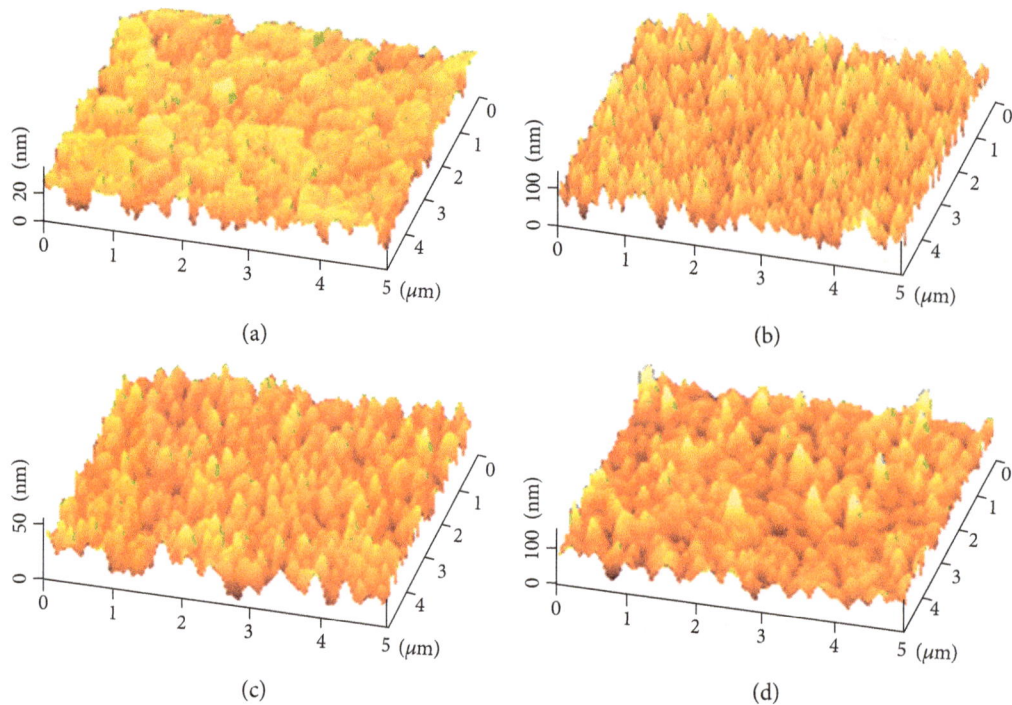

FIGURE 7: AFM images of ITO/glass (a), CuSCN/ITO/glass (b), $CH_3NH_3PbI_3$ (one-step)/CuSCN/ITO/glass (c), and $CH_3NH_3PbI_3$(two-step)/CuSCN/ITO/glass (d). The RMS roughness values are 4.6, 19.3, 7.6, and 17.0 nm, respectively. Reprinted with permission from [46].

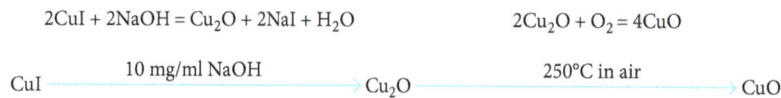

$$2CuI + 2NaOH = Cu_2O + 2NaI + H_2O \qquad\qquad 2Cu_2O + O_2 = 4CuO$$

$$CuI \xrightarrow{\quad 10\ mg/ml\ NaOH \quad} Cu_2O \xrightarrow{\quad 250°C\ in\ air \quad} CuO$$

FIGURE 8: Preparation process for Cu_2O and CuO films.

CuSCN/Au and achieved a PCE of 4.85% in 2013 [18]. Soon after, planar PSCs using CuSCN as a HTL received a PCE of 6.4%. There were reported low PCEs of PSCs based on CuSCN mainly because of the poor quality of $CH_3NH_3PbI_3$ active layer films on top of the CuSCN layers. Grätzel et al. enhanced the device PCE to 12.4% [45], via optimizing perovskite surface morphology by two times of iodide deposition. Then, the PCE of the CuSCN-based PSCs has already been improved to 16.6% [46, 47]. The latest announced PCE of PSCs using CuI as HTL has reached up to 18% [48]. In spite of the relatively low PCE compared to efficiency of PSCs based on organic HTMs, the low cost and air stability of CuSCN make it become a promising inorganic HTM.

The high PCE (16.6%) photovoltaic device used a one-step fast deposition method to fabricate high-quality perovskite films based on a rough CuSCN. $CH_3NH_3PbI_3$ prepared by a one-step faster deposition method with lower surface roughness and smaller interface contact resistance was compared to the perovskite films prepared by a conventional two-step deposition process, as showed in Figure 7 [46, 49].

Although cells using CuSCN as HTL have achieved considerable device efficiency, the CuSCN layer was fabricated via electrodeposition, which needs to be carefully compounded from a precursor solution containing potassium thiocyanate (KSCN), copper sulfate ($CuSO_4$), and ethylenediaminetetraacetic acid (EDTA). And the CuSCN films achieved by the above method are relatively rough and unshaped, which affect further improvements of PSCs [46].

2.4. Cu_2O and CuO. CuO and Cu_2O are well-known p-type semiconductors [27, 28, 50]. There is a simple low-temperature method to synthesize Cu_2O and CuO films and employ them as HTL for PSCs. Traditional methods for preparing Cu_2O film are thermal oxidation, sputtering, electrodeposition, and metal-organic chemical vapor deposition [51]. Cu_2O film can be obtained via in situ conversion of CuI film in aqueous NaOH solution, and CuO film is fabricated by heating Cu_2O film in the air, as showed in Figure 8 [52].

PSCs using Cu_2O and CuO as HTMs show observably increased V_{oc}, J_{sc}, and PCE. Recently, PSCs using Cu_2O and CuO as HTLs exhibited PCE of 13.35% and 12.16%, respectively [52]. Additionally, NH_4Cl was added into $CH_3NH_3PbI_3$ precursor to improve crystallinity of perovskite in Li et al. study. The increased V_{oc} mainly owns the VB of Cu_2O and CuO well matching with VB of the

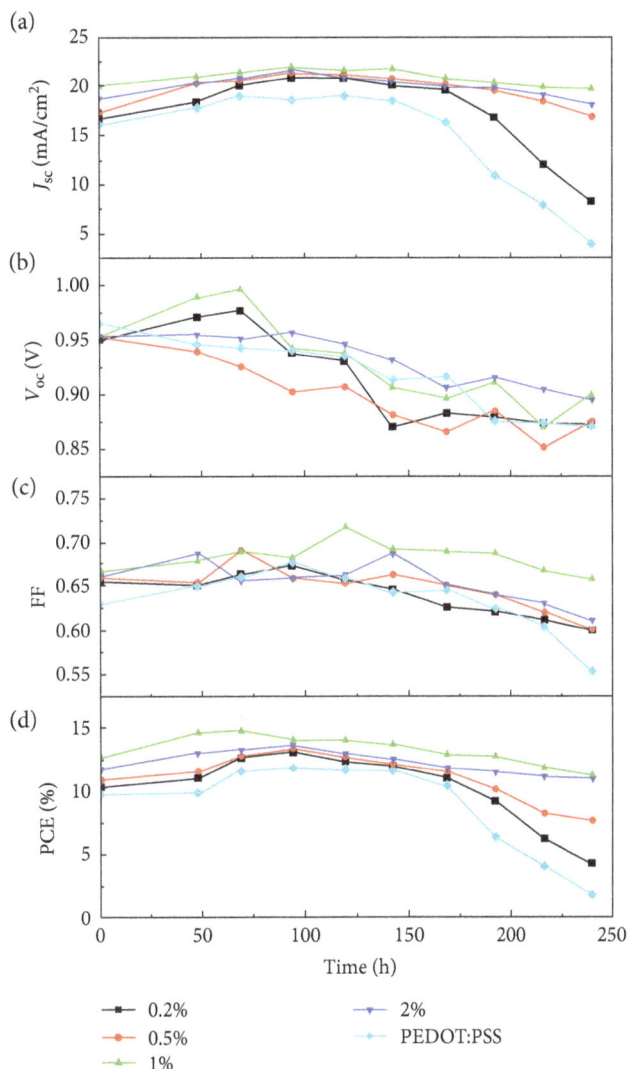

FIGURE 9: (a) J_{sc}, (b) V_{oc}, (c) FF, and (d) PCE values as a function of aging time of the devices with different MoO_3/PEDOT:PSS and pristine PEDOT:PSS HTLs. Reprinted with permission from [53].

perovskite and favorable crystallinity of perovskite on Cu_2O and CuO. What is more, well crystallinity enhanced the charge carrier transport and increased J_{sc}.

Compared to the PSCs employing NiO or Cu:NiO as HTLs, Cu_2O-based cells showed a superior property due to the higher mobility of Cu_2O, less energy loss, and favorable perovskite surface morphology on Cu_2O film [21].

2.5. MoO$_3$. MoO_3 is another potential HTM owing to the advantage of nontoxicity and air stability. However, PSCs with MoO_3 as HTL exhibit a low PCE mainly because of the poor quality of perovskite films deposited on MoO_3 [29, 31].

MoO_3 is a good HTM, but the poor perovskite films deposited on top of it limit further improvement. In order to solve the difficult problem, one simple solution would be to use an MoO_3/PEDOT:PSS composite film as the HTL in PSCs to take advantage of the ambient condition stability and favorable photovoltaic properties of MoO_3 and at the same time guarantee the admirable surface morphology of perovskite film [53].

Li et al. used a MoO_3/PEDOT:PSS bilayer structure as the HTL for PSCs. MoO_3 was prepared by a thermal decomposition of ammonium heptamolybdate $((NH_4)_6Mo_7O_{24} - 4H_2O)$ solution at 80°C. The PSCs have a structure of ITO/MoO_3/PEDOT:PSS/$CH_3NH_3PbI_3$/C60/Bphen/Ag. The application of an MoO_3 layer between ITO and PEDOT:PSS not only enhanced the hole extraction efficiency from perovskite to the ITO electrode but also avoided direct contact between rough MoO_3 and CH3NH3PbI$_3$. By this way, we can obtain a high-quality perovskite films deposited on top of PEDOT:PSS. These improvements contribute to the increase of stability and PCE of PSCs compared to the cells employing only PEDOT:PSS. Finally, the optimized PSCs exhibit a PCE of 14.87% [53]. What is more, the device stability increases when a MoO_3 layer is inserted between ITO and PEDOT:PSS as showed in Figure 9 [53].

2.6. VO$_x$. VO_x has become a favorable p-type semiconductor material primarily due to its higher work function (WF) and stability, and it can be prepared by low-temperature

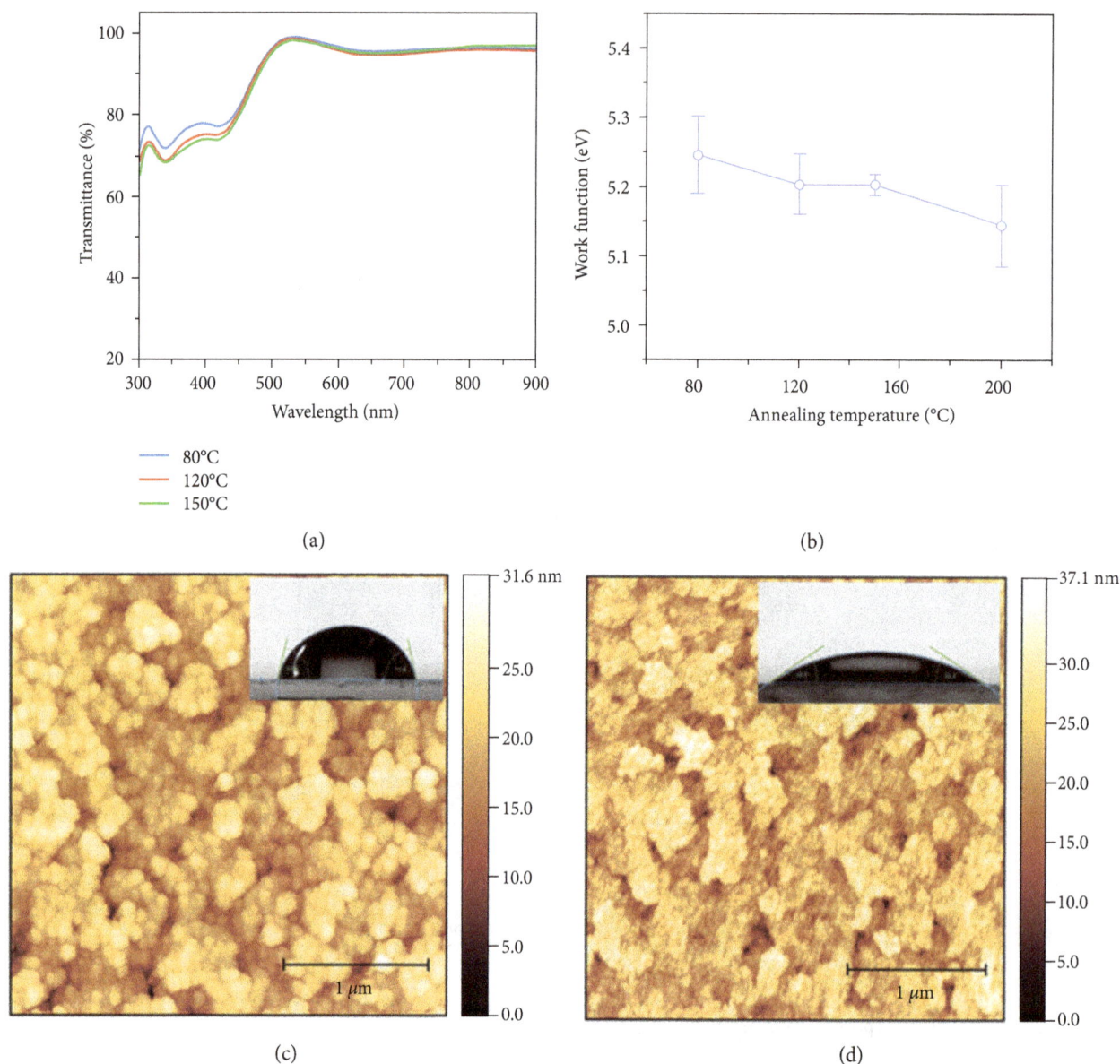

FIGURE 10: (a) Transmittance spectra of VO_x layers annealed at different temperatures. (b) Work function of VO_x layers annealed at different temperatures on ITO substrates. (c) AFM images of a bare ITO substrate. (d) AFM images of a VO_x-coated ITO substrate. Reprinted with permission from [55].

solution-processed method [32, 54]. At the beginning, VO_x was used as HTL in organic solar cells (OSCs). Recently, VO_x occurred as HTL for PSCs and achieved a relatively high PCE of 14.23% [55].

In general, VO_x films are fabricated by spin coating and annealed sol-gel precursor solution of VO_x. And the measurements of the VO_x layer exhibit high transmittance and well-quenching efficiency (Figure 10(a)). The value of x in VO_x was calculated at about 2.428 via X-ray photoelectron spectroscopy (XPS). High WF (Figure 10(b)) of VO_x not only extremely benefits cells containing high ionization potential donor materials but also reduces losses in V_{oc} and series resistance (Rs) [55]. However, VO_x is still faulty for it poor surface morphology result in harsh

deposition of perovskite films, as showed in Figures 10(c) and 10(d) [24, 45].

3. Conclusion

Inorganic semiconductor materials can be employed as hole-selective materials for PSCs due to their advantages of high hole mobility, wide band gap, and low cost, and they could be obtained by solution-processed method, showing promising respect of inorganic HTMs. What is more, the application of inorganic HTMs can enormously increase the stability and reduce cost of cells, which is very significant for PSCs. However, the reported device performance of most of the inorganic hole conductor-based PSCs is still much lower than that of cells with organic HTMs, which may result in the poor quality of perovskite films on top of the inorganic HTL.

Conflicts of Interest

The authors declare that there is no conflict of interest regarding the publication of this paper.

Acknowledgments

The authors would like to thank the National Natural Science Foundation of China (21271064 and 61306016), China Postdoctoral Science Foundation (2015M582179), and The Program for Changjiang Scholars and Innovative Research Team in the University (PCS IRT1126) of Henan University.

References

[1] A. Kojima, K. Teshima, Y. Shirai, and T. Miyasaka, "Organometal halide perovskites as visible-light sensitizers for photovoltaic cells," *Journal of the American Chemical Society*, vol. 131, no. 17, pp. 6050–6051, 2009.

[2] H. Zhou, Q. Chen, G. Li et al., "Interface engineering of highly efficient perovskite solar cells," *Science*, vol. 345, no. 6196, pp. 542–546, 2014.

[3] W. S. Yang, J. H. Noh, N. J. Jeon et al., "High-performance photovoltaic perovskite layers fabricated through intramolecular exchange," *Science*, vol. 348, no. 6240, pp. 1234–1237, 2015.

[4] N. J. Jeon, J. H. Noh, W. S. Yang et al., "Compositional engineering of perovskite materials for high-performance solar cells," *Nature*, vol. 517, no. 7535, pp. 476–480, 2015.

[5] Z. Yu and L. Sun, "Recent progress on hole-transporting materials for emerging organometal halide perovskite solar cells," *Advanced Energy Materials*, vol. 5, no. 12, 2015.

[6] S. Albrecht, M. Saliba, J. P. Baena et al., "Monolithic perovskite/silicon-heterojunction tandem solar cells processed at low temperature," *Energy & Environmental Science*, vol. 9, no. 1, pp. 81–88, 2016.

[7] H. S. Jung and N.-G. Park, "Perovskite solar cells: from materials to devices," *Small*, vol. 11, no. 1, pp. 10–25, 2015.

[8] "NREL efficiency chart," June 2016, http://www.nrel.gov/ncpv/images/efficiency_chart.jpg.

[9] G. Chen, J. Seo, C. Yang, and P. N. Prasad, "Nanochemistry and nanomaterials for photovoltaics," *Chemical Society Reviews*, vol. 42, no. 21, pp. 8304–8338, 2013.

[10] M. A. Green, A. Ho-Baillie, and H. J. Snaith, "The emergence of perovskite solar cells," *Nature Photonics*, vol. 8, no. 7, pp. 560–514, 2014.

[11] M.-H. Li, P.-S. Shen, K.-C. Wang, T. F. Guo, and P. Chen, "Inorganic p-type contact materials for perovskite-based solar cells," *Journal of Materials Chemistry A*, vol. 3, no. 17, pp. 9011–9019, 2015.

[12] W. Yan, S. Ye, Y. Li et al., "Hole-transporting materials in inverted planar perovskite solar cells," *Advanced Energy Materials*, vol. 6, no. 17, 2016.

[13] T. Leijtens, G. E. Eperon, N. K. Noel, S. N. Habisreutinger, A. Petrozza, and H. J. Snaith, "Stability of metal halide perovskite solar cells," *Advanced Energy Materials*, vol. 5, no. 20, article 1500963, 2015.

[14] Y. Rong, L. Liu, A. Mei, X. Li, and H. Han, "Beyond efficiency: the challenge of stability in mesoscopic perovskite solar cells," *Advanced Energy Materials*, vol. 5, no. 20, article 1501066, 2015.

[15] J. Liu, Y. Wu, C. Qin et al., "A dopant-free hole-transporting material for efficient and stable perovskite solar cells," *Energy & Environmental Science*, vol. 7, no. 9, pp. 2963–2967, 2014.

[16] J. You, L. Meng, T.-B. Song et al., "Improved air stability of perovskite solar cells via solution-processed metal oxide transport layers," *Nature Nanotechnology*, vol. 11, no. 1, pp. 75–81, 2016.

[17] J. A. Christians, R. C. M. Fung, and P. V. Kamat, "An inorganic hole conductor for organo-lead halide perovskite solar cells. Improved hole conductivity with copper iodide," *Journal of the American Chemical Society*, vol. 136, no. 2, pp. 758–764, 2014.

[18] S. Ito, S. Tanaka, H. Vahlman, H. Nishino, K. Manabe, and P. Lund, "Carbon-double-bond-free printed solar cells from TiO$_2$/CH$_3$NH$_3$PbI$_3$/CuSCN/au: structural control and photoaging effects," *ChemPhysChem*, vol. 15, no. 6, pp. 1194–1200, 2014.

[19] A. S. Subbiah, A. Halder, S. Ghosh, N. Mahuli, G. Hodes, and S. K. Sarkar, "Inorganic hole conducting layers for perovskite-based solar cells," *Journal of Physcal Chemistry Letters*, vol. 5, no. 10, pp. 1748–1753, 2014.

[20] S. Ito, S. Tanaka, K. Manabe, and H. Nishino, "Effects of surface blocking layer of Sb$_2$S$_3$ on nanocrystalline TiO$_2$ for CH$_3$NH$_3$PbI$_3$ perovskite solar cells," *The Journal of Physical Chemistry C*, vol. 118, no. 30, pp. 16995–17000, 2014.

[21] L. Hu, J. Peng, W. Wang et al., "Sequential deposition of CH$_3$NH$_3$PbI$_3$ on planar NiO film for efficient planar perovskite solar cells," *ACS Photonics*, vol. 1, no. 7, pp. 547–553, 2014.

[22] X. Yin, Z. Yao, Q. Luo et al., "High Efficiency Inverted Planar Perovskite Solar Cells with Solution-Processed NiOx Hole Contact," *ACS Applied Materials & Interfaces*, vol. 9, no. 3, pp. 2439–2448, 2017.

[23] H. Tian, B. Xu, H. Chen, E. M. J. Johansson, and G. Boschloo, "Solid-state perovskite-sensitized p-type mesoporous nickel oxide solar cells," *ChemSusChem*, vol. 7, no. 8, pp. 2150–2153, 2014.

[24] J. H. Kim, P.-W. Liang, S. T. Williams et al., "High-performance and environmentally stable planar heterojunction perovskite solar cells based on a solution-processed copper-doped nickel oxide hole-transporting layer," *Advanced Materials*, vol. 27, no. 4, pp. 695–701, 2015.

[25] Z. Zhu, Y. Bai, T. Zhang et al., "High-performance hole-extraction layer of sol–gel-processed NiO nanocrystals for inverted planar perovskite solar cells," *Angewandte Chemie*, vol. 126, no. 46, pp. 12779–12783, 2014.

[26] W. Chen, Y. Wu, J. Liu et al., "Hybrid interfacial layer leads to solid performance improvement of inverted perovskite solar cells," *Energy & Environmental Science*, vol. 8, no. 2, pp. 629–640, 2015.

[27] L. C. Chen, C. C. Chen, K. C. Liang et al., "Nano-structured CuO-Cu$_2$O complex thin film for application in CH$_3$NH$_3$PbI$_3$ perovskite solar cells," *Nanoscale Research Letters*, vol. 11, no. 1, p. 402, 2016.

[28] S. Chatterjee and A. J. Pal, "Introducing Cu$_2$O thin films as a hole-transport layer in efficient planar perovskite solar cell structures," *Journal of Physical Chemistry C*, vol. 120, no. 3, pp. 1428–1437, 2016.

[29] Y. Zhao, A. M. Nardes, and K. Zhu, "Effective hole extraction using MoO_x-Al contact in perovskite $CH_3NH_3PbI_3$ solar cells," *Applied Physics Letters*, vol. 104, no. 21, p. 213906, 2014.

[30] C. Liu, Z. Su, W. Li et al., "Improved performance of perovskite solar cells with a TiO2/MoO3 core/shell nanoparticles doped PEDOT:PSS hole-transporter," *Organic Electronics*, vol. 33, pp. 221–226, 2016.

[31] Z.-L. Tseng, L.-C. Chen, C.-H. Chiang, S. H. Chang, C. C. Chen, and C. G. Wu, "Efficient inverted-type perovskite solar cells using UV-ozone treated MoO_x and WO_x as hole transporting layers," *Solar Energy*, vol. 139, pp. 484–488, 2016.

[32] M. Xiao, M. Gao, F. Huang et al., "Efficient perovskite solar cells employing inorganic interlayers," *ChemNanoMat*, vol. 2, no. 3, pp. 182–188, 2016.

[33] L. Alibabaei, H. Luo, R. L. House, P. G. Hoertz, R. Lopez, and T. J. Meyer, "Applications of metal oxide materials in dye sensitized photoelectrosynthesis cells for making solar fuels: let the molecules do the work," *Journal of Materials Chemistry a*, vol. 1, no. 13, pp. 4133–4145, 2013.

[34] M. D. Irwin, B. Buchholz, A. W. Hains, R. P. H. Chang, and T. J. Marks, "*p*-type semiconducting nickel oxide as an efficiency-enhancing anode interfacial layer in polymer bulk-heterojunction solar cells," *Proceedings of the National Academy of Sciences of the United States of America*, vol. 105, no. 8, pp. 2783–2787, 2008.

[35] J.-Y. Jeng, K.-C. Chen, T.-Y. Chiang et al., "Nickel oxide electrode interlayer in $CH_3NH_3PbI_3$ perovskite/PCBM planar-heterojunction hybrid solar cells," *Advanced Materials*, vol. 26, no. 24, pp. 4107–4133, 2014.

[36] W. Chen, Y. Wu, Y. Yue et al., "Efficient and stable large-area perovskite solar cells with inorganic charge extraction layers," *Science*, vol. 350, no. 6263, pp. 944–948, 2015.

[37] S. Seo, I. J. Park, M. Kim et al., "An ultra-thin, un-doped NiO hole transporting layer of highly efficient (16.4%) organic–inorganic hybrid perovskite solar cells," *Nanoscale*, vol. 8, no. 22, pp. 11403–11412, 2016.

[38] K. X. Steirer, J. P. Chesin, N. E. Widjonarko et al., "Solution deposited NiO thin-films as hole transport layers in organic photovoltaics," *Organic Electronics*, vol. 11, no. 8, pp. 1414–1418, 2016.

[39] G. A. Sepalage, S. Meyer, A. Pascoe et al., "Copper(I) iodide as hole-conductor in planar perovskite solar cells: probing the origin of *J - V* hysteresis," *Advanced Functional Materials*, vol. 25, no. 35, pp. 5650–5661, 2015.

[40] W.-Y. Chen, L.-L. Deng, S.-M. Dai et al., "Low-cost solution-processed copper iodide as an alternative to PEDOT:PSS hole transport layer for efficient and stable inverted planar heterojunction perovskite solar cells," *Journal of Materials Chemistry A*, vol. 3, no. 38, pp. 19353–19359, 2015.

[41] W. Sun, S. Ye, H. Rao et al., "Room-temperature and solution - processed copper iodide as the hole transport layer for inverted planar perovskite solar cells," *Nanoscale*, vol. 8, no. 35, pp. 15954–15960, 2016.

[42] C. Chappaz-Gillot, S. Berson, R. Salazar et al., "Polymer solar cells with electrodeposited CuSCN nanowires as new efficient hole transporting layer," *Solar Energy Materials and Solar Cells*, vol. 120, pp. 163–167, 2014.

[43] B. Li, L. Wang, B. Kang, P. Wang, and Y. Qiu, "Review of recent progress in solid-state dye-sensitized solar cells," *Solar Energy Materials and Solar Cells*, vol. 90, no. 5, pp. 549–574, 2006.

[44] P. Pattanasattayavong, N. Yaacobi-Gross, K. Zhao et al., "Hole-transporting transistors and circuits based on the transparent inorganic semiconductor copper(I) thiocyanate (CuSCN) processed from solution at room temperature," *Advanced Materials*, vol. 25, no. 10, pp. 1504–1509, 2013.

[45] P. Qin, S. Tanaka, S. Ito et al., "Inorganic hole conductor-based lead halide perovskite solar cells with 12.4% conversion efficiency," *Nature Communications*, vol. 5, article 3834, 2014.

[46] S. Ye, S. W. Sun, Y. Li et al., "CuSCN-based inverted planar perovskite solar cell with an average PCE of 15.6%," *Nano Letters*, vol. 15, no. 6, pp. 3723–3728, 2015.

[47] V. E. Madhavan, I. Zimmermann, C. Roldan-Carmona et al., "Copper thiocyanate inorganic hole-transporting material for high-efficiency perovskite solar cells," *ACS Energy Letters*, vol. 1, no. 6, pp. 1112–1117, 2016.

[48] M. Jung, Y. C. Kim, N. J. Jeon et al., "Thermal stability of CuSCN hole conductor-based perovskite solar cells," *ChemSusChem Communications*, vol. 9, no. 18, pp. 2592–2596, 2016.

[49] P. Pattanasattayavong, G. O. Ngongang Ndjawa, K. Zhao et al., "Electric field-induced hole transport in copper(I) thiocyanate (CuSCN) thin-films processed from solution at room temperature," *Chemical Communications*, vol. 49, no. 39, pp. 4154–4156, 2013.

[50] B. K. Meyer, A. Polity, D. Reppin et al., "Binary copper oxide semiconductors: from materials towards devices," *Physica Status Solidi B*, vol. 249, no. 8, pp. 1487–1509, 2012.

[51] L.-C. Chen, "Review of preparation and optoelectronic characteristics of Cu_2O-based solar cells with nanostructure," *Materials Science in Semiconductor Processing*, vol. 16, no. 5, pp. 1172–1185, 2013.

[52] C. Zuo and L. Ding, "Solution-processed Cu_2O and CuO as hole transport materials for efficient perovskite solar cells," *Small*, vol. 11, no. 41, pp. 5528–5532, 2015.

[53] F. Hou, Z. Su, F. Jin et al., "Efficient and stable planar heterojunction perovskite solar cells with an MoO_3/PEDOT:PSS hole transporting layer," *Nanoscale*, vol. 7, no. 21, pp. 9427–9432, 2015.

[54] P. Li, C. Liang, Y. Zhang, F. Li, Y. Song, and G. Shao, "Polyethyleneimine High-Energy Hydrophilic Surface Interfacial Treatment toward Efficient and Stable Perovskite Solar Cells," *ACS Applied Materials & Interfaces*, vol. 8, no. 47, pp. 32574–32580, 2016.

[55] H. Sun, X. Hou, Q. Wei et al., "Low-temperature solution-processed p-type vanadium oxide for perovskite solar cells," *Chemical Communications*, vol. 52, no. 52, pp. 8099–8102, 2016.

16

Observer-Based Load Frequency Control for Island Microgrid with Photovoltaic Power

Chaoxu Mu,[1] Weiqiang Liu,[1] Wei Xu,[2,3] and Md. Rabiul Islam[4]

[1]*Tianjin Key Laboratory of Process Measurement and Control, School of Electrical and Information Engineering, Tianjin University, Tianjin 300072, China*
[2]*State Key Laboratory of Advanced Electromagnetic Engineering and Technology, School of Electrical and Electronic Engineering, Huazhong University of Science and Technology, Wuhan 430074, China*
[3]*College of Mechanical and Electrical Engineering, Huanggang Normal University, Huanggang 438000, China*
[4]*Department of Electrical and Electronic Engineering, Rajshahi University of Engineering and Technology, Rajshahi 6204, Bangladesh*

Correspondence should be addressed to Wei Xu; weixu@hust.edu.cn

Academic Editor: Zofia Stasicka

As renewable energy is widely integrated into the power system, the stochastic and intermittent power generation from renewable energy may cause system frequency deviating from the prescribed level, especially for a microgrid. In this paper, the load frequency control (LFC) of an island microgrid with photovoltaic (PV) power and electric vehicles (EVs) is investigated, where the EVs can be treated as distributed energy storages. Considering the disturbances from load change and PV power, an observer-based integral sliding mode (OISM) controller is designed to regulate the frequency back to the prescribed value, where the neural network observer is used to online estimate the PV power. Simulation studies on a benchmark microgrid system are presented to illustrate the effectiveness of OISM controller, and comparative results also demonstrate that the proposed method has a superior performance for stabilizing the frequency over the PID control.

1. Introduction

With the technological innovation of modern power systems, renewable energy has been widely incorporated to the power system, such as wind energy and solar energy. Although renewable energy causes less pollution and is energy saving, the large-scale integration of renewable energy would have a significant impact on the power system since renewable energy is not ideal for power generation [1–3].

Load frequency stability is regarded as an indispensable factor when considering the stability of power systems [4]. For a microgrid system, if the imbalance between load consumption and power generation frequently happens, then a load frequency controller is required to have an adequate ability to quickly damp the frequency oscillation [5, 6]. The traditional load frequency model for the power system is approximated as a linear model near an operation point without the generator dynamics being involved, and the most commonly used frequency control strategy for this kind of linear model is the PID control, as reported in [7]. After the offline tuning of parameters, the PID controller can have a good frequency damping performance within a certain range around the designed operation point [8]. With the nonlinear characteristics of power generation for renewable energy, it may be less effective to the microgrid system, especially when the microgrid system is far away from the set operation point. Besides, several advanced control methods have been used to the LFC problem in the last decade, such as fuzzy logic control [9–11], adaptive control [12, 13], and robust control [14]. Although the design of nonlinear control approaches is relatively complex, the research of advanced nonlinear control approaches on smart grid has been paid great attention since nonlinear control is more close to the characteristics of a power system itself [3, 15–18].

In this paper, we investigate the integral sliding mode (ISM) control method for load frequency control of an island

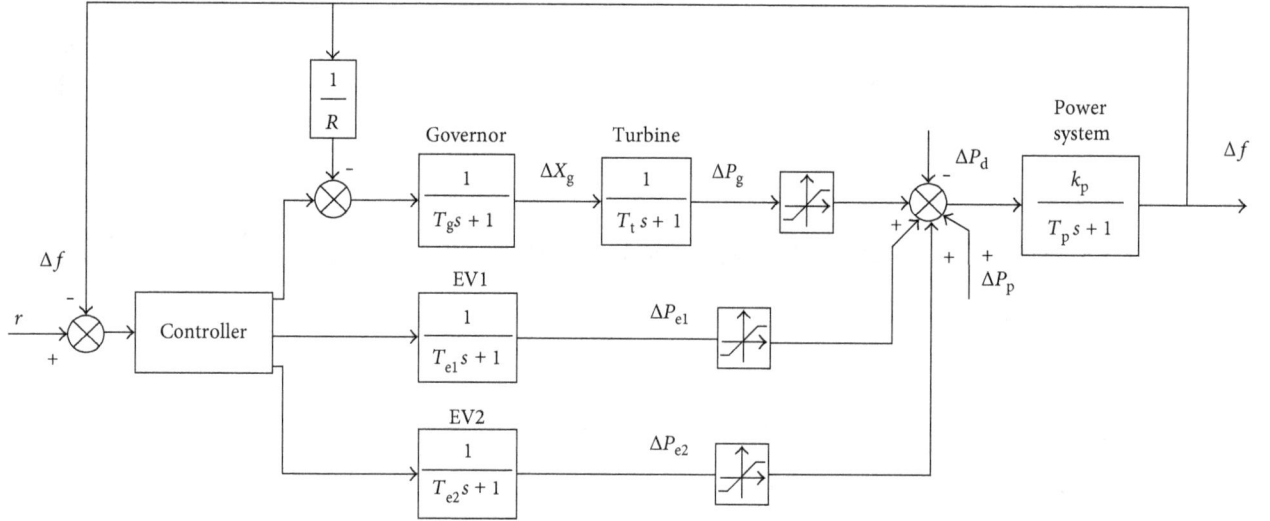

FIGURE 1: The schematic diagram of the benchmark system.

microgrid system with PV power integration, and a neural network observer is designed for an online observation of the PV power generation. Meanwhile, with the improvement of vehicle-to-grid (V2G) technique, EVs are also integrated into the microgrid as distributed storage devices to support the frequency regulation. The major contribution of this paper is as follows. First, an ISM controller for the microgrid with EVs is designed. Concerning the distinguished pattern of PV power generation, an online OISM controller is proposed to regulate the frequency of the microgrid, and the relevant stability proof is rigorously analyzed. Second, the robustness of the proposed controller against system parameter uncertainties are verified, where the control performance of this proposed method is compared with the traditional PID control.

This paper is organized as follows. Section 2 introduces the studied microgrid system and formulates the frequency control problem. Section 3 designs the neural network observer and further presents the OISM controller method. Simulation is carried out on the benchmark microgrid, and all the results are presented and analyzed in Section 4. Section 5 concludes this paper.

2. Problem Formulation

In this paper, the studied benchmark system includes the following general parts: an equivalent microturbine (MT), PV arrays, two equivalent EV models with battery banks, and a demand side such as smart homes and loads. Both island mode and grid-connected mode are the possible operation modes [19]. In this paper, this benchmark power system is considered as a microgrid and operated in an island mode, where the system power flow is balanced by local loads and local power generation. It means that MTs and PV arrays provide active power to balance all local loads, and EV stations can be considered as distributed battery energy storages to compensate the unbalance between power generation and load demand.

The power system is usually considered to be nonlinear and dynamic, but there only exists small load change during its normal operation. As EVs integrated into the benchmark system, the schematic diagram of the benchmark system is presented in Figure 1.

Since the microgrid is in the island operating mode, the LFC capacity is required to be adequate to quickly damp the frequency oscillation. By incorporating the EVs into the microgrid, the system inertia can be increased and the frequency stability can be improved [20, 21]. To demonstrate the benefit from the technology of V2G, an active power disturbance is added; the frequency dynamics with EVs and without EVs is shown in Figure 2.

The load frequency controller is expected to maintain the command frequency level when the load disturbances appear and renewable energy is incorporated into the gird. In order to formulate the LFC control problem of this microgrid, some mathematical notations are defined as follows: Δf, ΔP_t, ΔX_g, ΔP_{e1}, and ΔP_{e2} are the change of the frequency, the turbine power, the governor position valve, the first EV power, and the second EV power, respectively. T_t, T_g, T_{e1}, T_{e2}, and T_p are the time constants of the turbine, the governor, the first EV, the second EV, and the power system, respectively. k_p and R are the gain of the power system and the speed regulation coefficient, respectively.

Based on Figure 1, the state vector $x(t)$ of the system is defined as

$$x(t) = [\Delta f(t), \Delta P_t(t), \Delta X_g(t), \Delta P_{e1}(t), \Delta P_{e2}]^T, \quad (1)$$

and the LFC model is formulated by the following differential equation:

$$\dot{x}(t) = Ax(t) + Bu(t) + F(\Delta P_p(t) + \Delta P_d(t)). \quad (2)$$

The system matrix A, the control matrix B, and the disturbance matrix F can be expressed as

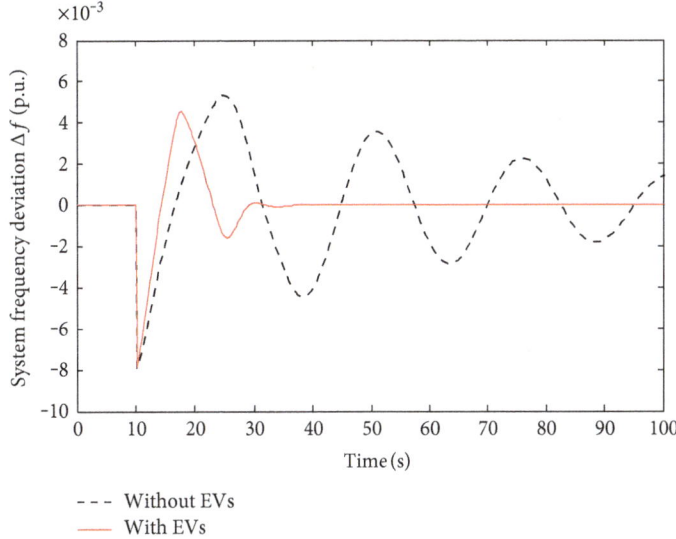

FIGURE 2: The frequency deviation with and without EVs.

$$A = \begin{bmatrix} -\dfrac{1}{T_p} & \dfrac{k_p}{T_p} & 0 & \dfrac{k_p}{T_p} & \dfrac{k_p}{T_p} \\ 0 & -\dfrac{1}{T_t} & \dfrac{1}{T_t} & 0 & 0 \\ -\dfrac{1}{RT_g} & 0 & -\dfrac{1}{T_g} & 0 & 0 \\ 0 & 0 & 0 & -\dfrac{1}{T_{e1}} & 0 \\ 0 & 0 & 0 & 0 & -\dfrac{1}{T_{e2}} \end{bmatrix},$$

$$B = \begin{bmatrix} 0 & 0 & 0 \\ 0 & 0 & 0 \\ \dfrac{1}{T_g} & 0 & 0 \\ 0 & \dfrac{1}{T_{e1}} & 0 \\ 0 & 0 & \dfrac{1}{T_{e2}} \end{bmatrix},$$

(3)

$$F = \begin{bmatrix} -\dfrac{k_p}{T_p} \\ 0 \\ 0 \\ 0 \\ 0 \end{bmatrix}.$$

$u(t)$ is the control vector, $u(t) = [\Delta u_1(t), \Delta u_2(t), \Delta u_3(t)]^T$, and $\Delta P_d(t)$ and $\Delta P_p(t)$ are the integrated disturbance from load change and the PV power, respectively. The following assumption is applied on the benchmark system.

Assumption 1. The induced norms of PV power disturbances and load change satisfy $\|\Delta P_p(t)\| \leq \alpha_1$ and $\|\Delta P_d(t)\| \leq \alpha_2$, respectively, where $\alpha_1 \geq 0$ and $\alpha_2 \geq 0$ represent upper bounds.

3. Controller Design

3.1. Integral Sliding Mode Controller. Compared with a large interconnected system, the microgrid system is more instable and easy to cause serious damage under parameter uncertainties and disturbances. Therefore, a controller with excellent robustness to maintain the microgrid system stable is required.

Sliding mode control is robust and systematic for matched disturbances and parameter variations [22–25]. To keep the frequency in the required level, ISM control is selected as the fundamental control method for the LFC problem of the benchmark system. For the microgrid system presented in (2), an ISM variable is designed as

$$s(t) = C_1 x(t) + \int_0^t C_2 x(t) dt, \tag{4}$$

where $C_1 = [c_1, c_2, c_3, c_4, c_5]$ and $C_2 = [c_6, c_7, c_8, c_9, c_{10}]$ are the coefficient vectors, and c_i meets that the two polynomials $c_5 p^4 + c_4 p^3 + c_3 p^2 + c_2 p + c_1$ and $c_{10} p^4 + c_9 p^3 + c_8 p^2 + c_7 p + c_6$ are Hurwitz.

The ISM controller is designed by adopting the reaching law $\dot{s}(x) = -\varepsilon \mathrm{sat}(s)$. With the idea of equivalent control, the ISM frequency controller is constituted as

$$u(t) = -(C_1 B)^{-1}(C_1 A x(t) + C_2 x(t) + \varepsilon \mathrm{sat}(s)), \tag{5}$$

where ε is the control gain and satisfies $\varepsilon > |C_1 F(\alpha_1 + \alpha_2)|$ and $\mathrm{sat}(s) = s(t)/\rho$ is the saturation function to eliminate the chattering in the control signal by setting a reasonable width ρ.

3.2. Neural Network Observer. In this subsection, the data of the PV power is measured by specific sensors and is stored in the control unit. Recently, intelligent algorithms have been developed for obtaining information from data [26–31]. Therefore, these data are available to design an observer to forecast the future PV power. In this paper, a three-layer neural

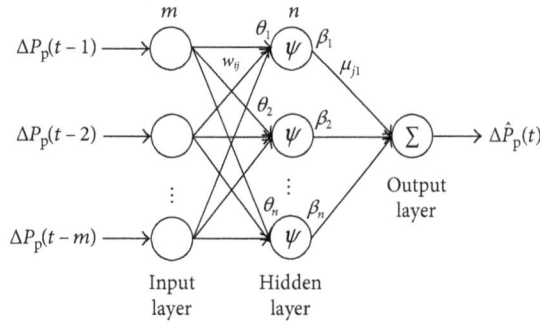

FIGURE 3: The structure of the neural network observer.

TABLE 1: The system parameters of the benchmark system.

Parameters	T_t	T_g	T_{e1}	T_{e2}	T_p	k_p	R
Values	10	0.1	1	1	10	1	0.5

TABLE 2: The disturbance from load change.

Time (s)	15	60	100	150
Disturbance (p.u.)	+0.2	−0.2	−0.4	+0.4

network with a hidden layer is used to realize the online observation of PV power, whose structure is shown in Figure 3.

From Figure 3, we can know that the neural network observer for the PV power is with m input-layer neurons, n hidden-layer neurons, and one output-layer neuron. Therefore, the inputs of the observer are m PV power values before time t, recorded as $I_t \in R^{1 \times m}$. The output $\Delta P_p(t)$ is the PV power value at time t. The rest of the variables in Figure 3 can be defined as

$$\theta_j(t) = \sum_{i=1}^{m} w_{ij}(t) \Delta P_p(t-i), \tag{6}$$

$$\beta_j(t) = \psi(\theta_j(t)) = \frac{1 - e^{-\theta_j(t)}}{1 + e^{-\theta_j(t)}}, \tag{7}$$

$$\Delta \widehat{P}_p(t) = \sum_{j=1}^{n} \mu_{j1}(t) \beta_j(t), \tag{8}$$

where i and j satisfy $i = 1,\ldots,m$ and $j = 1,\ldots,n$, respectively. $\theta_j(t)$ and $\beta_j(t)$ are input and output values of the jth hidden neuron, respectively. $w_{ij}(t)$ is the weight from the ith input neuron to the jth hidden neuron, and $\mu_{j1}(t)$ is the weights from the jth hidden node to the output node. $\psi(z) = (1 - e^{-z})/(1 + e^{-z})$ is used as the activation function.

The weight vectors $w_{ij}(t)$ and $\mu_{j1}(t)$ are randomly initialized in $[-1, 1]$. According to the calculation of forward propagation, the neural network outputs $\Delta \widehat{P}_p(t)$ as the estimation value of PV power at time t.

The weights of the neural network are updated by the error back-propagation algorithm. The difference between $\Delta \widehat{P}_p(t)$ and the measured real value of PV power is defined as the network error, which is

$$e_p(t) = \Delta P_p(t) - \Delta \widehat{P}_p(t). \tag{9}$$

If $e_p(t) = 0$, it represents the estimated PV power value which is completely equal to the real PV power value. In the back-propagation process, the aim is to minimize the objective function $E(t)$ associated with $e_p(t)$; that is,

$$E(t) = \frac{1}{2} e_p^2(t). \tag{10}$$

The gradient descent method is used to update the weights of the neural network observer in the back-propagation [26–28, 32]. Define the input-to-hidden weight vector $w_1(t)$ and the hidden-to-output weight vector $w_2(t)$ as

$$w_1(t) = \left\{ w_{ij}(t) \mid i = 1, 2, \ldots, m, j = 1, 2, \ldots, n \right\},$$
$$w_2(t) = \left\{ \mu_{j1}(t) \mid j = 1, 2, \ldots, n \right\}, \tag{11}$$

where $w_1(t) \in R^{m \times n}$ and $w_2(t) \in R^{n \times 1}$, respectively. Then (8) can be further simplified as

$$\Delta \widehat{P}_p(t) = \sum_{j=1}^{n} \mu_{j1}(t) \psi \left(\sum_{i=1}^{n} w_{ij}(t) \Delta P_p(t-i) \right) = w_2(t) \psi(w_1). \tag{12}$$

According to the chain derivation rule, the weights of the neural network observer are updated by

$$\dot{w}_2(t) = -\tau \frac{\partial E(t)}{\partial \Delta \widehat{P}_p(t)} \frac{\partial \Delta \widehat{P}_p(t)}{\partial w_2(t)} = \tau e_p(t) \psi(w_1). \tag{13}$$

Specifically, the weights $w_{ij}(t)$ and $\mu_{j1}(t)$ are regulated by

$$\dot{w}_{ij}(t) = -\tau \frac{\partial E(t)}{\partial \Delta \widehat{P}_p(t)} \frac{\partial \Delta \widehat{P}_p(t)}{\partial \alpha_j(t)} \frac{\partial \alpha_j(t)}{\partial \theta_j(t)} \frac{\partial \theta_j(t)}{\partial w_{ij}(t)}, \tag{14}$$

$$\dot{\mu}_{j1}(t) = -\tau \frac{\partial E(t)}{\partial \Delta \widehat{P}_p(t)} \frac{\partial \Delta \widehat{P}_p(t)}{\partial \mu_{j1}(t)}, \tag{15}$$

where $\tau > 0$ is the learning rate. The estimated PV power can be used as an input signal into the ISM controller to eliminate the influence of integrated PV power.

3.3. Observer-Based Integral Sliding Mode Controller. According to the universal approximation property of neural networks [32, 33], the output of the observer can approximate the real PV power with an allowable error. In other words, there exists a weight vector w to make the error $e_p(t)$ reach the minimum, which is

$$\left| e_p(t) \right| \leq \xi, \tag{16}$$

where ξ is an arbitrary enough small positive constant. It represents the maximal absolute difference between the

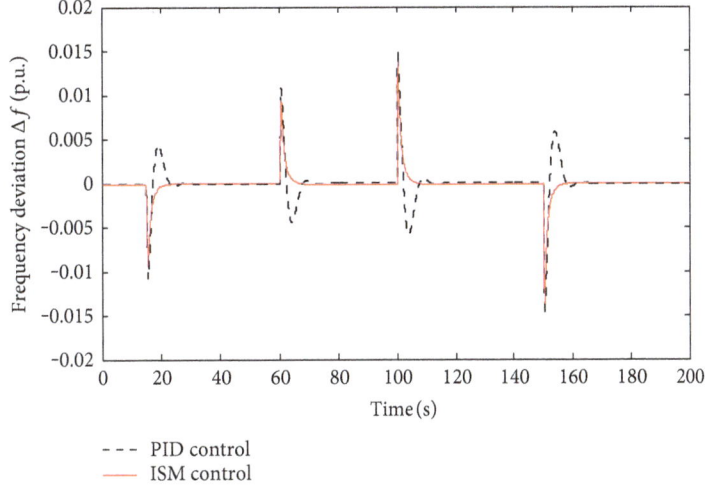

FIGURE 4: The frequency deviation without power output constraints.

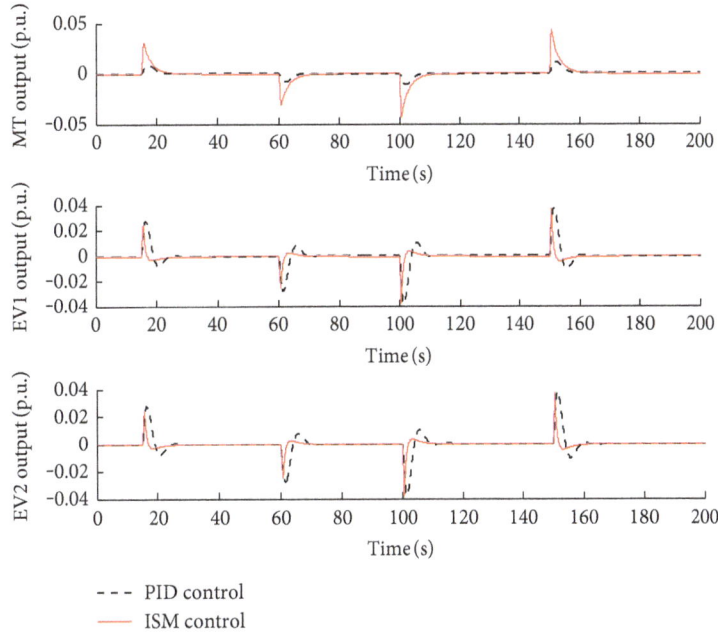

FIGURE 5: The output power of MT and EVs without power output constraints.

estimated value and the real PV power. By introducing the neural network observer, the OISM controller is constructed for the microgrid system with PV power integration. Thus, we have the following theorem.

Theorem 1. *For system (1), if the weights of the neural network observer are updated by (13), and the OISM controller is designed as*

$$\mu(t) = -\left(C_1 B\right)^{-1} \left(C_1 A x(t) + C_2 x(t) + \varepsilon \mathrm{sat}(s) + C_1 F \Delta \widehat{P}_p(t)\right),$$
(17)

then the microgrid system is asymptotically stable.

Proof. We define the Lyapunov function for system (1) with control law (17) as

$$V(t) = \frac{1}{2}s^2(t) + \frac{1}{2\tau}\tilde{w}_2^2(t).$$
(18)

It is obvious that $V(t) \geq 0$. Set the weight error vector as

$$\tilde{w}_2(t) = w_2^*(t) - w_2(t),$$
(19)

where $w_2^*(t)$ is the ideal weight vector and $w_2(t)$ is actually weight vector in the observer. By differentiating $\tilde{w}_2(t)$ with respect to time t, we could obtain

$$\dot{\tilde{w}}_2(t) = -\dot{w}_2(t).$$
(20)

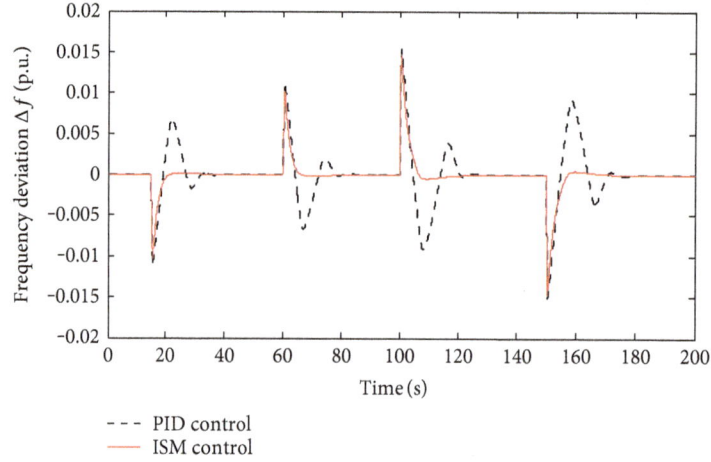

FIGURE 6: The frequency deviation with power output constraints.

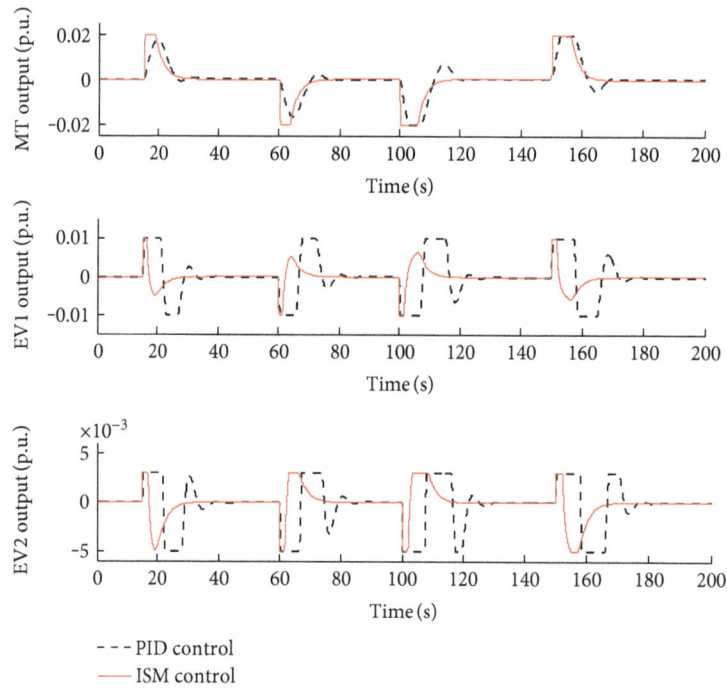

FIGURE 7: The output power of MT and EVs with power output constraints.

TABLE 3: The sum of frequency deviation.

Condition	PID	ISM	PID + constraints	ISM + constraints
E_{iae}	0.1309	0.066	0.3251	0.1212

By combining with (20), the derivative of $V(t)$ can be obtained as

$$\dot{V}(t) = s(t)\dot{s}(t) + \tau^{-1}\tilde{w}_2(t)\tilde{w}_2(t)$$
$$= s(t)\dot{s}(t) - \tau^{-1}\tilde{w}_2(t)\dot{w}_2(t). \tag{21}$$

According to (2) and (4), then $\dot{V}(t)$ can be deduced as

$$\dot{V}(t) = s(t)\left(C_1\dot{x}(t) + C_2 x(t)\right) - \tau^{-1}\tilde{w}_2(t)\dot{w}_2(t). \tag{22}$$

The weights of the neural network observer are updated according to (13); then $\dot{V}(t)$ can be derived as

$$\dot{V}(t) = s(t)\Big(C_1 A x(t) + \left(-C_1 A x(t) - C_2 x(t) - \varepsilon\text{sat}(s) - C_1 F\Delta\hat{P}_s(t)\right)$$
$$+ C_1 F(\Delta P_s(t) + \Delta P_{\text{d}}(t)) + C_2 x(t)\Big)$$
$$+ \tau^{-1}\tilde{w}_2(t)\dot{w}_2(t). \tag{23}$$

We simplify (23) and substitute $\tilde{w}_2(t) = w_2^*(t) - w_2(t)$; then $\dot{V}(t)$ becomes

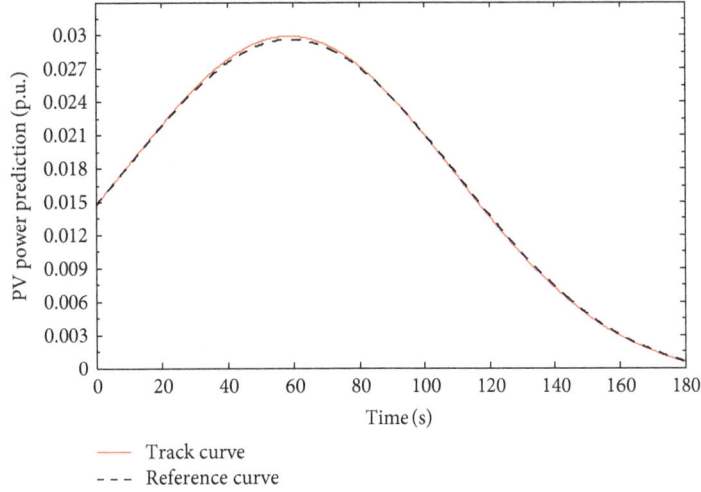

FIGURE 8: The predictive PV power by the neural network observer.

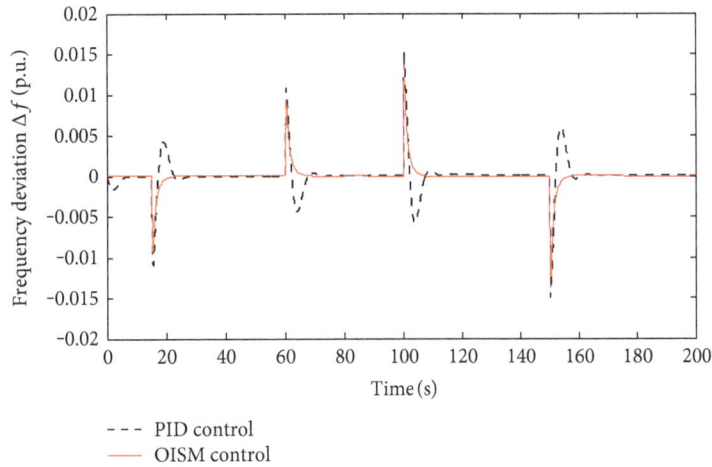

FIGURE 9: The frequency performance without power output constraints.

$$\dot{V}(t) = s(t)\left(-\varepsilon\mathrm{sat}(s) + C_1 F(e_\mathrm{p}(t) + \Delta P_\mathrm{d}(t))\right)$$
$$- \left(w_2^* - w_2\right)(t)\,\psi\left(w_1(t)I_t\right)e_\mathrm{p}(t). \qquad (24)$$

According to the universal approximation property of neural network and *Assumption* 1, that is, $|e_\mathrm{p}(t)| < \xi, |\Delta P_\mathrm{p}(t)| \le \alpha_1$, and $|\Delta P_\mathrm{d}(t)| \le \alpha_2$, then we can obtain that the sliding mode gain ε is required to satisfy

$$\varepsilon \ge |C_1 F(\alpha_1 + \alpha_2)|. \qquad (25)$$

After a sufficient learning, the error of the neural network observer is reasonable to be small enough, which means $\xi \le \alpha_1$. Therefore, it is easy to deduce

$$-\varepsilon \le C_1 F(e_\mathrm{p}(t) + \Delta P_\mathrm{d}(t)) \le \varepsilon. \qquad (26)$$

With $w_2^*\psi(w_1) = \Delta P_p(t)$ and $w_2\psi(w_1) = \Delta\widehat{P}_p(t)$, therefore we have the following:

$$\dot{V}(t) = s(t)\left(-\varepsilon\mathrm{sat}(s) + C_1 F\left(e_\mathrm{p}(t) + \Delta P_\mathrm{d}(t)\right)\right) - e_\mathrm{p}^2(t). \qquad (27)$$

It means

$$\dot{V}(t) \le -e_\mathrm{p}^2(t) \qquad (28)$$

for both $s(t) > 0$ and $s(t) < 0$.

Observing inequalities (18) and (28), controller (17) can ensure $V(t) \ge 0$ and $\dot{V}(t) \le 0$ for system (1). According to the Lyapunov stability theorem, system (1) is asymptotically stable. It means the frequency derivation of system (1) is regulated to zero under the designed OISM controller.

4. Simulation and Analysis

In this section, we apply the proposed frequency control method to the studied benchmark microgrid, which is a typical benchmark system. The parameters of a benchmark

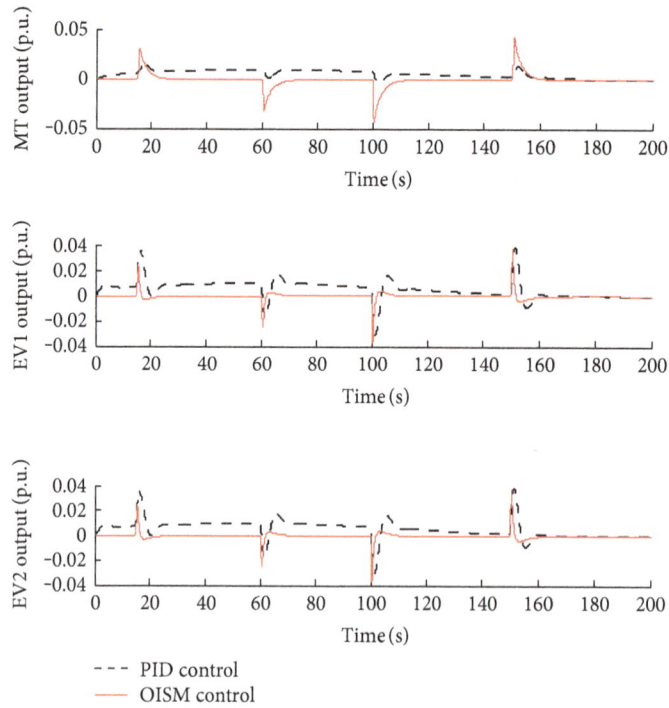

FIGURE 10: The output power of MT and EVs without power output constraints.

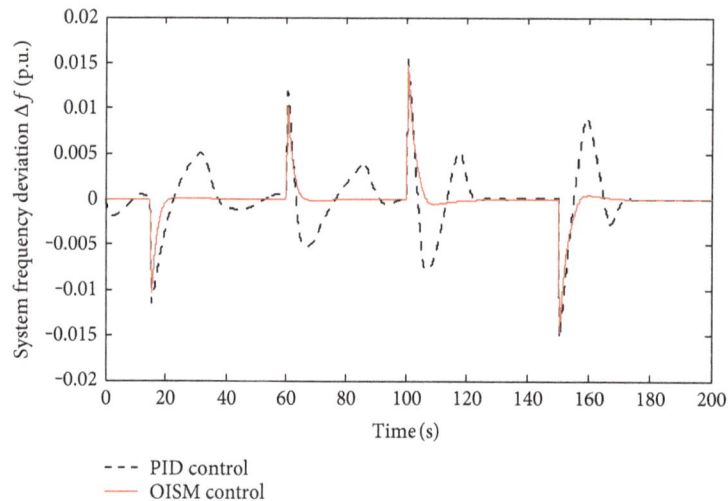

FIGURE 11: The frequency performance with power output constraints.

microgrid system are given in Table 1 by referring to [19, 34, 35]. The PID controller is also used to damp the frequency deviation of the benchmark system as a competitive method.

In the PID controller, it takes the frequency deviation $\Delta f(t)$ as the input, and the parameters of proportional, integral, and derivative gains are set as $K_p = 4$, $K_i = 1$, and $K_d = 0.6$, respectively. For the proposed ISM control method, the control gain is $\varepsilon = 10$, and the coefficient vectors C_1 and C_2 are defined as $C_1 = [0.65, 0.05, 0.05, 0.05, 0.05]$ and $C_2 = [0.15, 0.01, 0.01, 0.01, 0.01]$, respectively. In order to

validate the performance of designed controllers, we have the following cases.

4.1. Dynamic Response for Load Change.
In this case, the frequency of the microgrid system is initially stable; four sequential active power disturbances caused by load change are applied to the microgrid system, which is shown in Table 2. Without integrating the PV power, the ISM controller in (5) is used to suppress the frequency oscillation.

When these sequential disturbances are added on the system, the performance of the integral sliding mode controller

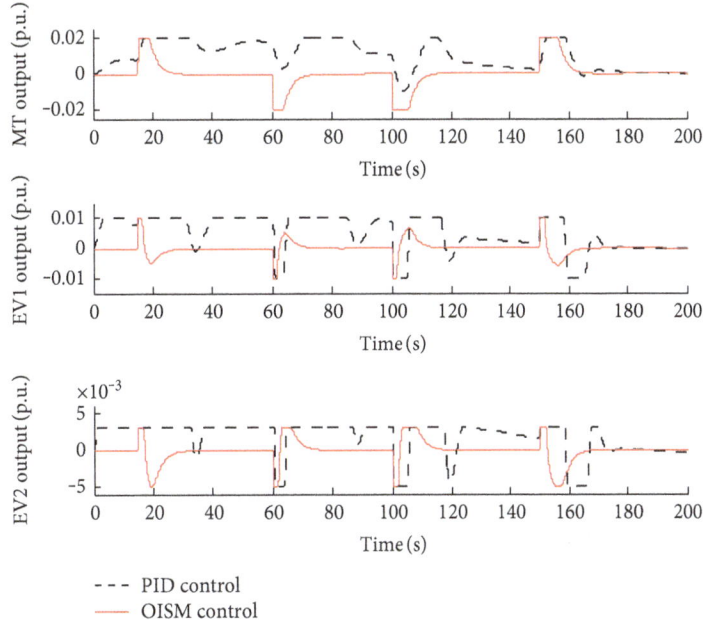

FIGURE 12: The frequency performance with power output constraints.

TABLE 4: The sum of frequency deviation.

	PID	OISM	PID + constraints	OISM + constraints
E_{iae}	0.1459	0.0679	0.4102	0.1230

is presented in Figure 4. As a competitive result, the regulation result of PID controller is also presented in Figure 4. The power outputs of MT and EVs are shown in Figure 5.

4.2. Dynamic Response to Load Disturbances and Power Output Constraints.
Consider the output power limits of MT and two EVs as [36]. The output power values of MT and two EVs have output limits which are [−0.02, +0.02], [−0.01, +0.01], and [−0.005, +0.003] for the output power ranges of MT, EV1, and EV2, respectively. The same disturbances given in Table 2 are added on this benchmark system. In this case, the performance of system frequency deviation is shown in Figure 6, including the PID control and the ISM control. The outputs of MT and EVs are shown in Figure 7.

From Figures 4 and 6, we can observe that every disturbance generates an obvious frequency deviation from the prescribed value, and the deviation depends on the amplitude of disturbance. Compared with the PID controller, the ISM controller has a faster response ability as well as smaller transient amplitudes. Moreover, without the power output constraints, the regulation of both PID and ISM controller has better performance than the two controllers under the condition of power output constraints. Figures 5 and 7 show the power outputs of MT and EVs without and with constraints.

The performance index based on the integral value of total frequency deviation in the whole regulation process is defined as

$$E_{iae} = \int_0^t |\Delta f|\, dt. \tag{29}$$

For the above two cases, the performance indexes of two controllers are presented in Table 3. Comparatively speaking, the ISM controller has a better regulation ability than the PID control.

4.3. Dynamic Response to Both Load Disturbances and PV Power.
In this case, the PV power is integrated into the microgrid system based on case 1. The PV power and four sequential power disturbances are all applied to the microgrid. Specifically, four sequential power disturbances from load changes are given in Table 2.

The parameters of PID controller and ISM controller are the same as the values in case 1. In order to eliminate the impact of PV power, we change the ISM controller to the OISM controller, which is designed in (17). The parameters of the neural network observer are set as follows: input layer nodes $m = 35$, hidden layer nodes $n = 6$, and the learning rate $\tau = 0.01$. It means that the neural network learns from 35 previous values before time t and then estimates the PV power at time t. The output of the neural network observer is shown in Figure 8, compared with the real PV power. From Figure 8, we can know that the neural network observer is able to predict the PV power well by learning the previous PV power data.

After integrating the PV power into the system and applying these sequential disturbances, the system frequency deviation curves with PID controller and OISM controller are presented in Figure 9, and the output power of MT and EVs are shown in Figure 10.

4.4. Dynamic Response to Load Disturbances, PV Power, and Power Output Constraints. Considering the output power limits of MT and two EVs, the output constraints of MT and EVs use the same settings as case 2. The applied disturbances are also shown in Table 2. By using the OISM controller, the system frequency deviation is shown in Figure 11, and the outputs of MT and EVs are shown in Figure 12. The performance of PID controller has been provided as the comparison.

From Figures 9 and 11, the frequency deviates from the command value under the disturbances from PV power and load change. However, the OISM controller is more effective than the PID controller, which can provide the control performance with faster regulation and smaller oscillation in the whole regulation process. From Figures 10 and 12, we can know that the PID controller hardly adjusts the deviation of the output power to zero; it means that the frequency is difficult to be regulated to the specified value. Comparing the control performance with the PID controller, the OISM controller still keeps fast regulation and strong robustness with an observed PV power. Similar to case 2, the performance of the PID controller becomes worse when the output power is constrained. From Figures 11 and 12, we can observe that it is hard for the PID controller to regulate the frequency to the required value under disturbances. On the contrary, the OISM controller still has a superior damping performance.

The performance index E_{iae} is also used to evaluate the control performance. The index values of the controllers are presented in Table 4. It can be seen that the performance of the OISM controller is obviously better than that of the PID controller for the whole regulation process.

5. Conclusions

This paper has addressed the LFC problem for the island microgrid with EVs and PV power integration. The OISM control strategy is adopted to regulate the frequency derivation, where the neural network observer is designed to predict the PV power disturbance. The theoretical analysis is presented for the OISM system in the sense of Lyapunov stability. Simulations have been executed on a benchmark system with different conditions, including load change, PV disturbances, and power output limits. Comparative studies with the PID controller are provided to demonstrate the superior performance of the OISM controller. In the future work, there are several significant topics that need to be intensively addressed. For example, integrating the diversified sources of energy into the island microgrid will be considered. Also, we will consider the coordination control of diversified energies and the frequency stability of interconnected microgrids.

Conflicts of Interest

The authors declare that they have no conflicts of interest.

Acknowledgments

This work was supported in part by the National Natural Science Foundation under Grants 51377065, 61304018, 61301035, and 6141130160, in part by Hubei Province Science and Technology Supporting Program under Grant 2014BAA035, and in part by Tianjin Natural Science Foundation under Grant 14JCQNJC05400.

References

[1] J. A. P. Lopes, F. J. Soares, and P. M. R. Almeida, "Integration of electric vehicles in the electric power system," *Proceedings of the IEEE*, vol. 99, no. 1, pp. 168–183, 2011.

[2] Y. Lei, W. Xu, C. Mu, Z. Zhao, H. Li, and Z. Li, "New hybrid damping strategy for grid-connected photovoltaic inverter with LCL filter," *IEEE Transactions on Applied Superconductivity*, vol. 24, no. 5, pp. 1–8, 2014.

[3] C. Mu, Y. Tang, and H. He, "Observer-based sliding mode frequency control for micro-grid with photovoltaic energy integration," *2016 IEEE Power and Energy Society General Meeting (PESGM)*, pp. 1–5, 2016.

[4] C. T. Pan and C. M. Liaw, "An adaptive controller for power system load-frequency control," *IEEE Transactions on Power Systems*, vol. 4, no. 1, pp. 122–128, 1989.

[5] X. Li, Y.-J. Song, and S.-B. Han, "Frequency control in microgrid power system combined with electrolyzer system and fuzzy PI controller," *Journal of Power Sources*, vol. 180, no. 1, pp. 468–475, 2008.

[6] I. Kamwa, R. Grondin, and Y. Hebert, "Wide-area measurement based stabilizing control of large power systems-a decentralized/hierarchical approach," *IEEE Transactions on Power Systems*, vol. 16, no. 1, pp. 136–153, 2001.

[7] P. Kundur, *Power System Stability and Control*, McGraw-Hill Education, New York, 1st edition, 1994.

[8] Y. Tang, J. Yang, J. Yan, and H. He, "Intelligent load frequency controller using GrADP for island smart grid with electric vehicles and renewable resources," *Neurocomputing*, vol. 170, pp. 406–416, 2015.

[9] M. H. Ali, T. Murata, and J. Tamura, "Transient stability enhancement by fuzzy logic-controlled SMES considering coordination with optimal reclosing of circuit breakers," *IEEE Transactions on Power Systems*, vol. 23, no. 2, pp. 631–640, 2008.

[10] E. Cam and I. Kocaarslan, "Load frequency control in two area power systems using fuzzy logic controller," *Energy Conversion and Management*, vol. 46, no. 2, pp. 233–243, 2005.

[11] I. Kocaarslan and E. Cam, "Fuzzy logic controller in interconnected electrical power systems for load-frequency control," *International Journal of Electrical Power & Energy Systems*, vol. 27, no. 8, pp. 542–549, 2005.

[12] M. Zribi, M. Al-Rashed, and M. Alrifai, "Adaptive decentralized load frequency control of multi-area power systems," *International Journal of Electrical Power & Energy Systems*, vol. 27, no. 8, pp. 575–583, 2005.

[13] A. Rubaai and V. Udo, "An adaptive control scheme for load-frequency control of multiarea power systems part I.

Identification and functional design," *Electric Power Systems Research*, vol. 24, no. 3, pp. 183–188, 1992.

[14] S. Vachirasricirikul and I. Ngamroo, "Robust LFC in a smart grid with wind power penetration by coordinated V2G control and frequency controller," *IEEE Transactions on Smart Grid*, vol. 5, no. 1, pp. 371–380, 2014.

[15] S. Saxena and Y. V. Hote, "Load frequency control in power systems via internal model control scheme and model-order reduction," *IEEE Transactions on Power Systems*, vol. 28, no. 3, pp. 2749–2757, 2013.

[16] X. Liu, X. Kong, and K. Y. Lee, "Distributed model predictive control for load frequency control with dynamic fuzzy valve position modelling for hydro-thermal power system," *IET Control Theory and Applications*, vol. 10, no. 14, pp. 1653–1664, 2016.

[17] K. Vrdoljak, N. Peric, and I. Petrovic, "Sliding mode based load-frequency control in power systems," *Electric Power Systems Research*, vol. 80, no. 5, pp. 514–527, 2010.

[18] Y. Tang, C. Mu, and H. He, "SMES-based damping controller design using fuzzy-GrHDP considering transmission delay," *IEEE Transactions on Applied Superconductivity*, vol. 26, no. 7, pp. 1–6, 2016.

[19] D. Qian, S. Tong, H. Liu, and X. Liu, "Load frequency control by neural-network-based integral sliding mode for nonlinear power systems with wind turbines," *Neurocomputing*, vol. 173, Part 3, pp. 875–885, 2016.

[20] S. D. G. Jayasinghe, D. M. Vilathgamuwa, and U. K. Madawala, "Direct integration of battery energy storage systems in distributed power generation," *IEEE Transactions on Energy Conversion*, vol. 26, no. 2, pp. 677–685, 2011.

[21] L. A. de Souza Ribeiro, O. R. Saavedra, S. L. de Lima, and J. G. de Matos, "Isolated microgrids with renewable hybrid generation: the case of Lencois Island," *IEEE Transactions on Sustainable Energy*, vol. 2, no. 1, pp. 1–11, 2011.

[22] B. Bandyopadhyay, F. Deepak, and K.-S. Kim, *Sliding Mode Control Using Novel Sliding Surfaces*, Springer-Verlag, Berlin Heidelberg, 2009.

[23] V. I. Utkin, *Sliding Modes in Control and Optimization*, Springer-Verlag, Berlin Heidelberg, 1992.

[24] C. Mu, W. Xu, and C. Sun, "On switching manifold design for terminal sliding mode control," *Journal of the Franklin Institute*, vol. 353, no. 7, pp. 1553–1572, 2016.

[25] C. Mu, Z. Ni, C. Sun, and H. He, "Air-breathing hypersonic vehicle tracking control based on adaptive dynamic programming," *IEEE Transactions on Neural Networks and Learning Systems*, vol. 28, no. 3, pp. 584–598, 2017.

[26] L. Cheng, Z.-G. Hou, and M. Tan, "Adaptive neural network tracking control for manipulators with uncertain kinematics, dynamics and actuator model," *Automatica*, vol. 45, no. 10, pp. 2312–2318, 2009.

[27] D. Wang, D. Liu, C. Mu, and H. Ma, "Decentralized guaranteed cost control of interconnected systems with uncertainties: a learning-based optimal control strategy," *Neurocomputing*, vol. 214, pp. 297–306, 2016.

[28] D. Wang, C. Li, D. Liu, and C. Mu, "Data-based robust optimal control of continuous-time affine nonlinear systems with matched uncertainties," *Information Sciences*, vol. 366, pp. 121–133, 2016.

[29] B. Gu, V. S. Sheng, K. Y. Tay, W. Romano, and S. Li, "Incremental support vector learning for ordinal regression," *IEEE Transactions on Neural Networks and Learning Systems*, vol. 26, no. 7, pp. 1403–1416, 2015.

[30] B. Gu, V. S. Sheng, Z. Wang, D. Ho, S. Osman, and S. Li, "Incremental learning for v-support vector regression," *Neural Networks*, vol. 67, pp. 140–150, 2015.

[31] B. Gu, X. Sun, and V. S. Sheng, "Structural minimax probability machine," *IEEE Transactions on Neural Networks and Learning Systems*, 2016, in press.

[32] D. Wang, C. Mu, and D. Liu, "Data-driven nonlinear near-optimal regulation based on iterative neural dynamic programming," *Acta Automatica Sinica*, vol. 43, no. 3, pp. 366–375, 2017.

[33] K. Hornik, "Approximation capabilities of multilayer feedforward networks," *Neural Networks*, vol. 4, no. 2, pp. 251–257, 1991.

[34] Y. Mi, Y. Fu, C. Wang, and P. Wang, "Decentralized sliding mode load frequency control for multi-area power systems," *IEEE Transactions on Power Systems*, vol. 28, no. 4, pp. 4301–4309, 2013.

[35] C. Wang, Y. Mi, Y. Fu, and P. Wang, "Frequency control of an isolated micro-grid using double sliding mode controllers and disturbance observer," *IEEE Transactions on Smart Grid*, 2016, in press.

[36] J. Yang, Z. Zeng, Y. Tang, J. Yan, H. He, and Y. Wu, "Load frequency control in isolated micro-grids with electrical vehicles based on multivariable generalized predictive theory," *Energies*, vol. 8, no. 3, pp. 2145–2164, 2015.

17

Recent Developments of Photovoltaics Integrated with Battery Storage Systems and Related Feed-In Tariff Policies

Angel A. Bayod-Rújula,[1] **Alessandro Burgio,**[2] **Zbigniew Leonowicz,**[3] **Daniele Menniti,**[2] **Anna Pinnarelli,**[2] **and Nicola Sorrentino**[2]

[1]*Department of Electrical Engineering, University of Zaragoza, Zaragoza, Spain*
[2]*Department of Mechanical, Energy and Management Engineering, University of Calabria, Rende, Italy*
[3]*Wroclaw University of Science and Technology, Wroclaw, Poland*

Correspondence should be addressed to Alessandro Burgio; alessandro.burgio@unical.it

Academic Editor: Cheuk-Lam Ho

The paper presents a review of the recent developments of photovoltaics integrated with battery storage systems (PV-BESs) and related to feed-in tariff policies. The integrated photovoltaic battery systems are separately discussed in the regulatory context of Germany, Italy, Spain, United Kingdom, Australia, and Greece; the attention of this paper is focused on those integrated systems subject to incentivisation policies such as feed-in tariff. Most of the contributions reported in this paper consider already existing incentive schemes; the remaining part of the contributions proposes interesting and novel feed-in tariff schemes. All the contributions provide an important resource for carrying out further research on a new era of incentive policies in order to promote storage technologies and integrated photovoltaic battery systems in smart grids and smart cities. Recent incentive policies adopted in Germany, Italy, Spain, and Australia are also discussed.

1. Introduction

In the last decade, the incentivisation of renewable source generation systems has received significant appreciation. The International Energy Agency (IEA) [1] calculates that subsidies granted to renewable energy worldwide amount to $135 billion in 2014 with an average growth rate of 25% in 2008. In 2014, Germany, the United States, and Italy hold 50% of the total of subsidies and 85% of the first 10 countries. After such a significant economic effort, from 2006 to 2015, the world total RES installed power doubled.

Figure 1 shows the global historical trend of power from installed renewable sources in the world in the last decade; the lower line indicates the percentage of power from photovoltaic sources. Figure 2 shows how power from renewable sources is divided between different continents. It is worth noting that Eurasia includes Armenia, Azerbaijan, Georgia, the Russian Federation, and Turkey. As clearly explained by IRENA in [2], smart grid technologies have facilitated the exploitation of renewable energy sources mainly by increasing grid flexibility. Furthermore, smart grid technologies have also favoured the integration of distributed renewable generation in transmission and distribution electric grids, simultaneously reducing the investment needs for operating existing infrastructures. The profound implications of smart grid technologies on transmission and distribution electric grids have a relevant and strategic importance given that IRENA (International Renewable Energy Agency 2013) estimates that these electric grids will account for almost half of the power sector investment until 2035.

The financial support for renewable source generation systems has significantly reduced the costs of renewable energy; however, since these costs continue, it is hoped that the renewables industry will soon survive without subsidies. It is not reasonable to think that an incentivisation policy can last forever; in fact, it should create and support the market in its initial life phase and then the market should be self-sustaining.

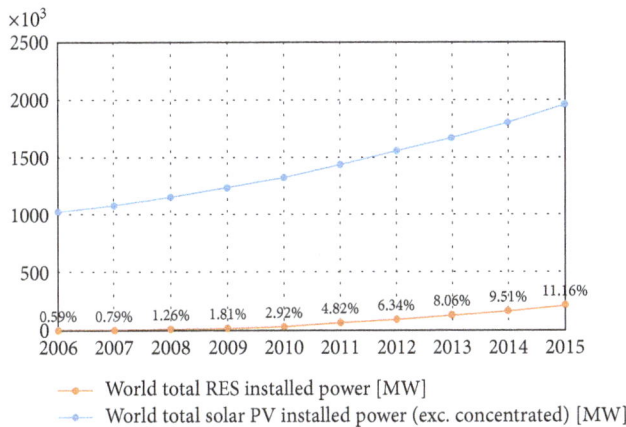

FIGURE 1: World total RES and solar PV installed power in MW in 2006–2015, IRENA statistic 2016.

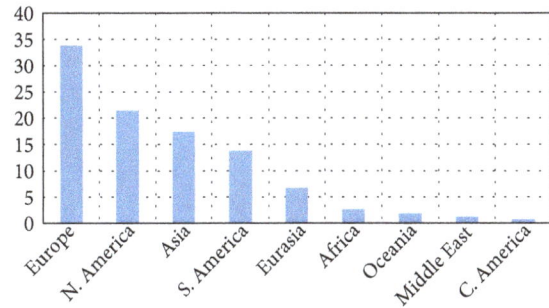

FIGURE 2: Percentage of world total RES installed power in 2006–2015, IRENA statistic 2016.

The tariff of an incentivisation scheme for the exploitation of renewable energy sources should be opportunely determined and, above all, be reasonable, given that financial support for renewable technologies is substantially paid by hardworking families and businesses via their electricity bills. In this connection, Pyrgou et al. [3] discuss the future of the feed-in tariff scheme in Europe; in particular, they exclusively study the case of photovoltaics and examine the regulatory and policy framework of the feed-in tariff scheme, specifically its effect on both the electricity pricing and the local and European renewable energy source markets.

In many countries, the value of the incentivising tariff is typically determined by the National Government considering the type of renewable source (solar, wind, and biomass) and the technical and economic parameters (size, investment, and cost) in order to guarantee certain repayment for those who operate the investment. It is evidently a political issue, and, as often is the case, opinions relating to it can be distinctly contrasting, as in the case of the economist and expert Darwall who states in [4] that renewable subsidies destroyed the UK electricity market. Considered by many people as a climate sceptic, Darwall believes that when politicians decided to impose renewables, they intentionally did not wanted to consider the entirely predictable destruction of the electricity market as a consequence of their policies. On the contrary, the politicians were convinced that the world should have adapted to their preferred generation technology.

Criticism of the effects of social welfare of policies enacted by countries to support the exploitation of renewable sources is not, however, the object of this study. Instead, this study simply seeks to address the case study of integrated battery storage photovoltaic systems. Therefore, in this study, the two technologies for smart grids, namely smart inverters and end user level distributed storage, and photovoltaic plants for the distributed generation of energy from renewables converge.

The contributions of the different authors mentioned in the present work discuss PV-BESs such as those illustrated in Figure 3 in the regulatory context of Germany, Italy, Spain, the United Kingdom, Australia, and Greece. More precisely,

only contributions where PV-BESs are the object of an incentivisation policy, such as a feed-in tariff, are considered.

The cases of these five countries are discussed separately in the order of the total solar PV power installed; as can be seen in Figure 4, the first case is Germany as it has an installed power of 214,000 MW, and the last case is Greece with an installed power of 10,200 MW. Figure 5 reports the production of electrical energy from solar PV sources. It is interesting to note that the order of Figure 5 is identical to that of Figure 4 with the exception of the UK and Australia; the latter has a production (13,656 GWh) that is almost double to that of the UK production (7718 GWh) even though the installed solar PV power in Australia is approximately 16% lower than that in the UK.

Most of the contributions reported in this paper consider already existing incentive schemes; therefore, these schemes constitute the input data for the techno-economic-financial analyses together with the solar radiation, the user load profiles, and the average electricity prices. Some of these contributions propose new feed-in tariff schemes in order to promote storage technologies in general as well as integrated photovoltaic battery systems for grid-connected end users.

2. A Brief Reference to Sizing and Integration of Photovoltaic and Batteries Systems in Distribution Grids

In the imminent future, renewable energy sources will certainly play a key role in electricity generation; solar and wind energy sources undoubtedly have the potential to meet the energy crisis to relevant extent. Solar and wind power plants have already created favourable conditions to switch the electricity generation from large-centralized facilities to small-decentralized units, exploiting the technological development and the increasing market competitiveness in the renewable energy sectors. These small-decentralized units are the implementation of the distributed generation as cultural concept, as feasible solution for a sustainable development extended also to remote areas, and as an effective environmental impact reduction.

In the current thinking, PV power generation on rooftop is the entry-level technology suitable for a massive transition towards a low-carbon power generation, also in the buildings sector. The increase in the use of incentives

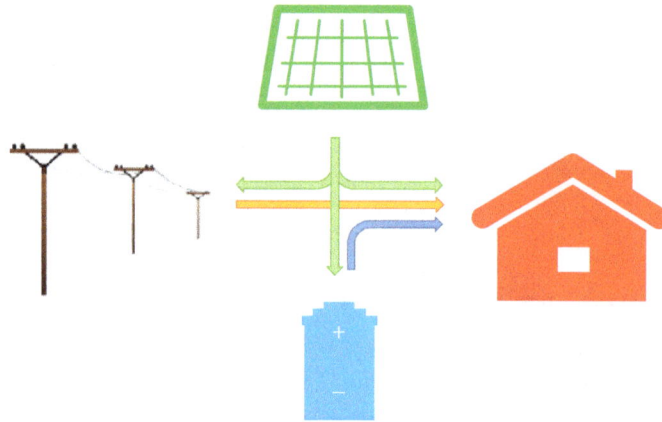

FIGURE 3: An integrated photovoltaic battery energy storage system (PV-BES).

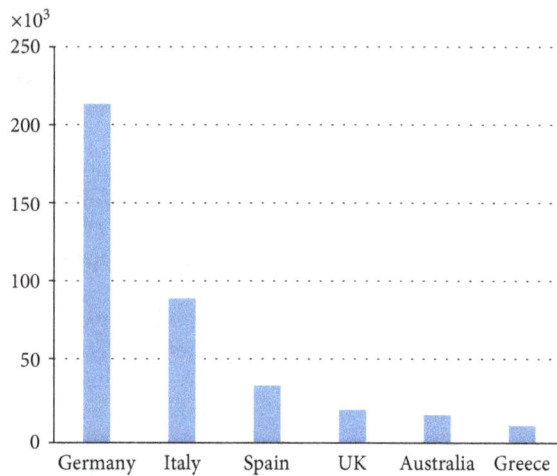

FIGURE 4: Solar PV installed power MW in 2006–2015, IRENA statistic 2016.

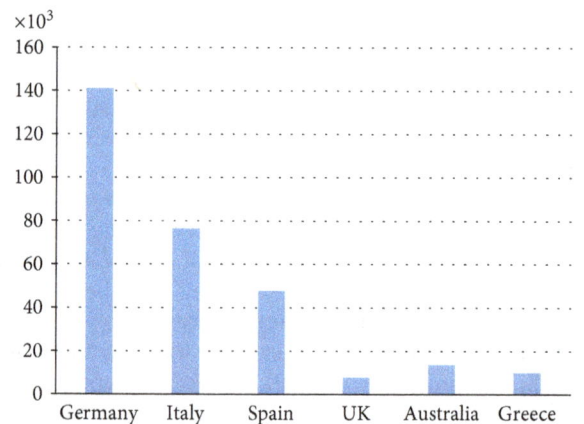

FIGURE 5: Solar PV production GWh in 2006–2015, IRENA statistic 2016.

for the construction of integrated photovoltaic systems in buildings has decisively established that such distributed generation, combined with the use of microinverters, is a viable option with great growth potential [5]. Undoubtedly, the wide spread of distributed generation strongly depended on government policies to support PV systems, by credit guarantee scheme in favour of the citizen and/or granting. Today, subsidies still exist in few countries but are subject to a fast reduction; so, self-consumption may be the only way to gain financial profit from distributed generation at residential and building level. Indeed, in those countries where there are no government policies to support PV systems, no credit guarantee schemes in favour of the citizen, and no grants, the sole incentive is self-consumption; the lower is the export price for PV energy with respect the retail price, the higher is the attractiveness of self-consumption. Thanks to self-consumption and the rapid decrease of the costs of modules and inverters, covering the roofs with PV modules remains an attractive investment, even in the absence of subsidies.

Performance of PV technology cannot be accounted for solely in terms of cost targets and energy efficiency. In this assessment, a holistic view is lacked; therefore, benefits as the peak demand reduction, the improved network stability, and loss reduction are not accounted [6]. Actually, a comprehensive and holistic analysis of how the combination of influencing factors determines the economics of rooftop modules is still missing [7]. Citizens who have installed PV modules on roofs are an example of distributed power generation at residential level, but they are also carriers of important benefits for the environment and the rest of the electrical system. If properly managed, prosumers may contribute to improving the system stability and reduce overall losses [8, 9]. On the other hand, the increasing electrification will put the reliability and stability of distribution grids under pressure. Integrating solar electricity generator at utility scaled into power networks may negatively affect the performance of the next generation of smart grids. In particular, the rapidly changing of power generated by many distributed generators is almost unpredictable; electrical storage systems like batteries are the technical trick to move problems from short term to mid term [10].

In the light of the above analyses, it is clear that distributed generation through PV plants on rooftops contributes to alleviate the overall load on LV distribution grids, the peak demand, and the power losses. Besides these benefits, new problems and issues arise. The LV grids, in fact, were originally designed to accommodate a unidirectional current flow, from the distribution network to consumers. With the high penetration of distributed generation, the current can reverse its direction; a possible consequence of reversal of the power flow is the occurrence of overvoltage. This is a new and unwelcome challenge for the distribution system operator, but it may be also a limitation to PV systems spread. A further limitation to such a spread may be the inability or high costs associated to the excessive generation with respect the demand; to face a surplus of generation, it is useful to mention the strengthening of the distribution networks and the widespread installation of small/limited storage systems at LV grid levels.

The examination of the hourly and intrahourly time series of the potential energy generation from distributed PV systems is recommended and necessary in order to obtain a proper understanding of the zonal energy balance [11]. The frequent variation of the solar radiation on PV modules and the frequent variation of speed and direction on a wind turbine make these two sources, that is, sun and wind, too intermittent and unreliable. Therefore, connecting these sources to the grid presents challenges in various technical aspects: power quality, protection, dispatching, control, and reliability. Levelling the output of PV systems and wind generators is necessary to maintain the grid stability [12]. In this regard, the adoption of energy storage systems is a feasible solution to compensate for the above-mentioned fluctuations and to meet the demand of energy during the night hours [13, 14]. Simultaneously, implementing demand response programmes can facilitate a resolution of this issue [15]. In addition to demand response, several research contributions examine the coordination of PV battery systems with demand side management. The total number of these research contributions is considered as limited; further comparative studies are required to achieve a complete overview of these technologies and their potential [16].

The use of battery energy storage systems, integrated with the PV modules, allows to decouple the generation and the consumption of electric energy. This deferred timeline is a simple and feasible solution, useful to reduce the grid load during peak hours so to avoid the strengthening of distribution networks. Novel control strategies for battery recharge management may be developed so to achieve benefits and savings for both the distribution system operators and prosumers [17]. Scheduling the battery operation installed at the prosumer level is strongly recommended because the battery degradation has a high cost and such a cost is mainly a function of the storage system operation [18].

The effects and the actual benefits of domestic integrated PV battery systems and the effects and real benefits on the distribution and transmission grids have to be identified and analyzed. The development of strategies for the coordinated management of the grid and storage systems is a necessary step to analyze and evaluate the real potential of domestic battery storage systems as a solution to mitigate the stress on the electric power system [19]. On the other hand, the use of these hybrid energy systems

looks promising because they are reliable and economical, thanks to the complementary nature of the two resources. However, at present, it is still unclear when and under what conditions the battery storage can be profitably used in residential photovoltaic systems without any political support [20–22].

As previously mentioned, self-consumption is a fundamental key for a proper evaluation of distributed generators when integrated to storage systems; scheduling of domestic loads such as washing machines, dryers, and dishwashers is an exciting opportunity to maximize self-consumption [23]. In this context, depending on the solar radiation, incentives, special discounts, and other factors, techniques for the demand side management can assist families in changing their usual habit in the electricity consumption [24]. Because these techniques use PV generation forecasting to determine the optimal loads scheduling, the greater the goodness of forecasting, the higher the savings. In case of inaccurate forecasts, errors are limited by using battery storage systems [25]. Because load profiles in commercial applications have a higher correlation with the daily solar radiation with respect to the residential applications, also the retail and commercial sectors are important sources of productivity gains and savings [26].

Finally, the introduction of electric vehicles in the home environment accentuates the "delay" between the load profile of the families and the generation profile of the PV modules [27]. Even in this case, the use of batteries seems a mandatory measure rather than an appropriate choice. Indeed, the spread of charging stations—which exploiting renewable energy sources—placed outside the home environment has increased in the recent years thanks to the use of distributed battery storage systems [28]. Although many contributions studied the integration of charging station and the utility grid, some researchers believe that none of these contributions has given due emphasis on the PV charger [29].

3. A Brief Reference to Incentive Polices and Implementation Schemes

In order to incentivise the adoption and use of renewable energy sources, different types of incentive mechanisms and schemes have been implemented in different countries and jurisdictions, such as feed-in tariff, quota obligation, green certificates, tendering system, and net metering.

The feed-in tariff (FIT) scheme is the most effective scheme in encouraging rapid and sustained spread of renewable energies because a feed-in tariff scheme significantly reduces the risks of investing in renewable energy technologies so that the market can grow rapidly.

There are different ways to structure an FIT scheme, each with its own strengths and weaknesses. In [30], Couture and Gagnon present seven different ways, which are not mutually exclusive, of structuring the payment of an FIT policy. These seven ways are divided into two large categories, namely those in which payment is dependent on the price of electrical energy and those that, on the contrary, remain independent of it.

Independent market FIT policies are policies that offer a purchase guarantee and either a fixed price or a minimum price for electrical energy from renewable sources fed into the grid. Instead, market-dependent FIT policies require renewable energy developers to provide their electricity to the market, so forcing developers to a competition with other suppliers to meet market demand. These policies offer a premium price or a feed-in premium, which comes on top of the market price.

In order to avoid windfall profits when the electricity market prices rise, governments can impose caps and floors on FIT premiums. From an analysis of the market-independent option, Couture and Gagnon highlight that the main advantage consists in the predetermined and guaranteed payment levels, which offer significant benefits including greater investment security, a more reliable and predictable revenue stream for developers, and lower overall risks. On the other hand, fixed price FITs distort competitive electricity prices because the purchase prices remain fixed regardless of the electricity market price; so, even if prices decline, RE producers will continue to receive the guaranteed prices. Moreover, fixed price FITs offer the same prices regardless of the time of day at which electricity is supplied, ignoring the prevailing electricity demand.

Couture and Gagnon analyzed the market-dependent option and highlights that the main advantage is encouraging demand sensitivity of RE producers, thus providing benefits to both grid operators and society. Indeed, producers are incentivised to supply electricity to the grid in times of high demand, when prices are the highest. In such a way, a more efficient electricity market is achieved because RE supply is encouraged at times when electricity is needed most. In addition, the premium price model could help meet peak demand especially when the spread between peak and off-peak prices is significant. However, the market-dependent option is not without disadvantages.

One disadvantage of the market-dependent option is the unpredictable electricity retail prices which create greater uncertainty for both investors and developers. This uncertainty is often an insurmountable obstacle because negative cash flows of RE projects are all concentrated at the beginning of the process to pay for technology and are amortized over periods of 15 years or more. Moreover, the reduction in market prices created by large increases in RE plants reduces appeal for investors and developers. The uncertainty of electricity retail prices is also a relevant obstacle for smaller investors who, in order to obtain project financing, need more stable and predictable revenue streams.

In [31] also, Sioshansi addresses the issue of how to incentivise the adoption of renewable energy sources and different types of incentive mechanisms commonly used in different jurisdictions. More interestingly, Sioshansi examines a problem of allocation of costs that arise following the use of distributed renewable energy (DRE) systems such as volumetric pricing.

As is well-known, residential customers pay a volumetric tariff to recover costs due to the providing energy service and due to the generation, transmission, and distribution capacity service; such a volumetric tariff mainly depends on energy consumption. Thus, for a customer who installs a PV plant, possibly integrated with a storage system, which drastically reduces electricity bought from the grid, then his payment to the utility company would be close to zero. However, the services mentioned above have to be installed and maintained to reliably serve the customer. Overall, volumetric costs determine the allocation of inefficient costs with DRE.

Furthermore, incentivisation mechanisms such as FIT worsen this inefficiency as the majority of these programmes offer incentives on the basis of energy generated by a DRE, without considering its effect on the capacity needs and costs.

4. The Case of Germany

Germany is notoriously one of the countries that firmly believed—and invested heavily—in the exploitation of renewables, also including technologies for energy storage on different scales. Since May 2013, the state-owned KfW bank has granted low-interest loans with an aggregate value of 163 million euros in order to promote PV-BESs with a PV peak power of up to 30 kW which should feed a maximum 60% of installed capacity into the grid. The Federal Ministry for Economic Cooperation and Development also covers 30% of the battery storage system costs. According to a study by RWTH University, almost 34,000 PV-BESs, with an average capacity of 6 kWh, was installed by 31 January 2016. Around 27% of the installed systems are mounted with lead-acid batteries, the rest with lithium batteries. When governmental support of PV-BES battery systems was introduced in May 2013, the budget was 60 million euros; a new programme started in March 2016 with a budget of 30 million euros will run until 2018. Since 2013, prices for lithium batteries in Germany have fallen by 18% per year.

4.1. PV-BESs in Commercial Applications. The study of integrated PV-BESs on the German FIT policy in commercial applications is addressed in 2016 in [26]; in the reference, Merei et al. focus their attention in this sector in that, in their opinion, there is a significant opportunity for economic savings. The reason for this is that commercial buildings usually have ample space on their roofs for the installation of photovoltaic panels and because their load profiles have a high correlation with the generated solar energy. Therefore, the authors study the real case of a supermarket in Aachen with a yearly electricity consumption of 238 MWh. For the cost calculation, the authors consider an import price of €20c/kWh, a cost of production of electrical energy from solar panels of about €8–12c/kWh, and a feed-in tariff of €10c/kWh.

The cost analysis for a PV system returns a cost reduction of 30% when the cheapest prices of PV systems of €1200/kWp and the lowest interest rates of 4% are considered. The cost analysis for the same PV systems discussed above but in combination with battery storage and a feed-in tariff of €10c/kWh returns no yearly cost reduction but rather leads to a rise in costs. In conclusion, battery storage increases self-consumption significantly, yet even unrealistic battery prices of less than €200/kWh cannot lead to an economic solution in the considered commercial application.

4.2. PV-BESs in Residential Applications. The study of integrated PV-BES on the German FIT policy in residential applications was addressed in 2009 in [32]; in this work, Braun et al. illustrate an integrated PV-BES developed in a French-German project called Sol-ion. Considering a storage system based on lithium-ion batteries and the German Renewable Energy Sources Act (EEG) which grants self-consumption in residential applications, the Sol-ion project proposes several models that are useful in analyzing energy flows and calculating the increase of PV self-consumption in residential PV-BESs installed in Germany and in France. The authors conclude that the adoption of the Sol-ion system is a profitable operation for a battery system price below €350/kWh.

In the reference, the authors introduce their study evaluating the reduction of the export price that occurred in 2009 from €43.01c/kWh to €25.01c/kWh for those customers that self-consume self-generated PV energy using battery storage systems. The difference of €18c/kWh is a cost which has to be compared with the import price. The German Federal Statistical Office calculated that the average electricity price in 2009 without VAT was approximately €19.45c/kWh; therefore, residential customers with a Sol-ion system could save €1.45c/kWh if 100% of the self-generated PV energy is self-consumed. Therefore, the authors calculated that residential customers with a 5 kWp PV plant could save electricity costs up to €73 per year if 100% of local generation is consumed locally. An increase of the profit margin is achieved when an increase of electricity prices of 4% per year is considered: the saved electricity costs rise to €5.66c/kWh and €283 per year. Additionally, the regression of the reimbursement tariffs leads to an increase of the profit margin; given depression rates of 8% in 2010 and 9% onwards, in 2012, residential customers with a Sol-ion system saved electricity costs of €7.71–8.61c/kWh and €385–430 per year. The case study considered by Braun et al. for the simulation results is a base scenario where the export price is €32.77c/kWh, the self-consumption feed is €19.05c/kWh, the annual increase of electricity prices is 4%, and the annual consumption is 5.5 MWh. Data refer to 2012. Based on these assumptions, a Sol-ion system using lithium-ion batteries with a capacity of 11.5 kWh and specific costs of €350/kWh is evaluated. The numerical results demonstrate that the Sol-ion system increases self-consumption by 82% compared to that of a conventional PV system without batteries; furthermore, the breakeven is reached between the 15th and the 20th year, depending on the installation and maintenance costs.

In 2012, the study of integrated PV-BESs on a current German feed-in tariff in residential applications is the focus of [33]; Mulder et al. who affirm that since 2012, the use of lead batteries up to 5 kWh is convenient even in the absence of subsidies, independent of an increase in the cost of electrical energy. Taking the progressive decrease of the cost of lithium batteries into consideration, the authors state that this type of technology will certainly soon be attractive. In particular, if the price of electrical energy increases by 4%, then 4 kWh lithium batteries will already be economically advantageous in 2017, even in the absence of subsidies. The authors also present a cost-benefit analysis based on 2012 market prices and best future expectations; financial indicators such as the net present value, the internal rate of return, and the payback period were used. The numerical results reported in the paper refer to 260 combinations obtained using data from 65 households and 4 PV systems in Belgium; data measured every 15 minutes over a year.

In 2013, [21] presented a study that is very similar to Braun et al.'s 2009 study and Mulder et al.'s 2012 study. In the reference, Hoppmann et al. conclude that investing in storage batteries in Germany was economically advantageous for small systems already in 2013, investing in storage batteries in Germany was economically advantageous for small PV systems even not considering feed-in type policies. However, a possible promotional policy for battery storage systems is a valid instrument, which is only necessary in the short term. The authors create a technical-economic model and use this model to assess the viability of battery storage under eight scenarios from 2013 to 2022; each scenario is generated by varying the PV costs and the electricity prices. For each scenario, the model generates and tests more than 1400 photovoltaic battery combinations and identifies the combination with the highest net present value; the final user receives no feed-in tariff or self-consumption premium for the electricity self-produced by using the PV system. The study concludes that the economic viability of the battery storage for residential PV is particularly high in a scenario where the excess of electricity cannot be sold on the wholesale market.

A year later, in 2014, the study of integrated PV-BESs on the German FIT policy was again addressed in [34]. Like many other academics, Weniger et al. also seek to identify the storage system price at which residential PV-BESs become economically sustainable. In actual fact, Weniger et al. question which factor mainly influences the break-even price. The results reveal that the main factor is the rate of interest, followed by the PV system price, the retail price of electricity, and the feed-in tariff. The authors consider a feed-in tariff of €0.12/kWh, an import price of €0.34c/kWh, and an interest rate of 4%. Based on these assumptions, investing in PV-BESs is economically interesting if the PV system price is €1500/kWp and the battery system price is below €1160/kWh. Nevertheless, if the PV system price drops to €1200/kWp, then installing a PV-BES is a profitable operation for a battery system price below €1500/kWh. Weniger et al. also verified that an increase in the retail electricity price of €0.08/kWh has a larger impact than a decrease of the feed-in tariff by the same magnitude. Therefore, they suggest evaluating the profitability of PV-BESs focusing on the future development of retail electricity prices instead of on the development of new feed-in tariffs. Moreover, in [35], Bergner et al. state that the integration of PV systems with batteries will be the most economical solution in the long-term scenario.

In 2015, Lissen et al. introduced a further contribution to the study of integrated PV-BESs on the German FIT policy [36]. In the reference, Linssen et al. conduct a techno-economic analysis of PV-BESs with particular attention on the influence of different consumer load profiles. Hence, the authors affirm that the use of realistic load and production profiles is mandatory in order to allow for reliable statements concerning both technical parameters and economic feasibility. Otherwise, the techno-economic analysis and cost optimization results might overestimate self-consumption and lead to an incorrect calculation of the total costs. Accordingly, the numerical results illustrated in the reference relate to temporal high-resolution consumer loads and PV production profiles; load profiles are three at all. Profile 1 is the German standard load profile used by German utilities as a representative load profile for consumer groups. Profile 2 is an average profile of five single family houses usually used as a reference profile for combined heat and power systems, whereas profile 3 is a synthetic load profile generated by a simulation tool.

The PV profile is the measured profile of a 5.8 kWp PV system in 2012 with a 5-minute resolution; the yearly electricity generation returned by such a plant is 1010 kWh/kWp representing a typical value for central Germany. The PV system price is assumed as being €1640/kWp. The storage system is mounted with lithium-ion batteries; the depth of discharge is 100%, the efficiency is 95%, and the cycle lifetime is 6200 with a degradation of 0.4% per year. The storage system price is assumed as being €1000/kWh, exclusive of VAT. Further economic parameters are a feed-in-tariff of €13.28c/kWh, an import price of €29c/kWh, an interest rate of 4%, and an electricity price increasing by 2.5% per year. Numerical results show that the optimal cost of the integrated PV-BES increases from profile 1 to profile 3 due to different self-consumption levels. In particular, profile 3, which is considered as being the most realistic one, leads to the lowest value of self-consumption. Moreover, the break-even price for the integrated system is about €900/kWh without a battery storage support scheme and about €1200/kWh when considering the German support scheme. Linssen et al. conclude their study underlining that the individual taxation of revenues can significantly lower the break-even costs.

5. The Case of Italy

In 2005, Italy introduced its first incentivisation policy for PV plants based on a FIT scheme by means of a ministerial decree, which made available a first incentivisation budget known as the "Primo Conto Energia." From 2005 to 2012, the incentivisation payment mechanism changed four times, and in 2012, a ministerial decree made the fifth and final incentivisation budget for PV systems available, the so-called "Quinto Conto Energia." In July 2013, after about eight years of incentives, there were approximately 500,000 working incentivised plants; 10% of the plants started functioning in June 2012. By the end of 2014, there was an installed power of approximately 18 million kW. Today in Italy, all residential users can install batteries. In order to do so, the user must simply make a new connection request of the storage system to the state-owned company GSE. The storage system can either be integrated or not with a production plant; it can feed energy stored in the battery pack to the grid and can charge the batteries from the grid. In the case of systems with batteries with a bidirectional convertor, where batteries can also charge using energy from the grid, these can be integrated with photovoltaic systems without losing the incentive as long as bidirectional electric meters have been installed. Nowadays, the only special term conceded to the installation of batteries is tax deduction equal to 50% of the investment costs to be spread over 10 years; it is allowed in the case of restructuring and building energy saving works.

An analysis of the costs/benefits of PV-BES for domestic users is reported in [37]; the analysis considers real electrical energy consumption data for approximately 400 domestic clients spread over the Italian territory as well as real data of photovoltaic production for Northern, Central, and Southern Italy. For each domestic client, the PV plant is dimensioned so that it generates the annual energy consumption of the client, which is approximately 3700 kWh; the battery storage system has a discharge yield of 80% and is dimensioned to maximize self-consumption. The result presented in the reference consists of a further saving of approximately 150 euros for an existing photovoltaic plant incentivised by the feed-in scheme and a saving of approximately 170 euros in the case of a new nonincentivised plant. The calculated annual benefit is estimated net of costs of the initial investment.

An analysis of the costs/benefits of PV-BES for an Italian Public Administration connected to medium voltage grid is reported in [38]; in the reference, Burgio et al. propose a novel scheme for an FIT policy to favour the adoption of PV-BESs by means of a constant tariff which exclusively rewards self-consumption. The generation price and the export price are both neglected. An optimization problem jointly scales the PV and the BESs considering actual data obtained measuring load consumptions each 15 minutes throughout 2011. The objective function is the sum of three terms, that is, saving, subsidy, and costs; saving is defined as the difference between the electric bill with and without the combined PV-BES, subsidy represents the money that the end user receives due to the FIT policy, and costs is the sum of the instalments for both the PV and the BESs. The feed-in tariff is in the set €0.05, €0.10, €0.15, €0.20, €0.25, and €0.30/kWh. The optimal solution consists of a PV plant of 30 kWp and a battery storage system of 25 kW/50 kWh. The cash flow analysis highlights that for feed-in tariffs lower than €0.2274c/kWh, there is a requirement for additional financial resources. The authors also evaluated the impact of the PV-BES on load profiles by calculating the frequencies of energy measurements during 2011. In particular, the recurrence of the measurement 0 kWh (i.e., full self-sufficiency) is 0% without the PV-BES; it increases to 7% when a PV system is adopted and it further increases to 19.02% when a storage system is also adopted.

The economic viability of a feed-in tariff scheme that solely rewards self-consumption to promote the use of PV-BES is studied in [39]; in the reference, Burgio et al. use an optimization problem so to determine the incentive and the size of the PV-BES. The incentive was calculated so that the yearly subsidy equals to the difference between the instalments paid for the PV-BES and the savings obtained from the electricity bill. The size was calculated so that the percentage of self-produced energy is at least 50% and the percentage of self-consumed energy is at least 80%. The optimization problem was applied to a real case; measured values of temperature, irradiation, energy consumption and electricity prices were considered from 2011 to 2015. The numerical results reported in the reference demonstrated that the feed-in tariff scheme for a PV-BES is feasible and advantageous: for the case study, the electricity bill in 2011 was reduced by 49.56%. Moreover, the yearly subsidy is lower than the instalments paid for the PV-BES; therefore, a positive socioeconomic impact is achieved. Because the movement in the electricity prices is a crucial point in the economic evaluation of a PV-BES, the optimal solution was studied in the years from 2012 to 2015 so to evaluate the possible consequences of the collapse in electricity peak-load prices occurred at the end of 2013. The numerical results show that the PV-BES allows a reduction of the electricity bill also in the presence of this radical change in electricity prices. In particular, the reduction equals to 44.98% when the PV-BES is adopted, whereas it equals to 33.65% when only the photovoltaic system is adopted.

6. The Case of Spain

In Spain, the PV support policy started in 1998 with the publication of the Royal Decree (RD) 2818/1998, and the conditions were improved some years after with the publication of the RD 436/2004 and RD 661/2007. The conditions and premium were greatly modified with the RD 1578/2008. The support mechanism for PV systems gave the producers the possibility to choose whether to sell the electricity under the FIT tariff or whether to sell the electricity in the free market, taking advantage of a premium above the market price. The FIT policy granted producers for an undefined number of years; a reduction was expected after 25 years [40, 41].

Furthermore, the Royal Decree was stated the impossibility of integrating the PV systems with any storage system; such a prohibition has remained valid up to 2015. In 2012, the PV support policy was suspended with the publication of the Royal Decree 1/2012; one year later, the Royal Decree-Law 9/2013 lowered the grants of FITs retroactively.

In 2015, the Royal Decree 900/2015 introduced a self-consumption policy. This Decree distinguishes two types of customer with self-consumption: Type 1—Supply with self-consumption, and Type 2—Generation with self-consumption. Type 1 includes facilities and end users with an installed power no larger than 100 kW; the surplus of generated electricity fed into the grid is not remunerated because the export price is set equal to zero. Type 2 includes production facilities signed in the "Registro administrativo de instalaciones de Produción de Energía" and with an installed power larger than 100 kW. The surplus of generated electricity fed into the grid is remunerated.

The Royal Decree 900/2015 also stated that all customers which adhere to any self-consumption policies, that is, all customers who adopt a PV system integrated with battery storage system device are subjected to a new fee, known in Spain as" Impuesto al sol" or "solar tax." Such a new fee is a distribution and transport grid access fee required in order to ensure technical and economic sustainability of the grid. For instance, the most of end users connected to the low-voltage distribution grid in Spain with an installed power no larger than 10 kW have to pay the 2.0 A tariff which consist of €0.049033 for each self-consumed kWh.

As expectable, the widespread opinion about solar tax in Spain is definitely negative. In particular, the Spanish solar PV association affirms that, far from encouraging self-consumption, distributed generation, and use of renewable energies, this fee discourages the development of electric self-consumption. Moreover, the association underlines that, due to this unjustified tax, the self-consumers will pay much money for the power system maintenance than the other users although the self-consumers use the power system the least.

7. The Case of the United Kingdom

The study of integrated PV-BESs on the United Kingdom FIT policy is addressed in [42]; McKenna et al. state that the PV-BES combination in the UK is not a convenient operation as, even if there is a feed-in tariff, there is no case of financial convenience in adopting lead batteries. Such a conclusion is valid even if considering ideal batteries without leaks and with an optimistic life expectancy.

Given an exchange rate GBP/EUR of 1.235, the UK feed-in tariff consists of a generation price of 21.0 p/kWh (about €25.93c/kWh) and an export price of 3.2 p/kWh (about €3.95c/kWh) paid for exported units. Assuming an import price of 11.8 p/kWh (about €14.57c/kWh), the export/import price ratio is 11.8/3.2 = 3.69. The authors used 5 minutes of recorded data on 37 domestic dwellings with installed PV modules; the sizes of PV systems range between 1.5 kW peak and 3.29 kW peak. The battery is charged using surplus PV generation and is discharged during the evening and at night. The authors calculated that benefits amount to about £30/year (approximately €37.05/kWh) for the larger combination PV-BES. Such a low value is mainly due to battery inefficiency. Indeed, bill savings increase up to £110/year (about €135.85/year) when assuming lossless batteries and maintenance and installation costs are ignored.

Considering a theoretical maximum benefit of 8.6 p/kWh (about €10.62c/kWh) of otherwise exported electricity and a modest discount factor of 4% over 20 years, the target up-front capital cost for the battery system to break even is £707 (about €873.14). Since the cheapest lead-acid battery system has an equivalent up-front capital cost of £3296 (about €4070.56), there is no economic case and specific commercial opportunity, even for idealised lossless batteries with optimistic lifetimes. The financial losses approach £1000/year (about €1235) for a 570 Ah battery storage system integrated to a 3.29 kWp PV system when realistic efficiencies and lifetimes are accounted for.

Lastly, McKenna et al. state that there is no case of economic convenience in adopting lead batteries even for the case of Germany and the Australian states of Queensland, Victoria, and Western Australia. The reason is that the solar resource in these countries is not dissimilar to that of the UK and the import/export price ratio for Germany (2.14) and for the Australian states (3.13) are lower than those for the UK (3.69).

8. The Case of Australia

Energy storage is one of the most critical topics facing energy utilities; currently, Australia is one of the top five distributed energy storage markets in the world. In 2015, the Australian Energy Storage Council (ESC) focused on laying the foundations for a successful long-term energy storage industry. Both the industry and the government worked together to develop standards and guidelines in order to ensure safe and high-quality storage products. They have also invested a considerable amount of time in providing clear information to consumers. Over 1.5 million dwellings in Australia have rooftop PVs; for those living in New South Wales, the generation price is $0.60c/kWh whereas the export price is about $0.06c/kWh. The Australian ESC has estimated that two-thirds of daytime solar generation are delivered to the grid. These plants are the expected results to incentive policies; today, these plants have paid off well. However, the incentive policies in New South Wales, Victoria, and South Australia are coming to an end, and therefore, the only right choice for residential customers is to install a battery to store excess generation during daytime and make it available at night-time.

The study of integrated PV-BESs along with two proposals of new FIT schemes in Australia is presented in [43, 44]; Ratnam et al. and Weller et al. proposed an optimization-based approach to scheduling residential battery storage with solar PV. The aim is to maximize the electricity generation in order to achieve financial benefits for residential customers and simultaneously alleviate the utility burden associated with peak demand and reverse power flow.

The financial benefits derived from a number of possible FIT schemes proposed by the authors are explored in detail; moreover, concrete examples of two FIT schemes commonly used in Australia are also investigated. One of the FIT schemes proposed by Ratnam et al. consist of a generous constant FIT of $0.4/kWh; such a value is higher than peak time-of-use billing but lower than the FIT offered in 2010 by North South Wales which paid a generation price of $60c. In the authors' study, each residential customer has a Home Energy Management system similar to the low-cost smartbox presented in [45]. One day in advance, the home energy management system forecasts the residential load and PV generation; it receives the electricity prices for energy delivered to and from the grid and receives existing additional incentives and then runs the optimization-based algorithms to schedule the battery storage.

Ratnam et al. study 145 residential users with a PV system and batteries; they considered measured load and generation profiles over one year. These residential customers were randomly selected from customers located in the low voltage Australian distribution network, operated by the Ausgrid distributor; the network includes load centers in Sydney and regional New South Wales. The battery capacity is initially

fixed to 10 kWh; subsequently, it varies within the range 0 kWh and 30 kWh. Thanks to the quadratic programme-based minimization of the energy supplied by, or to, the grid proposed by the authors, the PV-BES combination allows for an overall average saving of between $350/yr and $100/yr per residential customer.

9. The Case of Greece

There are thousands of Greek islands, yet only two hundred of them are inhabited; most of them are off-grid areas, powered by diesel generators. Therefore, Greek islands represent a unique opportunity for the integration of renewable energy sources and battery storage. The size of these generators must meet the peak demand, but their operation rarely generates the peak power; often, their diurnal operation is highly variable and fluctuates in accordance with the variable demand. Moreover, diesel generators require fuel imports by ship; naval transport is costly and leads to security risks. Therefore, energy storage systems might eliminate or drastically reduce reliance on diesel supplies.

The case of many Greek islands is studied in [46]; Krajačić et al. propose feed-in tariff schemes for different energy storage systems such as pumped hydro, hybrid wind pumped hydro, hydrogen, and combined PV-BESs. As an example, the techno-economic analysis of a FIT policy to promote the adoption of PV-BESs is performed considering the case study of the island of Corvo. The island has about 400 inhabitants; the yearly electricity demand is approximately 1086 MWh with a load peak of 204 kW. Now, two diesel generators of 120 kW and two of 160 kW serve the island. In order to calculate the subsidy for remunerating the adoption of batteries along with PV modules, the authors estimate the fuel savings achieved during operation thanks to the adoption of batteries. For a battery capacity up to 40 kWh mounted with a 4 kW inverter, the proposed remuneration scheme is a fixed tariff of €53.8/kWh multiplied by battery capacity. Further remuneration schemes are also proposed for higher values of battery and inverter capacity, at different penetration levels.

10. Conclusion

The paper presented a review on the recent developments of photovoltaics integrated with battery storage system and related feed-in tariff policies in the regulatory context of Germany, Italy, Spain, the United Kingdom, Australia, and Greece. The attention was focused on those integrated photovoltaic battery systems subject to incentivisation policies. The paper showed that the self-consumption is the key factor for the actual incentivisation policies; moreover, the paper confirmed that the feed-in tariff scheme is still the most effective and widely considered scheme for promoting the integration of storage batteries to existing or new photovoltaic systems. The contributions mentioned in the paper agree with each other about the adoption of integrated photovoltaic battery systems; these contributions show a positive scenario and a clear economic advantage in adopting such systems if the storage technology is lithium-ion batteries.

Germany confirms being one of the countries that firmly believes and invests in renewables and storage; a new programme for promoting the adoption of batteries started in March 2016 with a budget of 30 million euros. In the paper, studies of PV-BESs on the German FIT policy in commercial and residential applications were reported. For commercial applications, battery storage increases self-consumption significantly, yet even unrealistic battery prices of less than €200/kWh cannot lead to an economic solution. For the residential applications, the use of lead batteries up to 5 kWh is convenient since about four or five years, even in the absence of subsidies and independently of an increase in the cost of electrical energy. Other studies conclude that the adoption of PV-BESs is a profitable operation for a battery system price below €350/kWh when lithium-ion batteries substitute for the lead ones.

Italy seems weighed down by the significant incentive policy of systems for the exploitation of renewable sources started in 2005 and lasted eight years. To date, Italy only supports the installation of batteries with a special term consisting in a tax reduction of the 50% of the batteries costs to be spread in 10 years. The paper reported a costs/benefits analysis of PV-BES considering real electrical energy consumption data for approximately 400 domestic clients spread over the Italian territory. A saving of approximately 150 euros for existing PV plants already incentivised by the feed-in scheme is achieved thanks to the adoption of batteries; this saving increases to approximately 170 euros in the case of a new nonincentivised PV plants. The paper also reported the economic viability of a novel feed-in tariff scheme that solely rewards self-consumption; such a scheme is applied to an existing Italian Public Administration and allows a 50% self-generation, an 80% self-consumption, and a 45% reduction of the electricity bill.

Spain is still experiencing the legacies of past incentive policies but, with great strength and motivation, this country has recently introduced a self-consumption support policy. Unfortunately, the simultaneous introduction of the so-called "solar tax" leads to the wide disapproval of customers and industry trade associations in the recent Spanish self-consumption support policies. Due to solar tax, the most of end users connected to the low-voltage distribution grid with an installed power no larger than 10 kW have to pay €0.049 for each self-consumed kWh.

The United Kingdom has an unpromising legislative landscape at the moment; no incentives to companies and no subsidies to households are offered to install energy storage. But an optimistic feeling characterizes the UK storage industry; such a feeling contrasts the conclusions of the recent research contributions which affirm that, to date, the PV-BES combination in the UK is not a convenient operation as there is no case of financial convenience in adopting cost-effective and ideal lead-acid batteries.

Australia is one of the top five distributed energy storage markets in the world; the major effort in 2015 of the Australian Energy Storage Council has been lay the foundations for a successful energy storage industry in the long term. In June 2016, the Australian Capital Territory board launched a proposal for the Next Generation Energy Storage Grants; the government has allocated $2 million in funding five companies to install solar storage homes and commercial buildings in Canberra. This paper reported the financial benefits derived from a number of possible FIT schemes proposed for this country; among them, the most attractive scheme is a generous constant FIT, higher than peak time-of-use billing but lower than the FIT offered in 2010. A case study of 145 residential customers with a PV-BES demonstrates that the PV-BES combination allows for an overall average saving of between $100/yr and $350/yr per residential customer.

Greece is currently experiencing a phase of difficulties and uncertainties, a national economy devoted to address crucial issues such as employment and welfare; as a result, implementing policies aimed at encouraging the adoption of PV-BES is not a priority to date. Despite this, Greece maintains a high interest in PV-BESs because Greek islands are thousands and diesel generators power most of them. Therefore, battery storage represents a unique opportunity for the effective integration of renewable energy sources in these off-grid areas. The techno-economic analysis of a FIT policy to promote the adoption of PV-BESs in the island of Corvo was presented in the paper.

Conflicts of Interest

The authors declare that they have no conflicts of interest.

References

[1] International Energy Agency (IEA), *World Energy Outlook 2015*, 2015.

[2] International Renewable Energy Agency (IRENA), *Smart Grids and Renewables. A Cost-Benefit Analysis Guide for Developing Countries*, 2015.

[3] A. Pyrgou, A. Kylili, and P. A. Fokaides, "The future of the feed-in tariff (FiT) scheme in Europe: the case of photovoltaics," *Energy Policy*, vol. 95, pp. 94–102, 2016.

[4] R. Darwall, *How Renewable Subsidies Destroyed the UK Electricity Market. Central Planning with Market Features*, 2015.

[5] H. P. Ikkurti and S. Saha, "A comprehensive techno-economic review of microinverters for building integrated photovoltaics," *Renewable and Sustainable Energy Reviews*, vol. 47, pp. 997–1006, 2015.

[6] S. F. Baborska-Narozny and F. J. Ziyad, "User learning and emerging practices in relation to innovative technologies: a case study of domestic photovoltaic systems in the UK," *Energy Research & Social Science*, vol. 13, pp. 24–37, 2016.

[7] T. Lang, E. Gloerfeld, and B. Girod, "Don't just follow the sun – a global assessment of economic performance for residential building photovoltaics," *Renewable and Sustainable Energy Reviews*, vol. 42, pp. 932–951, 2015.

[8] A. Colmenar-Santos, S. Campíñez-Romero, C. Pérez-Molina, and M. Castro-Gil, "Profitability analysis of grid-connected photovoltaic facilities for household electricity self-sufficiency," *Energy Policy*, vol. 51, pp. 749–764, 2012.

[9] P. Kästel and B. Gilroy-Scott, "Economics of pooling small local electricity prosumers—LCOE & self-consumption," *Renewable and Sustainable Energy Reviews*, vol. 51, pp. 718–729, 2015.

[10] S. Koohi-Kamali, N. A. Rahim, and H. Mokhlis, "Smart power management algorithm in microgrid consisting of photovoltaic, diesel, and battery storage plants considering variations in sunlight, temperature, and load," *Energy Conversion and Management*, vol. 84, pp. 562–582, 2014.

[11] L. Ramirez Camargo, R. Zink, W. Dorner, and G. Stoeglehner, "Spatio-temporal modeling of roof-top photovoltaic panels for improved technical potential assessment and electricity peak load offsetting at the municipal scale," *Computers, Environment and Urban Systems*, vol. 52, pp. 58–69, 2015.

[12] S. Shivashankar and M. Karimi, "Mitigating methods of power fluctuation of photovoltaic (PV) sources – a review," *Renewable and Sustainable Energy Reviews*, vol. 59, pp. 1170–1184, 2016.

[13] M. Bortolini, M. Gamberi, and A. Graziani, "Technical and economic design of photovoltaic and battery energy storage system," *Energy Conversion and Management*, vol. 86, pp. 81–92, 2014.

[14] F. Marra and G. Yang, "Chapter 10 - decentralized energy storage in residential feeders with photovoltaics," *Energy Storage for Smart Grids*, vol. 2015, pp. 277–294, 2015.

[15] D. Vanhoudt, D. Geysen, B. Claessens, F. Leemans, L. Jespers, and J. Van Bael, "An actively controlled residential heat pump: potential on peak shaving and maximization of self-consumption of renewable energy," *Renewable Energy*, vol. 63, pp. 531–543, 2014.

[16] R. Luthander, J. Widén, D. Nilsson, and J. Palm, "Photovoltaic self-consumption in buildings: a review," *Applied Energy*, vol. 142, pp. 80–94, 2015.

[17] J. Li and M. A. Danzer, "Optimal charge control strategies for stationary photovoltaic battery systems," *Journal of Power Sources*, vol. 258, pp. 365–373, 2014.

[18] M. Gitizadeh and H. Fakharzadegan, "Battery capacity determination with respect to optimized energy dispatch schedule in grid-connected photovoltaic systems," *Energy*, vol. 65, pp. 665–674, 2014.

[19] J. Moshövel, K. P. Kairies, D. Magnor et al., "Analysis of the maximal possible grid relief from PV-peak-power impacts by using storage systems for increased self-consumption," *Applied Energy*, vol. 137, pp. 567–575, 2015.

[20] A. Bayod-Rújula, M. E. Haro-Larrodé, and A. Martínez-Gracia, "Sizing criteria of hybrid photovoltaic–wind systems with battery storage and self-consumption considering interaction with the grid," *Solar Energy*, vol. 98, pp. 582–591, 2013.

[21] J. Hoppmann, J. Volland, T. S. Schmidt, and V. H. Hoffmann, "The economic viability of battery storage for residential solar photovoltaic systems – a review and a simulation model," *Renewable and Sustainable Energy Reviews*, vol. 39, pp. 1101–1118, 2014.

[22] E. Waffenschmidt, "Dimensioning of decentralized photovoltaic storages with limited feed-in power and their impact on the distribution grid," *Energy Procedia*, vol. 46, pp. 88–97, 2014.

[23] J. Widén, "Improved photovoltaic self-consumption with appliance scheduling in 200 single-family buildings," *Applied Energy*, vol. 126, pp. 199–212, 2014.

[24] O. Motlagh, P. Paevere, T. S. Hong, and G. Grozev, "Analysis of household electricity consumption behaviours: impact of domestic electricity generation," *Applied Mathematics and Computation*, vol. 270, pp. 165–178, 2015.

[25] D. Masa-Bote, M. Castillo-Cagigal, E. Matallanas et al., "Improving photovoltaics grid integration through short time forecasting and self-consumption," *Applied Energy*, vol. 125, pp. 103–113, 2014.

[26] G. Merei, J. Moshövel, D. Magnor, and D. U. Sauer, "Optimization of self-consumption and techno-economic analysis of PV-battery systems in commercial applications," *Applied Energy*, vol. 168, pp. 171–178, 2016.

[27] J. Munkhammar, P. Grahn, and J. Widén, "Quantifying self-consumption of on-site photovoltaic power generation in households with electric vehicle home charging," *Solar Energy*, vol. 97, pp. 208–216, 2013.

[28] J. P. Torreglosa, P. García-Triviño, L. M. Fernández-Ramirez, and F. Jurado, "Decentralized energy management strategy based on predictive controllers for a medium voltage direct current photovoltaic electric vehicle charging station," *Energy Conversion and Management*, vol. 108, pp. 1–13, 2016.

[29] A. R. Bhatti, Z. Salam, M. Junaidi, B. A. Aziz, K. P. Yee, and R. H. Ashique, "Electric vehicles charging using photovoltaic: status and technological review," *Renewable and Sustainable Energy Reviews*, vol. 54, pp. 34–47, 2016.

[30] T. Couture and Y. Gagnon, "An analysis of feed-in tariff remuneration models: implications for renewable energy investment," *Energy Policy*, vol. 38, no. 2, pp. 955–965, 2010.

[31] R. Sioshansi, "Retail electricity tariff and mechanism design to incentivize distributed renewable generation," *Energy Policy*, vol. 95, pp. 498–508, 2016.

[32] M. Braun, K. Büdenbender, and D. Magnor, "Photovoltaic self-consumption in Germany using lithium-ion storage to increase self-consumed photovoltaic energy," in *Proc. 24th European Photovoltaic Solar Energy Conference and Exhibition*, Hamburg, Germany, 2009.

[33] G. Mulder, D. Six, B. Claessens, T. Broes, N. Omar, and J. V. Mierlo, "The dimensioning of PV-battery systems depending on the incentive and selling price conditions," *Applied Energy*, vol. 111, pp. 1126–1135, 2013.

[34] J. Weniger, T. Tjaden, and V. Quaschning, "Sizing of residential PV battery systems," *Energy Procedia*, vol. 46, pp. 78–87, 2014.

[35] J. Bergner, J. Weniger, and T. Tjaden, "Economics of residential PV battery Systems in the Self-Consumption Age," in *Proc. 29th European Solar Energy Conference and Exhibition (EUPVSEC)*, pp. 3871–3877, Amsterdam, Netherlands, 2014.

[36] J. Linssen, P. Stenzel, and J. Fleer, "Techno-economic analysis of photovoltaic battery systems and the influence of different consumer load profiles," *Applied Energy*, vol. 185, pp. 2019–2025, 2017.

[37] Ricerca sistema Energetico (RSE), *I Sistemi Di Accumulo Nel Settore Elettrico*, 2015, (In Italian).

[38] A. Burgio, G. Belli, G. Brusco, D. Menniti, A. Pinnarelli, and N. Sorrentino, "A novel scheme for a feed-in tariff policy to favorite photovoltaic and batteries energy storage systems for grid-connected end-user," *International Review on Modelling and Simulations*, vol. 6, no. 4, pp. 1123–1132, 2013.

[39] A. Burgio, G. Brusco, D. Menniti, A. Pinnarelli, and N. Sorrentino, "The economic viability of a feed-in tariff scheme that solely rewards self-consumption to promote the use of integrated photovoltaic battery systems," *Applied Energy*, vol. 183, pp. 1075–1085, 2016.

[40] L. M. Ayompe and A. Duffy, "Feed-in tariff design for domestic scale grid-connected PV systems using high resolution

household electricity demand data," *Energy Policy*, vol. 61, pp. 619–627, 2013.

[41] C. Gallego-Castillo and M. Victoria, "Cost-free feed-in tariffs for renewable energy deployment in Spain," *Renewable Energy*, vol. 81, pp. 411–420, 2015.

[42] E. McKenna, M. McManus, S. Cooper, and M. Thomson, "Economic and environmental impact of lead-acid batteries in grid-connected domestic PV systems," *Applied Energy*, vol. 104, pp. 239–249, 2013.

[43] E. L. Ratnam, S. R. Weller, and C. M. Kellett, "Residential load and rooftop PV generation: an Australian distribution network dataset," *International Journal of Sustainable Energy*, vol. 6451, pp. 1–20, 2015.

[44] S. R. Weller, E. L. Ratnam, and C. M. Kellett, "An optimization-based approach to scheduling residential battery storage with solar PV: assessing customer benefit," *Renewable Energy*, vol. 75, pp. 123–134, 2015.

[45] D. Menniti, G. Barone, G. Brusco, A. Burgio, A. Pinnarelli, and N. Sorrentino, "A smartbox as a low-cost home automation solution for prosumers with a battery storage system in a demand response program," in *Proc. 16th International Conference on Environment and Electrical Engineering (EEEIC)*, Florence, Italy, 2016.

[46] G. Krajačić, N. Duić, A. Tsikalakis et al., "Feed-in tariffs for promotion of energy storage technologies," *Energy Policy*, vol. 39, no. 3, pp. 1410–1425, 2011.

18

One-Pot Solid-State Reaction Approach to Synthesize Ag-Cu₂O/GO Ternary Nanocomposites with Enhanced Visible-Light-Responsive Photocatalytic Activity

Longfeng Li,[1,2] **Jing Zhang,**[1] **Xianliang Fu,**[1] **Peipei Xiao,**[1] **Maolin Zhang,**[1,2] **and Mingzhu Liu**[1,2]

[1]*School of Chemistry and Materials Science, Huaibei Normal University, Huaibei 235000, China*
[2]*Information College, Huaibei Normal University, Huaibei 235000, China*

Correspondence should be addressed to Maolin Zhang; zhangml@chnu.edu.cn and Mingzhu Liu; once1970@sina.com

Academic Editor: Laécio S. Cavalcante

A facile ball milling-assisted solid-state reaction method was developed to synthesize Ag-Cu₂O/graphene oxide (GO) nanocomposites. In the resultant complex heterostructures, Ag nanocrystals were mainly deposited on the surface of Cu₂O, while Ag-Cu₂O composites were anchored onto GO sheets. The resultant Ag-Cu₂O/GO nanocomposites exhibited excellent photocatalytic activity with 90% of methyl orange (MO) dye degradation efficiency after 60 min of visible-light irradiation, which was much higher than that of either Cu₂O or Ag-Cu₂O. This study opens a new avenue to fabricate visible-light-responsive photocatalyst with high performance for environmental pollution purification.

1. Introduction

Developing visible-light-responsive photocatalysts for the extensive application in environmental pollution purification has attracted considerable concerns in recent years. Of the concerned photocatalysts, cuprous oxide (Cu₂O), a p-type semiconductor with a small direct band gap (1.9–2.2 eV), has been proved to be the most promising candidate for visible-light-driven photocatalytic decontamination [1–4]. Unfortunately, the rapid recombination of photogenerated electron-hole pairs greatly decreases its quantum efficiency and limits its promising applications in photocatalysis [5]. To prevent the recombination of photogenerated electron-hole pairs, several strategies such as doping with elements [6, 7] and coupling with other semiconductors [8, 9] have been developed to design and synthesize the desired Cu₂O-based composite materials, since it has been demonstrated that semiconductor-based composite/hybrid materials can display new synergistic properties arising from the interaction between the different components to improve the performances of the original semiconductors [10].

Among various strategies, constructing noble metal-semiconductor binary nanohybrids is an effective method for improving the photocatalytic activity. It is believed that the Schottky barrier created between metal and semiconductor can effectively prevent the recombination of photogenerated charge carriers during the photocatalytic process, resulting in an enhanced photocatalytic activity [11, 12]. To date, several binary metal/Cu₂O heterogeneous nanostructures such as Cu/Cu₂O [13], Au/Cu₂O [14, 15], and Ag/Cu₂O [16, 17] have been successfully prepared and have shown expected enhanced photocatalytic activities in degradation of organic dyes like MO [13, 16, 17], methylene blue (MB) [14], and pyronine B [15] compared to pure Cu₂O.

Although dual-ingredient metal/Cu₂O composites display an enhanced photocatalytic performance for photocatalytic degradation of organic pollutants compared to pure Cu₂O, multicomponent nanomaterials are expected to provide much higher photoactivity and to promote the development of novel nanoarchitectures with extraordinary properties [18]. Recent studies [19–21] have evidenced that the combined effect of graphene and plasmonic metals on the semiconductor photocatalyst can effectively improve the photocatalytic activity via suppressing photogenerated electron-hole recombination and/or increasing the photoabsorption ability. For instance, Ahmad et al. [19] have introduced graphene to Mn-doped ZnO forming Mn-doped ZnO/graphene ternary nanocomposite photocatalysts using solvothermal method for photocatalytic degradation of MB under visible-light irradiation. They find that the as-synthesized Mn-doped ZnO/graphene ternary nanocomposite shows an impressive photocatalytic enhancement over Mn-doped ZnO or pure ZnO samples, which can be attributed to enhanced visible-light absorption, efficient charge separation, and fast transfer processes. Jin et al. [20] have fabricated Cu-P25-graphene ternary composite through hydrothermal method for photocatalytic degradation of MB. Compared to the pure P25 and the P25-graphene binary composite, the as-fabricated Cu-P25-graphene ternary composite exhibits extended visible-light absorption, good charge separation capability, and high degradation rate of MB under visible-light irradiation. Hsieh et al. [21] have constructed ternary Pt-TiO_2/graphene hybrids with higher visible-light-driven photocatalytic activities in degrading AO7 than TiO_2/graphene. To sum up, it will be beneficial to use the synergetic effect among graphene, metal, and semiconductor photocatalysts for improving visible-light absorption and photocatalytic activity.

Considering the advantages of the aforementioned types of photocatalysts, fabricating Ag-Cu_2O/graphene ternary composite may reveal some desirable properties for photocatalytic applications. However, at present, only one research [5] focuses on this kind of nanocomposite. And much work should be done to prepare Ag-Cu_2O/graphene ternary nanocomposites instead of binary nanomaterials for improving photocatalytic performance. On the other hand, a variety of methods such as solvothermal [19] and hydrothermal [20, 21] has been established to prepare metal-semiconductor/graphene ternary composites. These methods generally involve a multiple-step synthesis procedure, which is complicated and not suitable for mass production. Thus, a simple, high-yielding, and environmentally friendly method is highly desirable for the bulk production of this kind of ternary composites with high photocatalytic performance.

Compared to other methods for the preparation of nanomaterials, the ball milling-assisted solid-state reaction method has a simpler process, higher yield, and lower environmental impact, which has been reported in our previous work [4, 22]. Up to now, this method has not been used for the synthesis of Ag-Cu_2O/GO ternary nanocomposites. In this work, we first successfully synthesize Ag-Cu_2O/GO ternary nanocomposites via a facile ball milling-assisted solid-state reaction route. The photocatalytic performances of the as-synthesized Ag-Cu_2O/GO with various Ag loading amounts are investigated. The results indicate that all the ternary nanocomposites display much higher photocatalytic activity for the degradation of MO under visible-light irradiation than pure Cu_2O or Ag-Cu_2O binary composite. Our work may provide an alternative to synthesize visible-light-responsive ternary nanocomposites with high photocatalytic performance.

2. Experimental

2.1. Materials and Apparatus. $Cu_2(OH)_2CO_3$, $AgNO_3$, and $H_2C_2O_4 \cdot 2H_2O$ used in the study were of analytical grade quality, and graphene oxide nanoplatelets (thickness: 0.55–1.2 nm) were purchased from Beijing DK nano technology Co. LTD and used as received. An X-ray powder diffractometer (XRD; Bruker D8 Advance, Germany) with Cu Kα radiation ($\lambda = 0.15418$ nm), an accelerating voltage of 40 kV, and an emission current of 40 mA was used to determine the crystal phase composition and the crystallite size of the synthesized samples. A field-emission scanning electron microscope (FE-SEM; Hitachi S-4800, Japan), a high-resolution transmission electron microscope (HRTEM; JEOL JEM-2100, Japan), and a transmission electron microscope (TEM; HT7700, Japan) with an energy-dispersive X-ray spectroscopy system (EDS; Oxford Instruments, UK) were employed to observe the microstructures, morphology, and elemental composition of the as-synthesized samples.

2.2. Synthesis Procedure. In a typical synthesis process, $Cu_2(OH)_2CO_3$, $AgNO_3$, $H_2C_2O_4 \cdot 2H_2O$, and GO were firstly mixed together in a certain molar ratio. Then, the resultant mixtures were allowed to have a ball milling reaction for 1 h using the planetary ball mill (QM-3SP04, China) with zirconium oxide tanks at a rotation speed of 480 rpm at room temperature. Finally, the as-obtained intermediate product was calcined at 350°C for 1 h under the protection of nitrogen to get the Ag-Cu_2O/GO ternary nanocomposite. Following the same procedure, various Ag-Cu_2O/GO nanocomposites were synthesized at an $AgNO_3$/$Cu_2(OH)_2CO_3$/$H_2C_2O_4 \cdot 2H_2O$ molar ratio of $m : 1 : (1 + m)$ and a $Cu_2(OH)_2CO_3$/GO mass ratio of 50 : 1 and were termed as mAg-Cu_2O/GO (m was in the range of 0.02~0.2). Also, 0.1Ag-Cu_2O was prepared by removing GO from the starting materials, and pure Cu_2O was obtained by removing $AgNO_3$ and GO.

2.3. Photocatalytic Activity Measurement. The photocatalytic degradation experiments were carried out in a photochemical reactor using a 300 W Xe lamp with a 420 nm UV cutoff filter, and the photocatalytic performances of the as-synthesized mAg-Cu_2O/GO nanocomposites with m values of 0.04, 0.06, 0.1, and 0.2 were evaluated by the photocatalytic degradation of the MO solution. The reaction temperature was kept at room temperature by cooling water to prevent any thermal catalytic effect. In a typical photocatalytic experiment, 0.25 g of the mAg-Cu_2O/GO powder was first added into 100 ml of the MO solution (20 mg l^{-1}). Then, prior to light irradiation, the resultant solution was magnetically stirred in the dark for approximately 30 min to ensure the adsorption-desorption equilibrium between the catalyst and the MO molecules. Next, the resultant solution was exposed to Xe lamp irradiation. Finally, a 5.0 ml reaction suspension including the photocatalyst and MO was withdrawn and centrifuged to separate the photocatalyst at 15 min intervals. The degradation rate of MO was evaluated by recording the intensity of absorption peak of MO (464 nm) relative to its initial intensity (c/c_0) using a spectrophotometer. The details were referred to our recent work [22]. Following the same

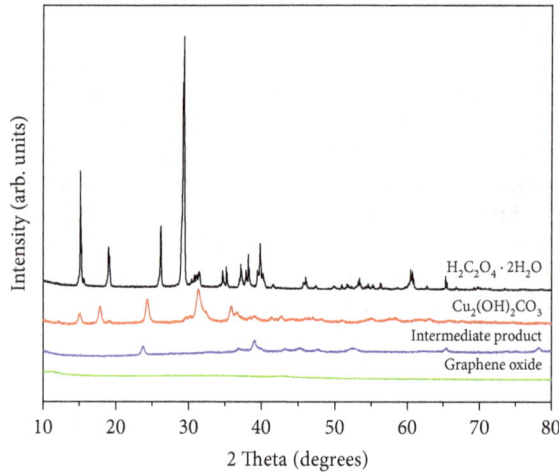

FIGURE 1: XRD patterns of the reactants and intermediate product.

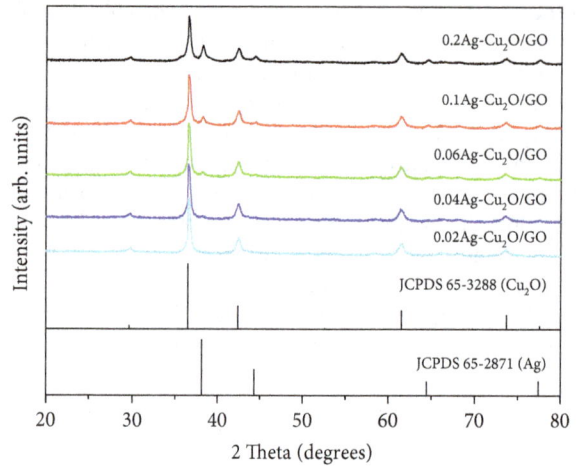

FIGURE 2: XRD patterns of mAg-Cu_2O/GO nanocomposites synthesized with different m values.

procedures, photocatalytic activities of pure Cu_2O, GO, and 0.1Ag-Cu_2O powders were also investigated for comparison with those of the mAg-Cu_2O/GO nanocomposites.

3. Results and Discussion

3.1. Principle of Synthesis Route. The synthesis of mAg-Cu_2O/GO nanocomposites is a two-stage process. At the first stage, the mixed reactants are ball milled for 1 h, which allows the solid-state reaction among $H_2C_2O_4 \cdot 2H_2O$, $Cu_2(OH)_2CO_3$, and $AgNO_3$, forming the intermediate product of $1/2$ $mAg_2C_2O_4 - Cu_2(OH)_{2n}(NO_3)_m(CO_3)_{1-n-m}(C_2O_4)_{1+1/2m}$ ($n = 0 \sim 1$) deposited on the surface of GO. At the second stage, the resultant intermediate product undergoes thermal decomposition reaction under the protection of nitrogen atmosphere at 350°C, which is an intramolecular redox reaction because of the reducibility of $C_2O_4^{2-}$ and the oxidizability of Cu^{2+}, to obtain Ag-Cu_2O/GO sample. The formation of Ag-Cu_2O/GO is represented by the following chemical equations:

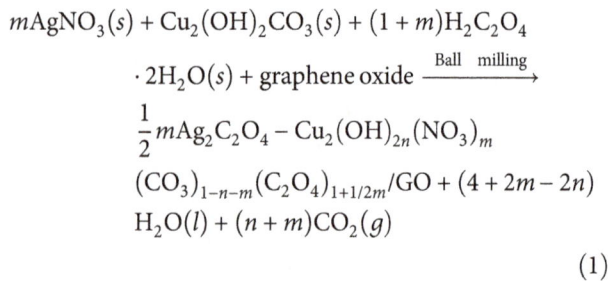

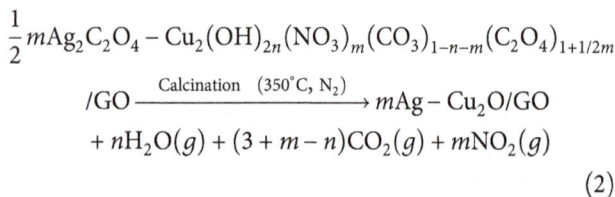

$$mAgNO_3(s) + Cu_2(OH)_2CO_3(s) + (1+m)H_2C_2O_4$$
$$\cdot 2H_2O(s) + \text{graphene oxide} \xrightarrow{\text{Ball milling}}$$
$$\frac{1}{2}mAg_2C_2O_4 - Cu_2(OH)_{2n}(NO_3)_m$$
$$(CO_3)_{1-n-m}(C_2O_4)_{1+1/2m}/GO + (4+2m-2n)$$
$$H_2O(l) + (n+m)CO_2(g)$$

$$(1)$$

$$\frac{1}{2}mAg_2C_2O_4 - Cu_2(OH)_{2n}(NO_3)_m(CO_3)_{1-n-m}(C_2O_4)_{1+1/2m}$$
$$/GO \xrightarrow{\text{Calcination } (350°C, N_2)} mAg - Cu_2O/GO$$
$$+ nH_2O(g) + (3+m-n)CO_2(g) + mNO_2(g)$$

$$(2)$$

The XRD patterns of the initial reactants and the intermediate product, which is obtained in the molar ratio of $0.1 : 1 : 1.1$ of $AgNO_3$, $Cu_2(OH)_2CO_3$, and $H_2C_2O_4 \cdot 2H_2O$

after 1 h of ball milling, are presented in Figure 1. It can be seen that the diffraction peaks of the initial reactants completely disappear in the intermediate product, and those of the new phases appear, indicating that the solid-state reaction of $H_2C_2O_4 \cdot 2H_2O$, $Cu_2(OH)_2CO_3$, and $AgNO_3$ is completed after 1 h of ball milling.

3.2. Phase Structures of mAg-Cu_2O/GO. Figure 2 shows the X-ray diffraction patterns of mAg-Cu_2O/GO nanocomposites synthesized with different molar ratios of $AgNO_3$ to $Cu_2(OH)_2CO_3$. It is found that the diffraction peaks of all the products match well with those of cubic Cu_2O (JCPDS 65-3288), and the characteristic diffraction peaks of Ag (JCPDS 65-2871) also appear in 0.06Ag-Cu_2O/GO, 0.1Ag-Cu_2O/GO, and 0.2Ag-Cu_2O/GO products, whose diffraction peak intensity gradually increases with an increase in the amount of $AgNO_3$. Besides, no signals of impurities are observed in all the products. Based on Scherrer's equation, the mean crystallite sizes of Cu_2O in mAg-Cu_2O/GO composites are calculated and are about 48, 44, 40, 36, and 31 nm corresponding to the m values of 0.02, 0.04, 0.06, 0.1, and 0.2, respectively, while those of Ag are about 5, 7, and 10 nm corresponding to the m values of 0.06, 0.1, and 0.2, respectively. The results indicate that, with the increase in the amount of $AgNO_3$, the crystallite sizes of Cu_2O in the composites gradually decrease, while those of Ag gradually increase.

3.3. Morphology and Microstructure. The morphology and microstructure of the 0.1Ag-Cu_2O/GO nanocomposite are examined by SEM and TEM as illustrated in Figure 3. It can be seen that spherical-like or nearly cubic-like Ag-Cu_2O particles with little aggregation are attached to the GO sheet, and most of the small Ag nanoparticles are directly deposited on the surface of Cu_2O particles with a size range of 45–70 nm (Figures 3(a) and 3(b)). The microstructure of 0.1Ag-Cu_2O/GO nanocomposite is further investigated by HRTEM. As shown in the HRTEM image (Figure 3(c)), Ag-Cu_2O interfaces are clearly observed and the observed interplanar spacings of 0.25 and 0.24 nm will correspond to

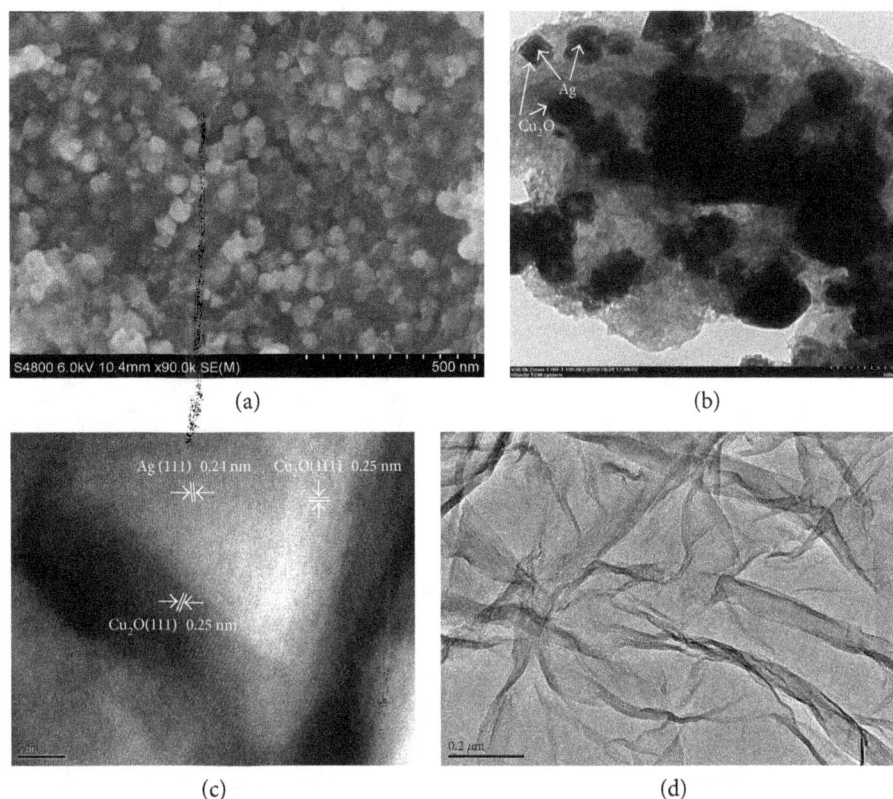

(a)

(b)

(c)

(d)

FIGURE 3: SEM (a), TEM (b), and HRTEM (c) images of 0.1Ag-Cu$_2$O/GO and TEM (d) image of GO.

the (111) lattice planes of Cu$_2$O and Ag, respectively, confirming that Ag nanocrystals are successfully anchored on the surfaces of Cu$_2$O.

To further observe the composition and the distribution of various elements, the EDS analyses of 0.1Ag-Cu$_2$O/GO are also performed. EDS analyses in the scanning transmission electron microscopy (STEM) mode have shown the presence of Ag, Cu, O, and C elements in the Ag-Cu$_2$O/GO composites, as shown in Figure 4. The EDS spectrum in Figure 4(a2) indicates that the marked area in Figure 4(a1) contains 57.7 wt% C, 11.9 wt% O, 27.9 wt% Cu, and 2.5 wt% Ag. That is, the atomic ratio of Cu to Ag is about 18.94 : 1, which is in agreement with the theoretical value of 20 : 1 in 0.1Ag-Cu$_2$O/GO. In contrast, another selected area for EDS analysis is located at the Ag deposition region on an individual Ag-Cu$_2$O nanoparticle (Figure 4(b1)) and the EDS spectrum indicates that the selected area contains 20 wt% C, 3.9 wt% O, 30.8 wt% Cu, and 45.3 wt% Ag (Figure 4(b2)). Through data calculation conversion, the atomic ratio of Cu to Ag is about 1.15 : 1 in the selected area in Figure 4(b1). It is seen that, compared with that in Figure 4(a1), the selected area in Figure 4(b1) is rich of Ag, which provides further evidence that the small Ag nanocrystalline is directly deposited on the surface of Cu$_2$O nanoparticles. In order to analyze the distribution of each element in the sample, EDS image scanning is performed on the framed area in Figure 4(c1) and the elemental maps of Cu, O, C, and Ag are shown in Figures 4(c2), 4(c3), 4(c4), and 4(c5), respectively. As can be seen, C element signals cover the entire selected area (Figure 4(c4)), suggesting that they are mainly from the GO.

The distribution of Cu, O, and Ag elements, in particular, that of Cu and Ag elements, basically corresponds to the STEM image of 0.1Ag-Cu$_2$O, suggesting that they are mainly from the 0.1Ag-Cu$_2$O. Therefore, combining the results of XRD, SEM, TEM, and EDS, it can be concluded that the ternary Ag-Cu$_2$O/GO nanocomposites have been successfully synthesized.

3.4. Photocatalytic Activities and Mechanism.

The photocatalytic activities of mAg-Cu$_2$O/GO with different Ag loading amounts (m = 0.04, 0.06, 0.1, and 0.2) are evaluated in terms of the degradation rate of MO in aqueous solution under visible-light irradiation, and experimental results are shown in Figure 5. For comparison, the test results of pure Cu$_2$O, GO, and 0.1Ag-Cu$_2$O are also displayed in Figure 5. Obviously, the photolysis of MO is negligible in the absence of the photocatalyst and pure GO exhibits 9.7% photocatalytic degradation efficiency for MO after 60 min of visible-light irradiation, suggesting its low photocatalytic activity. Besides, the photocatalytic degradation efficiency of pure Cu$_2$O is up to 64%, while 0.1Ag-Cu$_2$O is up to 78%. Acting as electron sinks, Ag nanoparticles deposited on Cu$_2$O can reduce the recombination of the photoinduced carriers, resulting in better photocatalytic activity of 0.1Ag-Cu$_2$O than pure Cu$_2$O. It is worth mentioning that all the Ag-Cu$_2$O/GO ternary nanocomposites exhibit excellent photocatalytic activities, which are much higher than those of either pure Cu$_2$O or Ag-Cu$_2$O binary composites. Moreover, the photocatalytic activities of the mAg-Cu$_2$O/GO increase slightly with increasing Ag loading amounts from 0.04

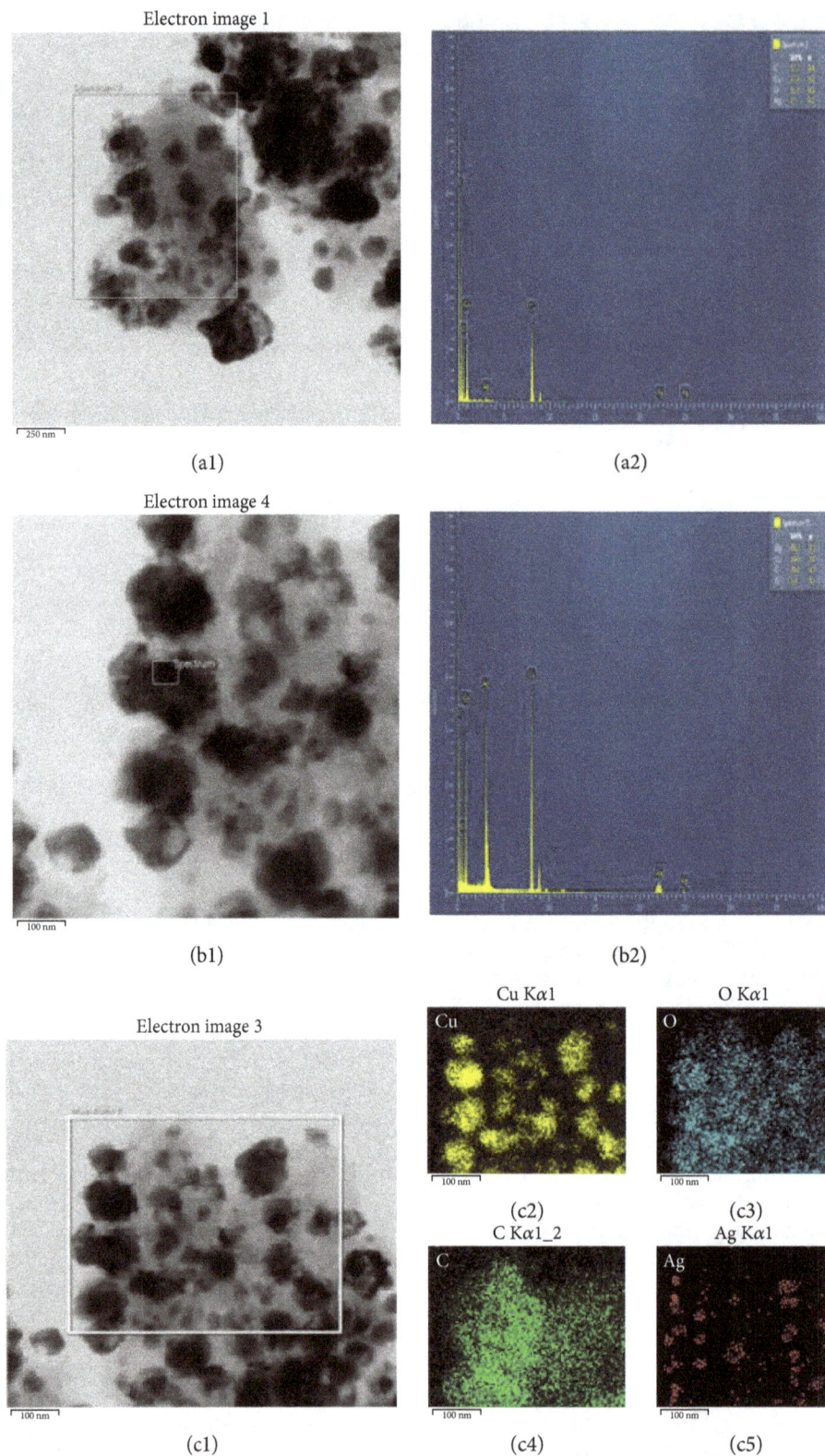

FIGURE 4: EDS analyses of 0.1Ag-Cu$_2$O/GO. STEM images (a1), (b1), and (c1), EDS spectra (a2) and (b2), and elemental maps (c2), (c3), (c4), and (c5).

to 0.1 of m values and the degradation percentage of MO follows the order of 0.1Ag-Cu$_2$O/GO (90%) > 0.06Ag-Cu$_2$O/GO (85%) > 0.04Ag-Cu$_2$O/GO (83%) after 60 min of visible-light irradiation. Interestingly, after the Ag loading amount reaches 0.2 of m value, the photocatalytic activity of 0.2Ag-Cu$_2$O/GO decreases slightly. Therefore,

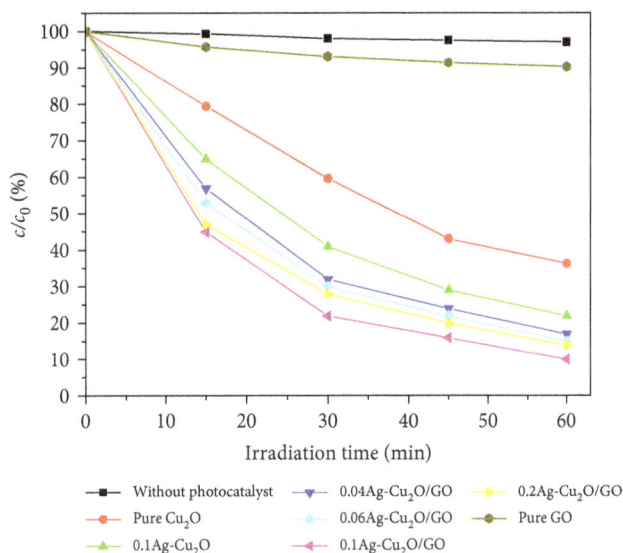

FIGURE 5: Photocatalytic degradation of MO under visible-light irradiation with different photocatalysts.

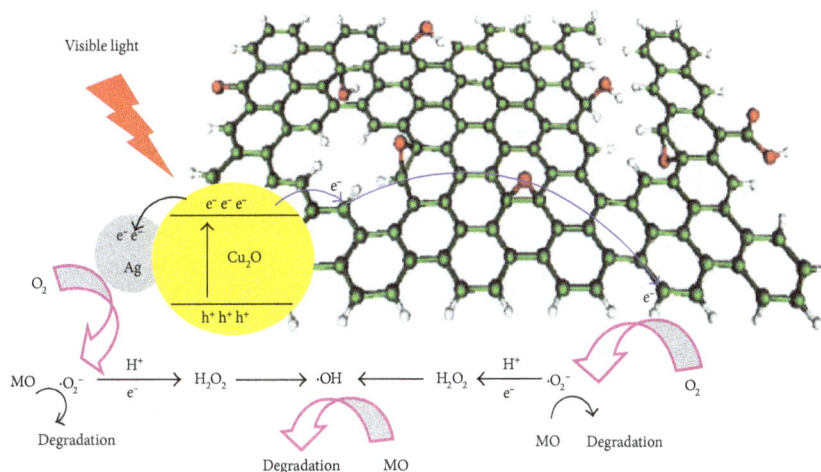

FIGURE 6: Schematic illustration of the charge transfer in Ag-Cu$_2$O/GO composite.

0.1 of m value is the optimal loading content of Ag and the higher content of Ag could be detrimental to the photocatalytic efficiency. It is due to its high content that Ag may serve as a new recombination center between photogenerated electrons and holes, thus reducing the photocatalytic activity of the photocatalyst [23]. In addition, the stability and reusability of 0.1Ag-Cu$_2$O/GO are evaluated over its multiple cycles for the photocatalytic degradation of MO. Stability tests are performed by repeating the reaction five times with the cycle length of 60 min using the recovered photocatalyst, and the degradation percentage of MO is about 89% at the end of the fifth cycle, which reveals that there is a negligible decrease in photocatalytic activity up to five cycles. So, 0.1Ag-Cu$_2$O/GO ternary nanocomposite prepared in the present work is a highly stable and reusable photocatalyst.

As expected, all the Ag-Cu$_2$O/GO nanocomposites show higher photocatalytic activities in the degradation of MO compared to the pure Cu$_2$O or the Ag-Cu$_2$O binary composite. It generally can be ascribed to a positive synergetic effect between various components. First of all, GO plays important roles in improving the visible-light-driven photocatalytic performance, which are described as follows: (1) the strong adsorption of GO to MO molecules resulting from the special π-conjugation and larger surface area of GO results in a high concentration of MO on the surface or in the vicinity of Ag-Cu$_2$O deposited on GO and thus increases the reaction rate. (2) GO with good electrical conductivity can act as an electron sink due to its higher work function than 4.42 eV of grapheme, and the existence of π-bands in GO facilitates charge transfer along the π-system [24], which accelerates electron transfer from Cu$_2$O to GO and further transfer to electronic receivers, leading to an improvement in the separation efficiency of the photoinduced carriers. Thereby, the higher photocatalytic performance of Ag-Cu$_2$O/GO ternary composites can be achieved. Apart from GO, Ag nanoparticles are also responsible for the higher photocatalytic performance of Ag-Cu$_2$O/GO ternary composites because they can also act as electron sinks to prevent the recombination of the photogenerated electrons and holes. And synergetic effects among different components for the superior photocatalytic activity of Ag-Cu$_2$O/GO ternary nanocomposites are illustrated in Figure 6. The valence band (VB) and the conduction band (CB) of Cu$_2$O with a band gap of about 1.94 eV in the composite are estimated to be about 0.5 eV and −1.44 eV versus normal hydrogen electrode (NHE) (pH = 0), respectively [25]. When Ag-Cu$_2$O/GO is under visible-light irradiation, the photoinduced electrons (e$^-$) will transfer from CB of Cu$_2$O to GO or Ag deposited on Cu$_2$O and then GO with superior conductivity will transfer these electrons along the π-system to the liquid/GO interface, while Ag nanoparticles with photoexcited plasmonic metallic nanostructure can extend the lifetime of electrons that arrived on its surface, which are crucial for the electron-dominated reduction reaction. At the liquid/GO interface or on the surface of Ag nanoparticles, the dissolved oxygen (O$_2$) in water acting as the electron scavenger can be reduced by trapping electrons to superoxide radical anions (O$_2^-$) and is further converted into hydrogen peroxide (H$_2$O$_2$) and hydroxyl radical (·OH). Then, hydroxyl radicals and superoxide radical anions acting as very powerful oxidants will oxidize MO molecules into small molecules such as CO$_2$ and H$_2$O. Therefore, combining photocatalytically active Cu$_2$O with GO acting as the electron sink and the adsorbent, as well as Ag nanoparticles acting as electron traps, can contribute to an effective photocatalytic degradation of MO under visible-light irradiation.

4. Conclusions

In this study, various Ag-Cu$_2$O/GO ternary nanocomposites as novel visible-light-responsive photocatalysts have been synthesized via a simple ball milling-assisted solid-state reaction route for the first time. Compared to the pure Cu$_2$O and the Ag-Cu$_2$O binary nanocomposite, the as-synthesized Ag-Cu$_2$O/GO ternary nanocomposites show superior visible-light-driven photocatalytic activity for MO degradation. The excellent photocatalytic activity of Ag-Cu$_2$O/GO can be attributed to the following facts: (1) the strong adsorption of GO to MO molecules results in increasing reaction rate of photocatalytic degradation; (2) good electrical conductivity and π-bands of GO are in favor of charge transfer, reducing the recombination of the photogenerated charge carriers; (3) Ag nanoparticles serve as electron sinks, accelerating the photoinduced charge carrier separation. The excellent performance of the Ag-Cu$_2$O/GO photocatalyst enables it to be used as a promising candidate in the field of environmental pollution purification. Also, this study provides a facile synthetic route for the synthesis of visible-light-responsive ternary nanocomposites with high photocatalytic performance.

Conflicts of Interest

The authors declare that they have no competing interest.

Acknowledgments

This work is financially supported by the Natural Science Foundation of Anhui Provincial Education Department (KJ2016A640, KJ2017A841, and KJ2013B249).

References

[1] X. Q. Liu, Z. Li, W. Zhao, C. X. Zhao, Y. Wang, and Z. Q. Lin, "A facile route to the synthesis of reduced graphene oxide-wrapped octahedral Cu$_2$O with enhanced photocatalytic and photovoltaic performance," *Journal of Materials Chemistry A*, vol. 3, no. 37, pp. 19148–19154, 2015.

[2] L. F. Yang, D. Q. Chu, L. M. Wang, H. L. Sun, and G. Ge, "Enhanced photocatalytic activity of porous cuprous oxide dodecahedron nanocrystals synthesized by solvothermal method," *Materials Letters*, vol. 159, pp. 172–176, 2015.

[3] X. H. Li, J. Q. Wang, Y. H. Zhang, and M. H. Cao, "Efficient visible-light photocatalytic performance of cuprous oxide porous nanosheet arrays," *Materials Research Bulletin*, vol. 70, no. 14, pp. 728–734, 2015.

[4] L. F. Li, W. X. Zhang, C. Feng, X. W. Luan, J. Jiang, and M. L. Zhang, "Preparation of nanocrystalline Cu$_2$O by a modified solid-state reaction method and its photocatalytic activity," *Materials Letters*, vol. 107, pp. 123–125, 2013.

[5] L. Xu, F. Y. Zhang, X. Y. Song, Z. L. Yin, and Y. X. Bu, "Construction of reduced graphene oxide-supported Ag-Cu$_2$O composites with hierarchical structures for enhanced photocatalytic activities and recyclability," *Journal of Materials Chemistry A*, vol. 3, no. 11, pp. 5923–5933, 2015.

[6] L. Xu, C. Srinivasakannan, J. H. Peng, M. Yan, D. Zhang, and L. B. Zhang, "Microfluidic reactor synthesis and photocatalytic behavior of Cu@Cu$_2$O nanocomposite," *Applied Surface Science*, vol. 331, pp. 449–454, 2015.

[7] L. Zhang, D. Jing, L. Guo, and X. Yao, "In situ photochemical synthesis of Zn-doped Cu$_2$O hollow microcubes for high efficient photocatalytic H$_2$ production," *ACS Sustainable Chemistry & Engineering*, vol. 2, no. 6, pp. 1446–1452, 2014.

[8] J. Zhang, W. X. Liu, X. W. Wang, X. Q. Wang, B. Hu, and H. Liu, "Enhanced decoloration activity by Cu$_2$O@TiO$_2$ nanobelts heterostructures via a strong adsorption-weak photodegradation process," *Applied Surface Science*, vol. 282, no. 2, pp. 84–91, 2013.

[9] J. Ma, K. Wang, L. Li, T. Zhang, Y. Kong, and S. Komarneni, "Visible light photocatalytic decolorization of Orange II on Cu$_2$O/ZnO nanocomposites," *Ceramics International*, vol. 41, no. 2, pp. 2050–2056, 2015.

[10] S. D. Sun, "Recent advances in hybrid Cu$_2$O-based heterogeneous nanostructures," *Nanoscale*, vol. 7, no. 25, pp. 10850–10882, 2015.

[11] P. Wang, B. B. Huang, X. Y. Qin et al., "Ag@AgCl: a highly efficient and stable photocatalyst active under visible light," *Angewandte Chemie, International Edition in English*, vol. 47, no. 41, pp. 7931–7933, 2008.

[12] F. P. Yan, Y. H. Wang, J. Y. Zhang, Z. Lin, J. S. Zheng, and F. Huang, "Schottky or ohmic metal–semiconductor contact: influence on photocatalytic efficiency of Ag/ZnO and Pt/ZnO model systems," *ChemSusChem*, vol. 7, no. 1, pp. 101–104, 2014.

[13] S. D. Sun, C. C. Kong, H. J. You, X. P. Song, B. J. Ding, and Z. M. Yang, "Facet-selective growth of Cu-Cu$_2$O heterogeneous architectures," *CrystEngComm*, vol. 14, no. 1, pp. 40–43, 2012.

One-Pot Solid-State Reaction Approach to Synthesize Ag-Cu2O/GO Ternary Nanocomposites with Enhanced...

189

[14] M. A. Mahmoud, W. Qian, and M. A. El-Sayed, "Following charge separation on the nanoscale in Cu_2O-Au nanoframe hollow nanoparticles," *Nano Letters*, vol. 11, no. 8, pp. 3285–3289, 2011.

[15] Z. H. Wang, S. P. Zhao, S. Y. Zhu, Y. L. Sun, and M. Fang, "Photocatalytic synthesis of M/Cu_2O (M = Ag, Au) heterogeneous nanocrystals and their photocatalytic properties," *CrystEngComm*, vol. 13, no. 7, pp. 2262–2267, 2011.

[16] X. L. Deng, C. G. Wang, E. Zhou et al., "One-step solvothermal method to prepare Ag/Cu_2O composite with enhanced photocatalytic properties," *Nanoscale Research Letters*, vol. 11, no. 1, p. 29, 2016.

[17] W. X. Zhang, X. N. Yang, Q. Zhu et al., "One-pot room temperature synthesis of Cu_2O/Ag composite nanospheres with enhanced visible-light-driven photocatalytic performance," *Industrial and Engineering Chemistry Research*, vol. 53, no. 42, pp. 16316–16323, 2014.

[18] N. Zhang, Y. H. Zhang, and Y. J. Xu, "Recent progress on graphene based photocatalysts: current status and future perspectives," *Nanoscale*, vol. 4, no. 19, pp. 5792–5813, 2012.

[19] M. Ahmad, E. Ahmed, W. Ahmed, A. Elhissi, Z. L. Hong, and N. R. Khalid, "Enhancing visible light responsive photocatalytic activity by decorating Mn-doped ZnO nanoparticles on graphene," *Ceramics International*, vol. 40, no. 7, pp. 10085–10097, 2014.

[20] Z. Jin, W. B. Duan, B. Liu, X. D. Chen, F. H. Yang, and J. P. Guo, "Fabrication of efficient visible light activated Cu–P25–graphene ternary composite for photocatalytic degradation of methyl blue," *Applied Surface Science*, vol. 356, pp. 707–718, 2015.

[21] S. H. Hsieh, W. J. Chen, and C. T. Wu, "Pt–TiO_2/graphene photocatalysts for degradation of AO7 dye under visible light," *Applied Surface Science*, vol. 340, pp. 9–17, 2015.

[22] M. M. Chen, Q. Y. Yang, L. F. Li, P. P. Xiao, M. Z. Liu, and M. L. Zhang, "Solid-state synthesis of $CuBi_2O_4$/MWCNT composites with enhanced photocatalytic activity under visible light irradiation," *Materials Letters*, vol. 171, pp. 255–258, 2016.

[23] W. W. Lu, S. Y. Gao, and J. J. Wang, "One-pot synthesis of Ag/ZnO self-assembled 3D hollow microspheres with enhanced photocatalytic performance," *Journal of Physical Chemistry C*, vol. 112, no. 43, pp. 16792–16800, 2008.

[24] C. Chen, W. M. Cai, M. C. Long et al., "Synthesis of visible-light responsive graphene oxide/TiO_2 composites with p/n heterojunction," *ACS Nano*, vol. 4, no. 11, pp. 6425–6432, 2010.

[25] X. Q. An, K. F. Li, and J. W. Tang, "Cu_2O/reduced graphene oxide composites for the photocatalytic conversion of CO_2," *ChemSusChem*, vol. 7, no. 4, pp. 1086–1093, 2014.

19

Feasibility Study of a Building-Integrated PV Manager to Power a Last-Mile Electric Vehicle Sharing System

Manuel Fuentes,[1] **Jesús Fraile-Ardanuy,**[2] **José L. Risco-Martín,**[3] **and José M. Moya**[4]

[1]*Renewable Energy Division, Energy Department, CIEMAT, Avda. Complutense 40, 28040 Madrid, Spain*
[2]*ETSI Telecomunicación, Universidad Politécnica de Madrid, Avda. Complutense 30, 28040 Madrid, Spain*
[3]*DACYA, Complutense University of Madrid, 28040 Madrid, Spain*
[4]*Integrated Systems Laboratory, Center for Computational Simulation, Universidad Politécnica de Madrid, Avda. Complutense 30, 28040 Madrid, Spain*

Correspondence should be addressed to Manuel Fuentes; manuel.fuentes@ciemat.es

Academic Editor: Leonardo Sandrolini

Transportation is one of the largest single sources of air pollution in urban areas. This paper analyzes a model of solar-powered vehicle sharing system using building-integrated photovoltaics (BIPV), resulting in a zero-emission and zero-energy mobility system for last-mile employee transportation. As a case study, an electric bicycle sharing system between a public transportation hub and a work center is modeled mathematically and optimized in order to minimize the number of pickup trips to satisfy the demand, while minimizing the total energy consumption of the system. The whole mobility system is fully powered with BIPV-generated energy. Results show a positive energy balance in e-bike batteries and pickup vehicle batteries in the worst day of the year regarding solar radiation. Even in this worst-case scenario, we achieve reuse rates of 3.8 people per bike, using actual data. The proposed system manages PV energy using only the batteries from the electric vehicles, without requiring supportive energy storage devices. Energy requirements and PV generation have been analyzed in detail to ensure the feasibility of this approach.

1. Introduction

Most countries around the world are trying to reduce their total fossil-fuel consumption with the main objective of reducing their greenhouse gas (GHG) emissions, which are mainly responsible for global warming, climate change, and deterioration of air quality in cities [1]. Solar energy is an abundant source of renewable/sustainable energy, which has an enormous potential in reducing the footprint of the greenhouse gases [2]. The integration of photovoltaic systems into buildings (BIPV), as it is shown in Figure 1, is proving to be an increasingly endorsed solution, and this is the market segment with the greatest growth potential in the photovoltaic industry [3], replacing a significant amount of the electricity that would otherwise be generated by burning fossil fuels [4]; BIPV will gain increasing attention of the building energy efficiency market in the twenty-first century [5].

Nowadays, 50% of people are living in cities (reaching 69% in the European Union) and it is estimated that over 60% of the total world's population will live in urban areas by 2030 [6], increasing the pressure in the urban environments. Bicycles can be used in urban areas to reduce air pollution, traffic congestion, noise emission, and energy consumption, allowing a personal healthier lifestyle [7]. For this reason, governments are promoting different initiatives to use bicycles as an alternative to private motor vehicles [8]. Different types of public bike sharing programs have emerged in many cities worldwide in recent years [9, 10]. Bikes are distributed in different bike stations across the city, and users can pick up a bicycle from any docking station, returning it to any other one located within the network. These systems must be managed and maintained by operators, who are responsible for monitoring the stations and moving bicycles from the most loaded stations to the emptiest ones. Usually, these operators use a diesel-powered vehicle to deliver bikes between different bike stations as it is shown in Figure 2. This management reduces problems that arise (a) when a particular user does not find a free bike when he/she wants to

FIGURE 1: Southeast view of building 42 before and after rehabilitation of the shell. (BIPV) © CIEMAT.

FIGURE 2: e-bike sharing system operators in Madrid (Spain).

use it in an empty station or (b) when he/she is not able to return the bike in a full station.

Bike sharing can be used in point-to-point trips, or it can be used combined with other transportation modes in cities, increasing the flexibility of public transport infrastructures. For example, one of the weakest points in a public transport system is the access (i.e., to reach public transport stations) and egress trips (i.e., from public transport stations to the final destination). Bikes can be used to cover these specific trips, reducing the door-to-door total travel time, making the combination of bicycle-public transport more competitive compared to private motor cars [11]. Cycling has also some disadvantages, like close passing traffic, exposure to weather conditions (heat, cold, rain, wind, etc.), difficulty to carry loads, and physical effort in hilly cities. Electric bicycles (e-bikes) are electric-powered-assisted bicycles that can

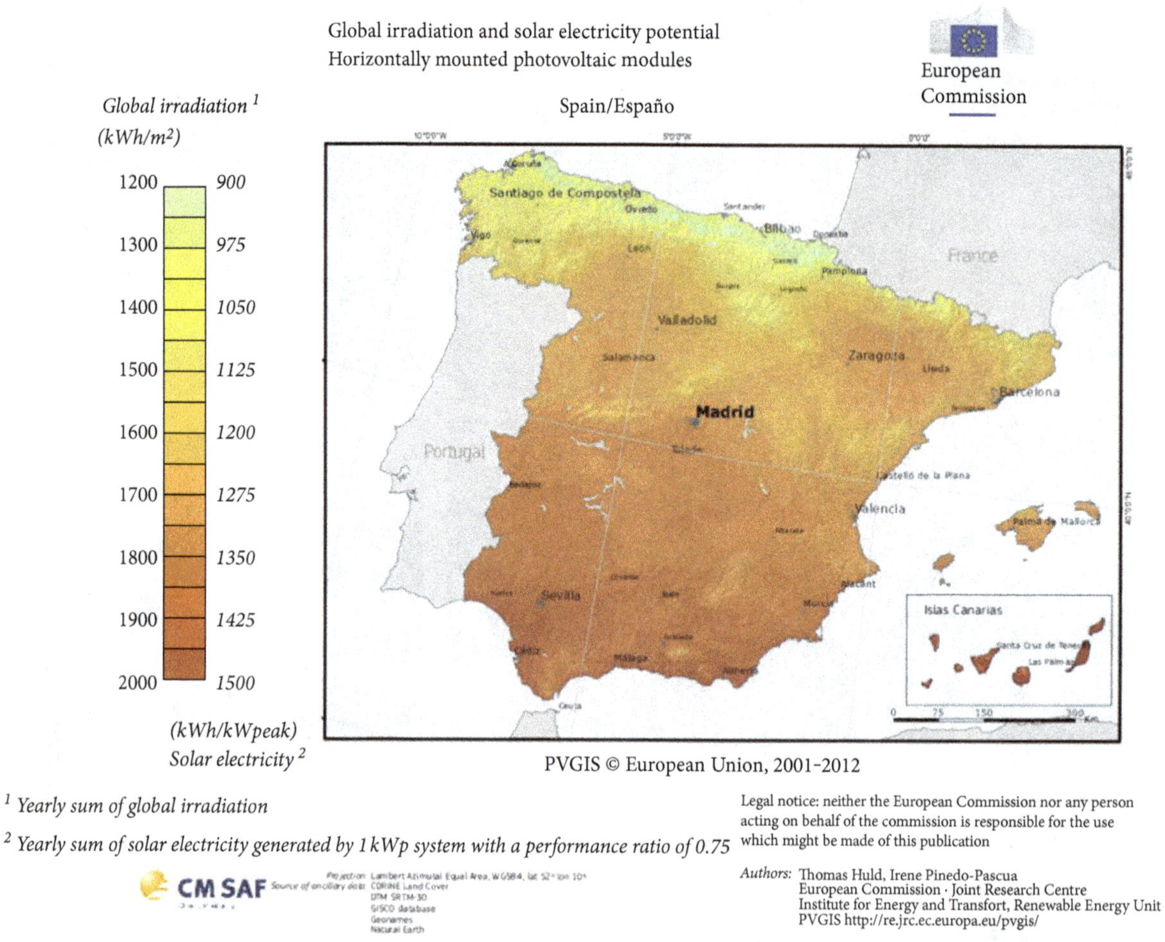

Global irradiation and solar electricity potential
Horizontally mounted photovoltaic modules

European
Commission

Global irradiation [1]   Spain/Españo
(kWh/m²)

PVGIS © European Union, 2001-2012

[1] *Yearly sum of global irradiation*

[2] *Yearly sum of solar electricity generated by 1 kWp system with a performance ratio of 0.75*

Legal notice: neither the European Commission nor any person acting on behalf of the commission is responsible for the use which might be made of this publication

Authors: Thomas Huld, Irene Pinedo-Pascua
European Commission · Joint Research Centre
Institute for Energy and Transfort, Renewable Energy Unit
PVGIS http://re.jrc.ec.europa.eu/pvgis/

FIGURE 3: Global irradiation and solar electricity potential on horizontally mounted PV modules in Spain [16].

eliminate some of these problems, keeping all the benefits of traditional cycling. e-bike users can go further and faster with less fatigue than users on conventional ones, increasing the daily covered range [7]. This type of bicycle can be pedalled when the battery runs out.

Previous works demonstrated the feasibility and the economic relevance of introducing electric vehicles in last-mile urban logistics operations [12]. In this paper, we propose an e-bike sharing mobility system fully powered by BIPV-generated energy as a last-mile transportation solution, for people to go from public transportation hubs to their final destinations. This approach has been validated with real data obtained from a pilot implementation in a large work center in Spain.

Some e-bike sharing programs have emerged in different urban environments: that is, university campuses, such as the University of Tennessee (UTK), Knoxville Campus in the US [13] and in different European cities. In these e-bike sharing systems, batteries can be directly charged when e-bikes are parked or they can be swapped for fully charged ones in the docking stations. In both cases, the charging process extracts electric energy from the grid. The gap between PV power output and vehicle charging demand is highly variable [14]. In this paper, a system has been designed to adapt the daily BIPV generation curve of a work center to

charge the batteries of the electric vehicles during working hours, which are also the period of maximum PV generation, to power the mobility system.

The remainder of this paper is organized as follows: Section 2 describes the case study, its characteristics, constraints, and models, including the estimation of PV generation and the energy demand of the mobility system; Section 3 describes and discusses the results of the energy-optimized system for this case study, resulting in a positive energy balance even in the worst day of the year; and finally, Section 4 summarizes the main conclusions of this work.

2. Case Study: Last-Mile e-Bike Sharing System for CIEMAT

As a case study, we analyze the feasibility of an electric bicycle sharing system between a public transportation hub and a research center in Madrid, Spain. This section provides a detailed analysis of this case, its constraints, and the estimation models for PV generation and total energy consumption of the mobility system.

2.1. CIEMAT Location. The Center for Advanced Research in Energy, Environment and Technology (Centro de Investigaciones Energéticas, Medioambientales y Tecnológicas—

CIEMAT) is a public research institution located at the very end of the Moncloa Campus in Madrid (Spain) and occupying 71,000 m². This integrated campus is shared with othereducational and research centers like Complutense University of Madrid, Technical University of Madrid, and some other partner institutions [15]. In Figure 3, spatial distribution of Spain cities is depicted against a backdrop of irradiation and PV theoretical energy.

The Moncloa Campus is situated in the west side of Madrid, covering an area of 2 square kilometers, and it is linked to the rest of the city public transport network by a single underground station called *Ciudad Universitaria* (line 6) [15], shown in Figure 4. The distance between CIEMAT offices and the nearest underground station is 1.1 km in straight and slightly sloping line.

2.2. Expected Solar Energy Generation.

To estimate PV generation, we use the online calculator from Photovoltaic Geographical Information System (PVGIS) [16].

CIEMAT research centers are distributed in more than 70 different buildings in Moncloa Campus. Among them, CIEMAT building 42, which is the headquarters of the Renewable Energies Division, has recently installed a peak power of 27.2 kWp of photovoltaic (PV) cells in the BIPV facades, occupying a total surface area of 176 m² [3]. These PV modules are integrated in the upper areas of the east, south, and west facades with 90° angle of inclination. There are 42 crystalline silicon modules installed in the east facade (13.7 kWp), 28 modules installed in the south facade (8.5 kWp), and 16 modules installed in the west facade (4.9 kWp). This west PV subsystem is not considered in this work. The schematic of the BIPV modules installed in the south and east facades is shown in Figure 5.

In order to estimate the solar electricity production of this PV, in a similar way to [17], we need to first answer the following questions: What is the relation of irradiation values to the power load on a daily basis, and how should BIPV be integrated in the e-bike sharing power system?

Table 1 presents the average daily electricity production (Ed) and the average monthly electricity production (Em), also in kWh, for the east and south facades. It is observed that the estimated annual energy generated by this PV system is 17.88 MWh/year, assuming an estimated loss due to temperature and low irradiance of 10.5%, using local ambient temperature, an estimated loss due to angular reflectance effects of 5.3%, and other losses (cables, inverters, etc.) of 14%.

The worst months for PV generation, due to the low daily irradiance, are December and January [16] shown in Figure 6. The estimated solar energy generated by the PV panels located at CIEMAT building 42 during a single day in December is presented in Figure 7.

2.3. Modeling and Predicting the Energy Consumption of the System

2.3.1. Users' Daily Working Schedule.

Most of the 1200 members of the CIEMAT staff at Moncloa Campus have a 40-hour working week, between 8:00 and 17:00, Monday to Friday, but they usually have a flexible work schedule. From

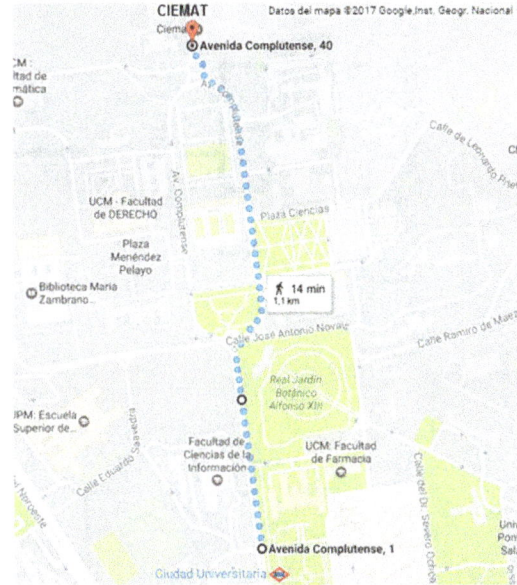

FIGURE 4: Moncloa Campus, Ciudad Universitaria underground station, and CIEMAT location.

different surveys [7], it has been pointed out that the rush hours are between from 7:00 to 10:00 in the morning and from 16:00 to 19:00 in the afternoon, and around 8% of the staff (95 people) would be interested in participating in the proposed eco-mobility program.

Analyzing the underground timetable for line 6, the interval between two consecutive trains from 7:00 to 10:00 is 4 minutes. Taking into account the arrival time information extracted from those surveys, it is assumed that CIEMAT employees will arrive according to the probability density distribution shown in Figure 8, which resembles a Poisson distribution and matches the real observed mobility behavior. Most people arrive to work early (from 7:08 to 8:20), and then there is a long tail showing employees who arrive to work later. A similar probability density function is used for the afternoon return trips.

2.3.2. Evaluating the Optimal Number of e-Bikes and Electric Vehicles to Fulfill the Mobility Constraints.

As aforementioned, the distance between the underground station and CIEMAT headquarters is 1.1 km. e-bikes are initially locked in the bike station located outside the underground station. During the morning trips, subscribers will release the e-bikes from this dock station and will return it over an empty dock station located at the CIEMAT.

Assuming an average speed of 15–20 km/hour per e-bike [18], the average time to unlock the e-bike, ride, travel, and finally lock it back at the CIEMAT dock station is around 6 minutes.

As soon as the underground bike station empties and the CIEMAT bike station fills up, an electric pickup truck will tow back these e-bikes to the initial station. The estimated time for this shuttle and swapping operation is around 3.75 minutes (see Figure 9). It is assumed 4 minutes for the whole operation of this pickup electric vehicle (EV),

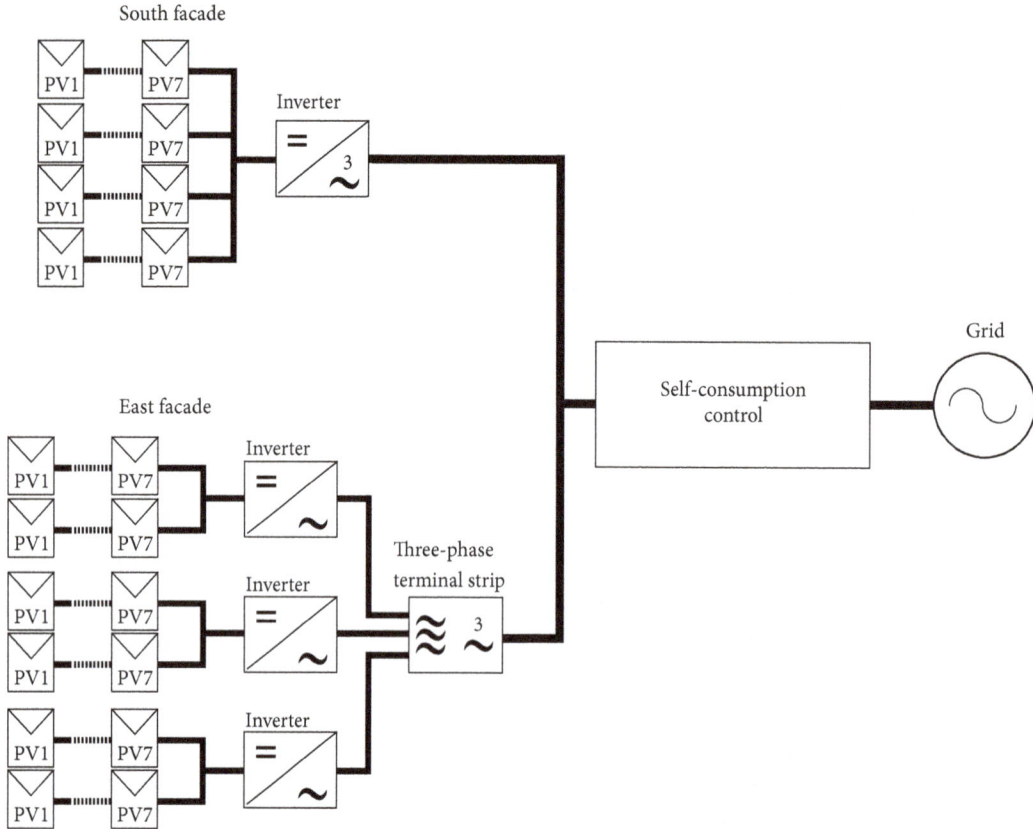

FIGURE 5: Schematic diagram of the BIPV modules installed in the south and the east facades.

TABLE 1: Average daily and monthly expected electricity production (kWh) from the PV modules installed on the two considered facades, obtained from the PVGIS software [16].

Month	East facade		South facade	
	Ed	Em	Ed	Em
Jan	24.10	748	13.80	429
Feb	28.00	785	20.00	560
Mar	27.90	864	28.30	878
Apr	21.20	635	31.00	929
May	16.40	507	34.70	1080
Jun	14.00	419	37.70	1130
Jul	15.00	466	38.40	1190
Aug	20.60	639	35.00	1090
Sep	26.50	795	29.10	872
Oct	28.00	867	21.50	667
Nov	24.90	748	14.80	445
Dec	24.10	749	12.60	392
Yearly average	22.50	685	26.50	805
Total for year		8220		9660
Total PV installation for a year				
		17880		

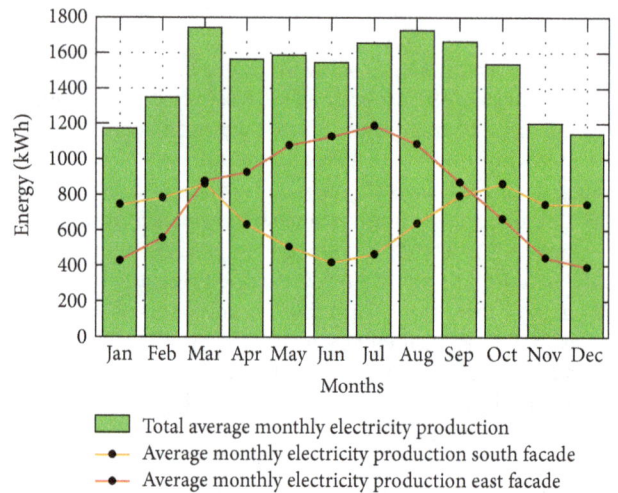

FIGURE 6: Average monthly expected electricity production.

these two locations. This is the upper-bound limiting factor of the whole e-bike sharing system.

2.3.3. Estimation of e-Bike Sharing System Energy Demand.

(1) Estimation of e-Bike Energy Consumption. The electric motor used is a 250 W brushless synchronous motor embedded into the rear wheel hub. The main electric specifications of the e-bike are presented in Table 2.

including loading and unloading e-bikes, which is based on actual measurements. In a single hour, only $60/4 = 15$ trips can be performed by this EV towing the e-bikes between

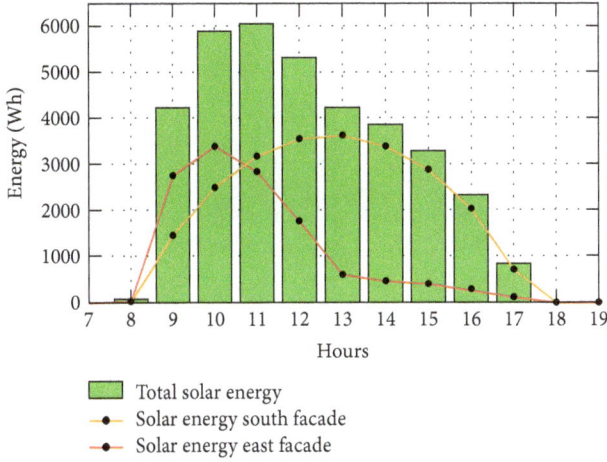

FIGURE 7: Estimated solar energy generated at December day.

To estimate the e-bike average consumption, two different tests were performed. Firstly, different users rode an e-bike during 3 months around the CIEMAT headquarters, measuring travelled distance, average vehicle speed, travel time, and energy consumed. From this data, energy consumption per kilometer was evaluated. Table 3 summarizes this information.

A second test was performed over the real route between the underground station and CIEMAT headquarters. Two different users, with very different weight (122 kg versus 75 kg), travelled across this route at the same time, riding two e-bikes. Table 4 shows the energy consumption in this situation. The energy consumption per kilometer and energy consumption per kilometer and kilogram were derived from this data. Taking into account these results, a conservative 7.7 Wh/km was used in this work.

(2) Estimation of the Towing Electric Vehicle. The EV considered in this analysis is the pickup vehicle Nissan e-NV200 with the same electric propulsion system with Nissan Leaf. This car is 100% EV powered by AC synchronous motor 80 kW-280 Nm, with a 24 kWh laminated lithium-ion battery. Its range is 170 km when tested according to the New European Driving Cycle (NEDC), which corresponds to 0.165 kWh/km [19]. According to different real field tests [20], the average real consumption for this vehicle in this study has been selected to 0.18 kWh/km.

2.4. Determining the Mobility Requirements. The objective of the designed mobility system is to satisfy the demand of CIEMAT employees between the underground station and work site in the morning and afternoon, minimizing the number of trips performed by the pickup EV trailer, defined by (1).The optimization problem is defined as a mixed integer linear programming problem given by (1)–(9). Table 5 presents the optimization model parameters with their associated value ranges. Table 6 shows integer variables used in the optimization model with their associated limits. In order to clarify the notation, underground station will be labeled as A, whereas the CIEMAT station will be labeled as B.

$$\min \sum_t \left(dk_{AB}(t) + dk_{BA}(t) \right), \tag{1}$$

subject to

$$
\begin{aligned}
nb_A(t) &= NB_{A0}, \ \text{if } t \le 0 \\
nb_A(t) &= 0, \ \text{if } t > 0, \ t \in BCTI \\
nb_A(t) &= nb_A(t-1) - D_{AB}(t) + D_{BA}\left(t - \Delta T_{BA}{}^{AEB}\right) \\
&\quad - Cb_{AB}(t) + Cb_{BA}(t - \Delta T_{BA}), \\
&\qquad \text{if } t > 0, \ t \notin BCTI,
\end{aligned} \tag{2}
$$

$$
\begin{aligned}
nb_B(t) &= NB_{B0}, \ \text{if } t \le 0 \\
nb_B(t) &= NB_{TOT}, \ \text{if } t > 0, \ t \in BCTI \\
nb_B(t) &= nb_B(t-1) - D_{BA}(t) + D_{AB}\left(t - \Delta T_{AB}{}^{AEB}\right) \\
&\quad - Cb_{BA}(t) + Cb_{AB}\left(t - \Delta T_{AB}{}^{K}\right)), \\
&\qquad \text{if } t > 0, \ t \notin BCTI,
\end{aligned} \tag{3}
$$

$$
\begin{aligned}
nk_A(t) &= NK_{A0}, \ \text{if } t \le 0 \\
nk_A(t) &= 0, \ \text{if } t > 0, \ t \in BCTI \\
nk_A(t) &= nk_A(t-1) - dk_{AB}(t) + dk_{BA}\left(t - \Delta T_{BA}{}^{K}\right), \\
&\qquad \text{if } t > 0, \ t \notin BCTI,
\end{aligned} \tag{4}
$$

$$
\begin{aligned}
nk_B(t) &= NK_{B0}, \ \text{if } t \le 0 \\
nk_B(t) &= NK_{TOT}, \ \text{if } t > 0, \ t \in BCTI, \\
nk_B(t) &= nk_B(t-1) - dk_{BA}(t) + dk_{AB}\left(t - \Delta T_{AB}{}^{K}\right), \\
&\qquad \text{if } t > 0, \ t \notin BCTI,
\end{aligned} \tag{5}
$$

$$
\begin{aligned}
Cb_{AB}(t) &= 0, \ \text{if } t \le 0 \text{ or } t \in BCTI \\
Cb_{AB}(t) &\le A_{EB} CT \cdot dk_{AB}(t), \ \text{if } t > 0, \ t \notin BCTI,
\end{aligned} \tag{6}
$$

$$
\begin{aligned}
Cb_{BA}(t) &= 0, \ \text{if } t \le 0 \text{ or } t \in BCTI \\
Cb_{BA}(t) &\le A_{EB} CT \cdot dk_{BA}(t), \ \text{if } t > 0, \ t \notin BCTI,
\end{aligned} \tag{7}
$$

$$
\begin{aligned}
db_{AB}(t) &= 0, \ \text{if } t \le 0 \text{ or } t \in BCTI \\
db_{AB}(t) &= db_{AB}(t-1) + D_{AB}(t-1) \\
&\quad - D_{AB}\left(t - \Delta T_{AB}{}^{AEB}\right), \\
&\qquad \text{if } t > 0, \ t \notin BCTI,
\end{aligned} \tag{8}
$$

$$
\begin{aligned}
db_{BA}(t) &= 0, \ \text{if } t \le 0 \text{ or } t \in BCTI \\
db_{BA}(t) &= db_{BA}(t-1) + D_{BA}(t-1) \\
&\quad - D_{BA}\left(t - \Delta T_{BA}{}^{AEB}\right), \\
&\qquad \text{if } t > 0, \ t \notin BCTI.
\end{aligned} \tag{9}
$$

Equation (1) defines the objective function to be minimized by the optimization algorithm. This equation evaluates

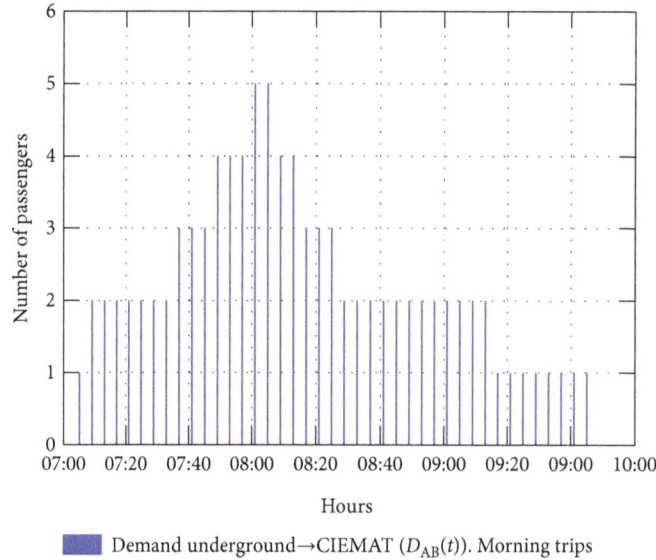

FIGURE 8: Temporal distribution of CIEMAT staff members' arrivals per time slot during the morning period.

the number of trips performed by the pickup EV between both nodes (from A to B) along the day at each time slot (given by $dk_{AB}(t) + dk_{BA}(t)$). Remember that these trips are done to maintain the balance of the e-bike fleet parked at each docking station. Energy required for e-bikes depends only on the demand, and there is no way to reduce it. The remaining energy requirements come from the pickup trips required to satisfy the demand with a certain number of e-bikes, which can be minimized.

Equations (2) and (3) define the general balance of the number of e-bikes in each docking station at each time slot. During the charging time interval (10:00 to 16:00), the e-bikes are charging in B. In normal operation, in (2), the number of e-bikes at A at time slot t, denoted by $nb_A(t)$, is equal to the number of e-bikes parked at this station in the previous time slot, $nb_A(t-1)$, minus the number of e-bikes used by the demand to travel from the A to B, denoted by $D_{AB}(t)$, adding the number of e-bikes which departed from B previously, at $t = t - \Delta T_{BA}{}^{AEB}$, denoted by $D_{BA}(t - \Delta T_{BA}{}^{AEB})$, minus the number of e-bikes transported by the pickup EV from A to B, denoted by $Cb_{AB}(t)$, and adding the number of e-bikes transported by the pickup EV from the B to the A, which departed at $t = t - \Delta T_{BA}{}^{K}$.

During the morning, the employees' demand to travel, using their e-bikes from CIEMAT to the underground station, $D_{BA}(t)$, will be zero, and also the number of e-bikes transported by the pickup trailer from the underground dock station to CIEMAT one, $Cb_{AB}(t)$, will be zero. During the afternoon trips, this behavior is reversed; therefore, $D_{AB}(t)$ and $Cb_{BA}(t)$ will be zero. The same procedure has been applied in (3).

Equations (4) and (5) model the behavior of the pickup EV in each docking station at each time slot.

Restrictions (6) and (7) limit the maximum number of e-bikes that can be transported in each pickup-EV trip from A to B and reversed. Equations (8) and (9) present the continuity equations for each e-bike movement.

IBM ILOG CPLEX Optimization Studio v. 12.5.1.0 was used for solving the defined optimization problem, and MATLAB® was later used for analyzing and plotting the results.

3. Results and Discussion

3.1. Mobility. It is assumed that a single pickup EV was used in this system. The optimal number of e-bikes in this fleet is then evaluated running the optimization model and checking its convergence. If a feasible solution is not obtained, the number of e-bikes is increased in one until convergence is reached.

With this procedure, the minimum number of e-bikes in the fleet that can fulfill the mobility requirements was fixed to 25, as it is presented in Table 5.

Once the minimum number of pickup EV and e-bikes were set in the sharing system, the optimization algorithm was run to determine the minimum number of trips from pickup EV required to balance the e-bike fleet. This value was 28 during each demand period (28 in the morning trips and 28 more in the afternoon).

The optimization model also determines the optimal moment to transport the e-bikes from CIEMAT dock station to the underground station, avoiding e-bike scarcity in this last dock station, and the exact number of bikes moved in each trips (it is an integer variable between 1 and 5). It is important to notice that at the end of the morning, all e-bikes and the pickup EV are located at the CIEMAT in order to be recharged before the afternoon trips.

3.2. Energy Demanded by the e-Bike Sharing System. In a similar way, controlling charging and discharging of lead-acid batteries is critical to extend the lifetime of microgrid systems [21]; our work has taken into account the optimum management of the state of charge (SOC) for lithium-polymer

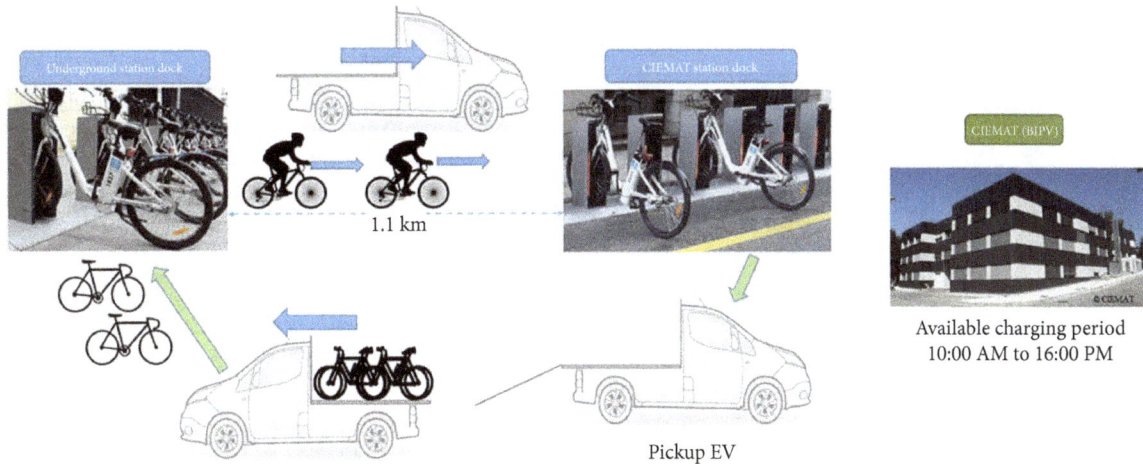

FIGURE 9: e-bike sharing system mobility description.

TABLE 2: e-bike specifications.

Battery tech.	Nominal voltage	Nominal capacity	Weight	Motor	Nominal power	Bike weight
Li-ion polymer	36 V	360 Wh	4.6 kg	Brushless	250 W	20.4 kg

TABLE 3: Electric consumption data from CIEMAT headquarters tests.

Distance	Time	Average speed	Energy consumed	Consumption
1450 km	108.5 h	13.6 km/h	11.62 kWh	8.01 Wh/km

batteries of the electric bicycles and lithium-ion batteries of the electric pickup vehicles, with the aim of extending the system lifetime.

From the previous mobility analysis, it is observed that during the morning period (from 7:00 to 10:00), the total number of e-bike trips is 95. The energy required per bike during this period is 8.459 Wh/e-bike, and the energy demanded by the e-bike fleet during the morning period will be 803.605 Wh. The total energy demanded by the e-bike fleet for the full day will be 1607.21 Wh.

The total number of trips by pick up EV trailer during the morning period is 28, consuming 5.544 kWh. The total energy demanded by the EV trailer for the full day will be 11.088 kWh, and the total amount of energy required by the proposed e-bike sharing system (e-bikes plus EV trailer) will be 12.7 kWh/day. Table 7 summarizes all energy mobility requirement of the proposed system.

There is a charging station located outside CIEMAT building 42, which is connected to the PV panels through three 10 kW single-phase MPPT inverters. This charging station is composed by an AC level 2 charging point, which operates at 3.7 kW (230 V/240 V-16 A) to charge the pickup EV trailer and 25 Schuko plugs type F (also known as CEE 7/4) protected by 5 single-pole 6A 230/240 V, 50 Hz, circuit breaker to charge the e-bikes.

The complete e-bike sharing system will be recharged during the midday period (10:00 to 16:00), when all CIEMAT employees are working and there is no demand for trips. At the beginning of the day, all e-bikes are parked at the underground docking station and the EV trailer is empty and parked at CIEMAT. It is assumed that the initial capacity of the e-bikes is 1 kWh (11.11% of the total e-bike capacity), and the initial capacity of the EV trailer is 8 kWh (33.33% of the EV trailer battery capacity). The solar resource is not available early in the morning; therefore, it is necessary to have energy in the batteries of the pickup and e-bikes to be able to perform the first morning trips. With these assumptions, we prove that, even in the worst day of the year, the remaining energy at the end of the day is even higher.

As soon as the employees start to arrive in the morning, they pick up an e-bike from the docking station, returning it to the EV trailer in the CIEMAT headquarters. When the EV trailer is full, this vehicle will carry back the e-bikes to the docking station located near the metro station. During these trips, all vehicles involved in this sharing system (EV trailer and e-bikes) will spend energy. Figure 10 shows the e-bike sharing system hourly energy consumption for a sample day.

The worst months for PV generation, due to the daily low irradiance are December and January [16]. The estimated solar energy generated by the PV panels located at CIEMAT building 42 during a single day in December is presented in Figure 11. In this figure, negative blue bars represent the total sharing system hourly energy consumption and the positive estimated solar generation is shown in red and yellow bars. e-bike sharing system will be recharged during the remainder period between 10:00 and 16:00 (highlighted with yellow bars).

TABLE 4: Electric consumption from two different users over the same specific route.

	Distance	Time	Average speed
	1.1 km	5 min	15 km/h
	Energy consumed	Energy consumed/km	Energy consumed/km/kg
User 122 kg	12.15 Wh	9.34 Wh/km	0.076 Wh/km/kg
User 75 kg	8.9 Wh	6.6 Wh/km	0.088 Wh/km/kg

TABLE 5: Optimization model parameters.

Symbol	Description	Value
(t)	Time interval index, from 7:00–19:00	[1,720] min
BCTI	Charging time interval for e-bikes, from 10:00–16:00	[180,540] min
$D_{AB}(t)$	Passenger demand from A (metro station) to B (CIEMAT)	$[D_{AB}]$
$D_{BA}(t)$	Passenger demand from B (CIEMAT) to A (metro station)	$[D_{BA}]$
$A_{EB}CT$	Pickup EV trailer capacity (number of e-bikes which can be transported)	5
$\Delta T_{AB}{}^{AEB}$	Travel time from A to B on e-bikes	6 min
$\Delta T_{BA}{}^{AEB}$	Travel time from B to A on e-bikes	5 min
$\Delta T_{AB}{}^{K}$	Travel time from A to B of pickup EV	4 min
$\Delta T_{BA}{}^{K}$	Travel time from B to A of pickup EV	4 min
NB_{TOT}	Total number of e-bikes available in the system	25
NB_{A0}	Initial number of e-bikes in the docking station A at 7:00 am	25
NB_{B0}	Initial number of e-bikes located at B at 7:00 am	0
NK_{TOT}	Total number of pick-up EV available in the system	1
NK_{A0}	Number of pick-up EV located at A at 7:00 am	0
NK_{B0}	Number of pick-up EV located at B at 7:00 am	1
P_A	e-bikes parking slots in A	25
P_B	e-bikes parking slots in B	25

TABLE 6: Integer variables of the optimization model.

Symbol	Description	Value
$nb_A(t)$	Number e-bikes at A	(0,...,25)
$nb_B(t)$	Number e-bikes at B	(0,...,25)
$Cb_{AB}(t)$	Number e-bikes on pickup EV from A to B	(0,...,5)
$Cb_{BA}(t)$	Number e-bikes on pickup EV from B to A	(0,...,5)
$nk_A(t)$	Number of pickup EV trailers at A	(0,1)
$nk_B(t)$	Number of pickup EV trailers at B	(0,1)
$dk_{AB}(t)$	Number of pickup EV trailers traveling from A to B	(0,1)
$dk_{BA}(t)$	Number of pickup EV trailers traveling from B to A	(0,1)
$db_{AB}(t)$	Number e-bikes traveling from A to B	(0,...,25)
$db_{AB}(t)$	Number e-bikes traveling from B to A	(0,...,25)

Figure 12 represents the charging process along the day. The EV trailer is charged using a level 1230 V-16 A charging station, absorbing 3.7 kWh during each hour. The rest of available energy generation is used to charge e-bike fleet. It is observed that the EV trailer is charged during the first 5 hours (from 10:00 to 14:00), and when the solar generation falls below 3.7 kW, the remaining generation capacity is only used to charge the e-bike fleet. Although from 16:00 there is still solar generation capacity available, the system interrupts the charging process because the employees are starting to come back home using this mobility system again. To get an idea of the evolution of the state of charge of the electric vehicle battery, we use the state of charge (SOC) metric [21]. Since in our case we decide to idealize the efficiencies of the electric power elements of the system, we use a simplified definition of the SOC function.

$$SOC = \frac{Ah_{int} - Ah_{consum}}{Ah_{rate}}. \tag{10}$$

Ah_{int} is the initial battery charging state. Ah_{consum} is the consumption of the battery. Ah_{rate} is the battery rating.

Figure 13 shows the battery state of charge of the pickup EV along the day. At the beginning of the day, the initial SOC is 33.3%, corresponding to 8 kWh. As soon as the employees start to arrive to the underground station in the morning, the pickup EV must tow the e-bikes between both locations, consuming energy. This consumption is observed from 7:00 to 10:00. Once all employees are working, the pickup EV is

TABLE 7: Energy demanded by the proposed e-bike sharing system (e-bikes plus pick up EV trailer).

Description	Value	Units
Pickup EV trailer consumption	0.18	kWh/km
Number of trips by EV trailer/day	56	
Total daily pickup EV trailer consumption	11.088	kWh
e-bike consumption	0.0077	kWh/km
Number of trips by e-bikes/day	190	
Total daily e-bike fleet consumption	1.607	kWh
Total daily e-bike sharing system consumption	**12.7**	kWh

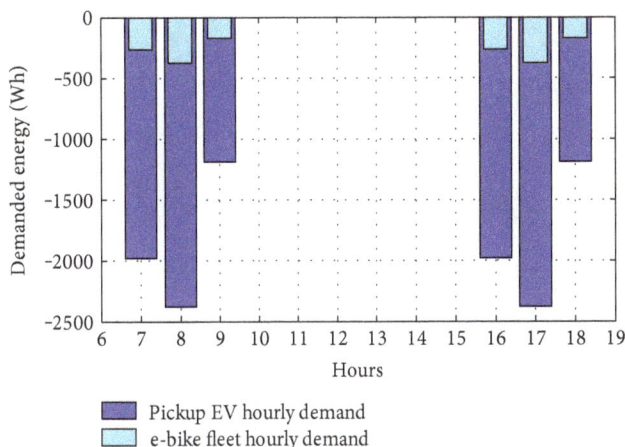

FIGURE 10: e-bike sharing system hourly energy consumption.

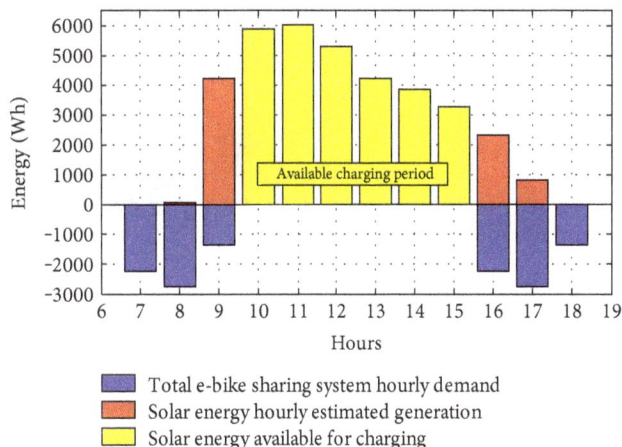

FIGURE 11: Total e-bike sharing system energy demand versus solar energy available for charging.

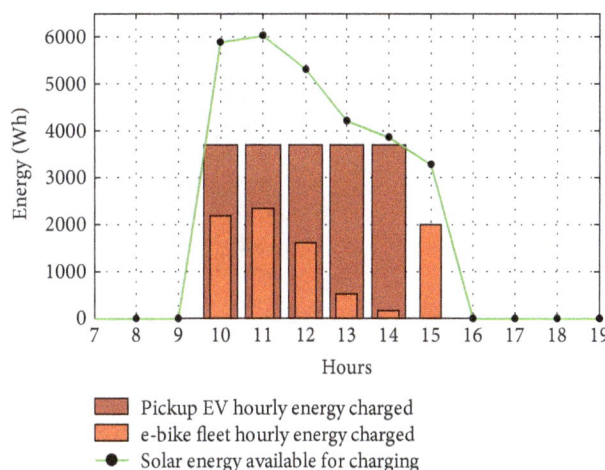

FIGURE 12: Charging process of the e-bike sharing system.

charged for 5 hours at CIEMAT building 42. The SOC is linearly increased reaching 87.32% (21 kWh) at 15:00. During the afternoon, the employees return to their homes, using this e-bike sharing system to reach again the underground station. One more time, the pickup EV will spend energy during this period (until 19:00), reaching a final SOC of 64.22% (15.4 kWh). This energy will be the initial SOC for the next day, assuming that the battery self-discharging rate is negligible.

Figure 14 shows the battery state of charge (SOC) of the e-bike fleet along the day. At the beginning of the day, the initial SOC is 11.11%, corresponding to 1 kWh. This consumption is observed from 7:00 to 10:00. Once all employees are working, the e-bike fleet is charged for 5 hours at CIEMAT building 42. The SOC is linearly increased reaching 100% (9 kWh) at 15:00. During the afternoon, the employees return to their homes, using this e-bike sharing system to reach again the underground station. One more time, the e-bike fleet will spend energy during this period (until 19:00), reaching a final SOC of 91.07% (8.2 kWh). This energy will be the initial SOC for the next day, assuming that the battery self-discharging rate is negligible.

4. Conclusions

This study concludes that photovoltaics have a huge potential to satisfy the energy demand of mobility systems for last-mile employee transportation. The proposed system manages PV energy using only the batteries from the electric vehicles, without requiring any supportive energy storage device.

Taking advantage of the existence of a building-integrated PV system currently available at the workplace, we have analyzed and optimized an e-bike sharing system fully powered by solar energy, providing a zero CO_2 emission and zero grid electricity consumption system. To determine the total daily electric demand of the e-bike sharing system, different tests were performed over real e-bike models, evaluating the e-bike consumption under several conditions. From the daily mobility requirements of several employees (traveling from the nearest public underground station to their common workplace), an optimization model was designed to size this e-bike sharing system, determining the optimal number of e-bikes and the minimum number of trips required by the pickup EV to keep the balance of e-bikes in both dock stations. In addition, the pickup EV consumption was estimated based on real consumption information.

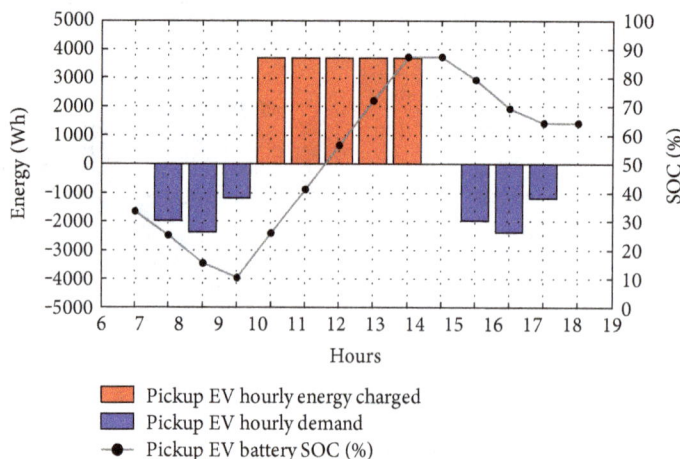

FIGURE 13: EV charging and discharging process and SOC evolution.

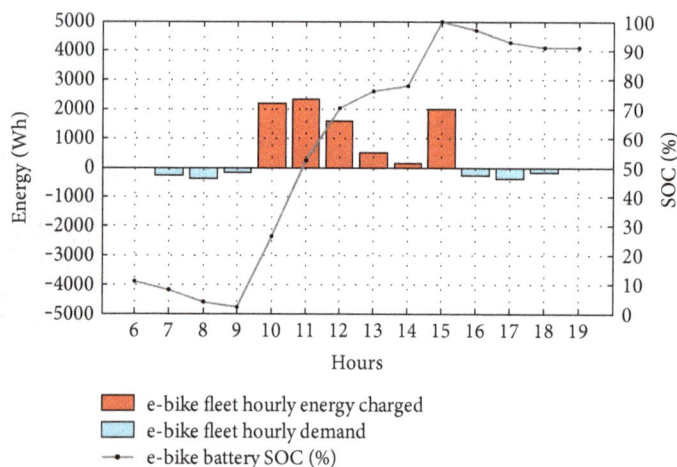

FIGURE 14: e-bike charging and discharging process and SOC evolution.

Related to PV generation, monthly and daily PV electricity production was estimated based on accurate solar radiation data, tilt and orientation data of the PV modules installed in each facade, and the inverter and solar module datasheets.

With all this information, it has been demonstrated that it is completely feasible to design a zero-emission e-bike sharing system to solve the last-mile problem, completing the public transport system. This healthy solution will also allow reducing the GHG emissions in urban areas.

Conflicts of Interest

The authors declare that they have no competing interests.

Acknowledgments

The research leading to these results has received funding from the European Union Seventh Framework Programme (FP7/2007-2013) under Grant Agreement no. 270833.

References

[1] EU, "Directive 2009/29/EC of the European Parliament and of the Council of 23 April 2009 amending directive 2003/87/EC so as to improve and extend the greenhouse gas emission allowance trading scheme of the community," *Journal of the European Union*, vol. 5, pp. 63–87, 2009.

[2] Q. K. Hassan, K. Mahmud Rahman, A. S. Haque, and A. Ali, "Solar energy modelling over a residential Community in the City of Calgary, Alberta, Canada," *International Journal of Photoenergy*, vol. 2011, Article ID 216519, p. 8, 2011.

[3] N. M. Chivelet, J. C. G. García, M. A. Abella, and F. C. Romero, "Integration of solar photovoltaic energy into the rehabilitation of CIEMAT building 42," *Vértices*, vol. 5, no. 25, pp. 28–32, 2016, http://www.ciemat.es/portal.do?IDM=226&NM=3.

[4] Y. C. Huang, C. C. Chan, S.-C. Kuan, S. J. Wang, and S. K. Lee, "Analysis and monitoring results of a building integrated photovoltaic façade using PV ceramic tiles in Taiwan," *International Journal of Photoenergy*, vol. 2014, Article ID 615860, p. 12, 2014.

[5] Y. J. Chiu and T. M. Ying, "A novel method for technology forecasting and developing R&D strategy of building integrated photovoltaic technology industry," *Mathematical Problems in Engineering*, Article ID 273530, p. 24, 2012.

[6] UN-Habitat, *State of the World's Cities 2012/2013*, 2016, https://sustainabledevelopment.un.org/content/documents/745habitat.pdf.

[7] M. Fuentes, *Impact of Energetic Management on the Development of LEVS as a Mode Alternative Transport Solutions Mobility to Urban/Metropolitan*, Technical Report CIEMAT, p. 126, 2011.

[8] K. Martens, "The bicycle as a feedering mode: experiences from three European countries," *Transportation Research Part D: Transport and Environment*, vol. 9, no. 4, pp. 281–294, 2004.

[9] S. Shaheen, S. Guzman, and H. Zhang, "Bikesharing in Europe, the Americas, and Asia: past, present, and future," *Transportation Research Record: Journal of the Transportation Research Board*, vol. 2143, pp. 159–167, 2010.

[10] P. DeMaio, "Bike-sharing: history, impacts, models of provision, and future," *Journal of Public Transportation*, vol. 12, no. 4, p. 3, 2009.

[11] H. Sayarshad, S. Tavassoli, and F. Zhao, "A multi-periodic optimization formulation for bike planning and bike utilization," *Applied Mathematical Modelling*, vol. 36, no. 10, pp. 4944–4951, 2012.

[12] P. Lebeau, C. De Cauwer, J. VanMierlo, C. Macharis, W. Verbeke, and T. Coosemans, "Conventional, hybrid, or electric vehicles: which technology for an urban distribution centre?" *The Scientific World Journal*, vol. 2015, Article ID 302867, p. 11, 2015.

[13] S. Ji, C. R. Cherry, L. D. Han, and D. A. Jordan, "Electric bike sharing: simulation of user demand and system availability," *Journal of Cleaner Production*, vol. 85, pp. 250–257, 2014.

[14] E. Blasius, E. Federau, P. Janik, and Z. Leonowicz, "Heuristic storage system sizing for optimal operation of electric vehicles powered by photovoltaic charging station," *International Journal of Photoenergy*, vol. 2016, Article ID 7134904, p. 12, 2016.

[15] Campus Moncloa, *Campus of International Excellence MONCLOA*, 2016, http://www.campusmoncloa.es/en/campus-moncloa/welcome.php.

[16] Joint Reseach Center, EU, "Photovoltaic geographical information system - interactive maps," 2016, http://re.jrc.ec.europa.eu/pvgis/apps4/pvest.php.

[17] J. Jurasz and J. Mikulik, "Investigating theoretical PV energy generation patterns with their relation to the power load curve in Poland," *International Journal of Photoenergy*, vol. 2016, Article ID 3789840, p. 7, 2016.

[18] Cycling London, *Cycle Mph Average Speed in London*, 2016, http://cycling-london.blogspot.com.es/2006/04/cycle-mph-average-speed-in-london.html.

[19] Nissan, *E-nv200. Range and Specifications*, 2016, https://www.nissan.co.uk/vehicles/new-vehicles/e-nv200/charging-range.html.

[20] M. Thwaite, *Review: 2014 Nissan e-NV200 Electric Van*, 2016, https://transportevolved.com/2015/04/18/2014-nissan-e-nv200-review/.

[21] B. G. Yu, "Design and experimental results of battery charging system for microgrid system," *International Journal of Photoenergy*, vol. 2016, Article ID 7134904, p. 6, 2016.

PERMISSIONS

LIST OF CONTRIBUTORS

S. Daliento and P. Guerriero
Department of Electrical Engineering and Information Technology (DIETI), University of Naples Federico II, Via Claudio 21, 80125 Naples, Italy

A. Chouder
Department of Génie Electrique, Faculty of Technologies, University of M'Sila, BP 166, Ichbelia, M'Sila, Algeria

A. Massi Pavan
Department of Engineering and Architecture, University of Trieste, Piazzale Europa 1, 34127 Trieste, Italy

A. Mellit
Renewable Energy Laboratory, Faculty of Sciences and Technology, Jijel University, Jijel, Algeria

R. Moeini and P. Tricoli
School of Electronic, Electrical and Systems Engineering, University of Birmingham, Gisbert Kapp Building, Edgbaston, Birmingham B15 2TT, UK

Abu Jahid
Department of Electrical, Electronic and Communication Engineering, Military Institute of Science and Technology, Dhaka 1216, Bangladesh

Abdullah Bin Shams
Department of Electrical and Electronic Engineering, Islamic University of Technology, Gazipur 1704, Bangladesh

Md. Farhad Hossain
Department of Electrical and Electronic Engineering, Bangladesh University of Engineering and Technology, Dhaka 1000, Bangladesh

Bulent Yaniktepe, Osman Kara and Coskun Ozalp
Engineering Faculty, Department of Energy Systems Engineering, University of Osmaniye Korkut Ata, Fakiusagi, 80000 Osmaniye, Turkey

Xiaodan Sun and Li Xiao
State Key Laboratory of Alternate Electrical Power System with Renewable Energy Sources, North China Electric Power University, Beijing 102206, China

Jia Xu
Beijing Key Laboratory of Energy Safety and Clean Utilization, North China Electric Power University, Beijing 102206, China

Jing Chen and Jianxi Yao
State Key Laboratory of Alternate Electrical Power System with Renewable Energy Sources, North China Electric Power University, Beijing 102206, China
Beijing Key Laboratory of Energy Safety and Clean Utilization, North China Electric Power University, Beijing 102206, China

Bing Zhang and Songyuan Dai
Beijing Key Laboratory of Energy Safety and Clean Utilization, North China Electric Power University, Beijing 102206, China
Beijing Key Laboratory of Novel Film Solar Cell, North China Electric Power University, Beijing 102206, China

Satoshi Koyasu, Daiki Atarashi, Etsuo Sakai, and Masahiro Miyauchi
School of Materials and Chemical Technology, Tokyo Institute of Technology, 2-12-1 Ookayama, Meguro-ku, Tokyo 152-8552, Japan

Shengjun Wu and Qingshan Xu
School of Electrical Engineering, Southeast University, No. 2 Sipailou, Nanjing 210096, China

Qun Li, Xiaodong Yuan and Bing Chen
Jiangsu Electric Power Research Institute, No. 1 Paweier, Nanjing 211103, China

Hashim A. Hussein, Ali H. Numan, and Ruaa A. Abdulrahman
Electromechanical Engineering Department, University of Technology, Baghdad, Iraq

Junjie Qian, Kaiting Li, Huaren Wu, Jianfei Yang, and Xiaohui Li
School of Electrical and Automation Engineering, Nanjing Normal University, Nanjing 210042, China

Md. Feroz Ali
Department of Electrical and Electronic Engineering (EEE), Pabna University of Science & Technology (PUST), Pabna 6600, Bangladesh

Md. Faruk Hossain
Electrical & Electronic Engineering, Rajshahi University of Engineering & Technology (RUET), Rajshahi 6204, Bangladesh

Md. Asaduzzaman, Md. Billal Hosen, Md. Karamot Ali and Ali Newaz Bahar
Department of Information and Communication Technology (ICT), Mawlana Bhashani Science and Technology University (MBSTU), Santosh, Tangail 1902, Bangladesh

Fei Cao
College of Mechanical and Electrical Engineering, Hohai University, Changzhou 213022, China
Sunshore Solar Energy Company Limited, Nantong 226300, China

Lei Wang and Tianyu Zhu
College of Mechanical and Electrical Engineering, Hohai University, Changzhou 213022, China

Ciaran Lyons, Pratibha Dev, Owen Byrne, Praveen K. Surolia, Pathik Maji, J. M. D. MacElroy, Niall J. English and K. Ravindranathan Thampi
SFI Strategic Research Cluster in Solar Energy Conversion, UCD School of Chemical and Bioprocess Engineering, University College Dublin, Dublin 4, Ireland

Neelima Rathi and Edmond Magner
SFI Strategic Research Cluster in Solar Energy Conversion, Department of Chemical Sciences and Bernal Institute, University of Limerick, Limerick, Ireland

Aswani Yella and Michael Grätzel
Laboratoire de Photonique et Interfaces (LPI), Ecole Polytechnique Fédérale de Lausanne, 1015 Lausanne, Switzerland

Peijie Lin, Zhicong Chen, Lijun Wu, Lingchen Chen and Shuying Cheng
Institute of Micro/Nano Devices and Solar Cells, College of Physics and Information Engineering, Fuzhou University, Fuzhou 350116, China

Yaohai Lin
College of Computer and Information Sciences, Fujian Agriculture and Forestry University, Fuzhou 350002, China

Pierluigi Guerriero, Renato Rizzo and Santolo Daliento
Department of Electrical Engineering and Information Technology (DIETI), University of Naples Federico II, Via Claudio 21, 80125Naples, Italy

Luigi Piegari
Department of Electronics, Information and Bioengineering (DEIB), Politecnico di Milano, Piazza Leonardo da Vinci 32, 20133 Milan, Italy

Pan-Pan Zhang, Zheng-Ji Zhou, Dong-Xing Kou and Si-Xin Wu
Key Laboratory for Special Functional Materials of Ministry of Education, Henan University, Kaifeng, Henan Province 475004, China
Collaborative Innovation Center of Nano Functional Materials and Applications, Henan University, Kaifeng, Henan Province 475004, China

Chaoxu Mu and Weiqiang Liu
Tianjin Key Laboratory of Process Measurement and Control, School of Electrical and Information Engineering, Tianjin University, Tianjin 300072, China

Wei Xu
State Key Laboratory of Advanced Electromagnetic Engineering and Technology, School of Electrical and Electronic Engineering, Huazhong University of Science and Technology, Wuhan 430074, China
College of Mechanical and Electrical Engineering, Huanggang Normal University, Huanggang 438000, China

Md. Rabiul Islam4
Department of Electrical and Electronic Engineering, Rajshahi University of Engineering and Technology, Rajshahi 6204, Bangladesh

Angel A. Bayod-Rújula
Department of Electrical Engineering, University of Zaragoza, Zaragoza, Spain

Alessandro Burgio, Daniele Menniti, Anna Pinnarelli and Nicola Sorrentino
Department of Mechanical, Energy and Management Engineering, University of Calabria, Rende, Italy

Zbigniew Leonowicz
Wroclaw University of Science and Technology, Wroclaw, Poland

Longfeng Li, Maolin Zhang and Mingzhu Liu
School of Chemistry and Materials Science, Huaibei Normal University, Huaibei 235000, China
Information College, Huaibei Normal University, Huaibei 235000, China

Jing Zhang, Xianliang Fu and Peipei Xiao
School of Chemistry and Materials Science, Huaibei Normal University, Huaibei 235000, China

Manuel Fuentes
Renewable Energy Division, Energy Department, CIEMAT, Avda. Complutense 40, 28040 Madrid, Spain

Jesús Fraile-Ardanuy
ETSI Telecomunicación, Universidad Politécnica de Madrid, Avda. Complutense 30, 28040 Madrid, Spain

José L. Risco-Martín
DACYA, Complutense University of Madrid, 28040 Madrid, Spain

José M. Moya
Integrated Systems Laboratory, Center for Computational Simulation, Universidad Politécnica de Madrid, Avda. Complutense 30, 28040 Madrid, Spain

Index

www.ingramcontent.com/pod-product-compliance
Lightning Source LLC
Chambersburg PA
CBHW080704200326
41458CB00013B/4961